The Coordination Chemistry of Metalloenzymes

The Role of Metals in Reactions Involving Water, Dioxygen, and Related Species

NATO ADVANCED STUDY INSTITUTES SERIES

Proceedings of the Advanced Study Institute Programme, which aims at the dissemination of advanced knowledge and the formation of contacts among scientists from different countries

The series is published by an international board of publishers in conjunction with NATO Scientific Affairs Division

A	Life Sciences	Plenum Publishing Corporation
B	Physics	London and New York
C	Mathematical and Physical Sciences	D. Reidel Publishing Company Dordrecht, Boston and London
D	Behavioural and Social Sciences	Martinus Nijhoff Publishers
E	Engineering and Materials Sciences	The Hague, London and Boston
F	Computer and Systems Sciences	Springer Verlag Heidelberg
G	Ecological Sciences	

Series C – Mathematical and Physical Sciences

Volume 100 – The Coordination Chemistry of Metalloenzymes

The Coordination Chemistry of Metalloenzymes

The Role of Metals in Reactions Involving Water, Dioxygen, and Related Species

Proceedings of the NATO Advanced Study Institute held at San Miniato, Pisa, Italy, May 28 - June 8, 1982

edited by

I. BERTINI
University of Florence, Italy

R. S. DRAGO
University of Florida, Gainesville, U.S.A.

and

C. LUCHINAT
University of Florence, Italy

D. Reidel Publishing Company

Dordrecht : Holland / Boston : U.S.A. / London : England

Published in cooperation with NATO Scientific Affairs Division

Library of Congress Cataloging in Publication Data

NATO Advanced Study Institute (1982 : San Miniato, Italy)
The coordination chemistry of metalloenzymes.

(NATO advanced study institutes series. Series C, Mathematical and physical sciences ; v. 100)
"Published in cooperation with NATO Scientific Affairs Division."
Includes index.
1. Metalloenzymes–Congresses. 2. Coordination compounds–Congresses. I. Bertini, I. (Ivano) II. Drago, Russell S. III. Luchinat, C. (Claudio), 1952- IV. Title. V. Series.
QP601.3.N37 1982 574.19'25 82-24083
ISBN-13: 978-94-009-7051-9 e-ISBN-13: 978-94-009-7049-6
DOI: 10.1007/978-94-009-7049-6

Published by D. Reidel Publishing Company
P.O. Box 17, 3300 AA Dordrecht, Holland

Sold and distributed in the U.S.A. and Canada
by Kluwer Boston Inc.,
190 Old Derby Street, Hingham, MA 02043, U.S.A.

In all other countries, sold and distributed
by Kluwer Academic Publishers Group,
P.O. Box 322, 3300 AH Dordrecht, Holland

D. Reidel Publishing Company is a member of the Kluwer Group

CONTENTS

PREFACE

Assembling a program in bioinorganic chemistry that is scientifically relevant, well defined, and self-consistent is not an easy task. In this attempt we decided to consider zinc enzymes, copper oxidases, cytochromes and cytochrome oxidase. The choice is in part due to the great attention that the current specialized literature devotes to these topics, which are now debated among chemists, biochemists, biophysicists, etc.. We believe that hydration reactions, hydrolytic and oxidative processes have much in common from the point of view of the reaction mechanisms, the comprehension of which represents a frontier of science.

For these reasons these topics have been the subject of the NATO-ASI held at San Miniato, Pisa, Italy, from May 28 to June 8, 1982. We hope we can transfer here the main conclusions of what (we believe) was a very stimulating scientific meeting.

We would like to thank the local saving bank, Cassa di Risparmio di San Miniato, for helping in many ways. The financial contribution from the European Research Office of the US Army, and from the Bruker Spectrospin s.r.l., Italy, is also acknowledged. The National Science Foundation of the United States has provided a travel grant to one of the participants from the U.S.A. We are grateful to the NATO Scientific Affairs Division which provided a grant to finance this Institute. Finally, we would like to mention that despite the presence of three editors, this book could be arranged in the present form only thanks to the work of Dr. Gianni Lanini, who also helped in the organization of the meeting at every stage.

Florence, June 1982

The Editors

I. Bertini, R. S. Drago, and C. Luchinat (eds.), The Coordination Chemistry of Metalloenzymes, xi.

PARTICIPANTS

Aalmo, K.M.	University of Trondheim, Department of Chemistry, NLHT-Rosenborg, N-7000 Trondheim, NORWAY
Alpoim, M.C.	Universidade de Coimbra, Departamento de Quimica da Faculdade de Ciências e Tecnologia, 3000 Coimbra, PORTUGAL
Aronsson, A.C.	University of Stockholm, Arrhenius Laboratory, Department of Biochemistry, S-106 91 Stockholm, SWEDEN
Avigliano, L.	Università di Roma, Istituto di Chimica Biologica, Piazzale Aldo Moro 2, 00185 Roma, ITALY
Banci, L.	Università di Firenze, Istituto di Chimica Generale e Inorganica, Via J. Nardi 39, 50132 Firenze, ITALY
Beltramini, M.	Biochemisches Institut der Universität Zürich, Zürichbergstrasse 4, CH-8028 Zürich, SWITZERLAND
Bencini, A.	Istituto per lo Studio della Stereochimica ed Energetica dei Composti di Coordinazione, C.N.R., Via F.D. Guerrazzi 27, 50132 Firenze, ITALY
Bertini, I.	Università di Firenze, Istituto di Chimica Generale e Inorganica, Via J. Nardi 39, 50132 Firenze, ITALY
Bianconi, A	Università di Roma, Istituto di Fisica "G. Marconi", Piazzale Aldo Moro 2, 00185 Roma, ITALY
Bill, E.	Universität des Saarlandes, Fachbereich 12.1, Werkstoffwissenschaften, D-6600 Saarbrücken, F.R.G.
Bordas, J.	EMBL-Outstation c/o DESY, Notkestrasse 85, D-2000, Hamburg 52, F.R.G.

Borges Coutinho, I.	Centro de Quimica Estrutural, Complexo Interdisciplinar, Instituto Superior Tecnico, 1096 Lisboa, Codex, PORTUGAL
Borghi, E.	Istituto per lo Studio della Stereochimica ed Energetica dei Composti di Coordinazione, C.N.R., Via F.D. Guerrazzi 27, 50132 Firenze, ITALY
Büther, H.	Universität Bremen, Institut für Biophysikalische Chemie, FB3/NW2, Achterstrasse, D-2800 Bremen 33, F.R.G.
Canti, G.	Università di Firenze, Istituto di Chimica Generale e Inorganica, Via G. Capponi 7, 50121 Firenze, ITALY
Castro, M.M.	Universidade de Coimbra, Departamento de Quimica da Faculdade de Ciências e Tecnologia, 3000 Coimbra, PORTUGAL
Chan, S.I.	California Institute of Technology, Division of Chemistry and Chemical Engineering, The Chemical Laboratories, Pasadena, California 91125, U.S.A.
Cinellu, M.A.	Università di Sassari, Istituto di Chimica Biologica, Via Muroni 23/A, 07100 Sassari, ITALY
Cocco, D.	Università di Roma, Istituto di Chimica Biologica, Piazzale Aldo Moro 2, 00185 Roma, ITALY
Dahl, K.H.	Universitetet i Oslo, Biokjemisk Institutt, Postboks 1041, Blindern Oslo 3, NORWAY
Dahlin, S.	Chalmers Tekniska Högskola och Göteborgs Universitet, Institutionen för Biokemi & Biofysik, S-412 96 Göteborg, SWEDEN
Dawson, J.H.	University of South Carolina, Department of Chemistry, Columbia, South Carolina 29208, U.S.A.
Desideri, A.	Università di Roma, Istituto di Chimica Biologica, Piazzale Aldo Moro 2, 00185 Roma, ITALY
Drago, R.S.	University of Florida, Department of Chemistry, Gainesville, Florida 32610, U.S.A.
Dreyer, J.L.	Université de Fribourg, Institut de Chimie Physiologique, CH-1700 Fribourg, SWITZERLAND

Dutler, H.	Laboratorium für Organische Chemie, ETH-Hönggerberg, CH-8093 Zürich, SWITZERLAND
Dutton, C.J.	University Chemical Laboratory, Lensfield Road, Cambridge CB2 1EW, UNITED KINGDOM
Frekel, A.	The Weitzmann Institute of Science, Department of Cell Biology, Rehovot, ISRAEL
Gatteschi, D.	Università di Firenze, Istituto di Chimica Generale, Via G. Capponi 7, 50121 Firenze, ITALY
Groves, J.T.	The University of Michigan, Department of Chemistry, Ann Arbor, Michigan 48109, U.S.A.
Haase, W.	Institut für Physikalische Chemie, Technische Hochschule Darmstadt, Petersenstrasse 20, D-6100 Darmstadt, F.R.G.
Harmer, H.R.	Royal Roads Military College, Department of Chemistry, FMO Victoria, B.C. VOS 1BO, CANADA
Hermes, C.	EMBL-Oustation, c/o DESY, Notkestrasse 85, D-2000 Hamburg 52, F.R.G.
Ianzini, F.	Istituto Superiore di Sanità, Laboratorio delle Radiazioni, Viale Regina Elena 299, 00161 Roma, ITALY
Jensen, P.	Chalmers Tekniska Högskola och Göteborgs Universitet, Institutionen för Biokemi & Biofisik, S-412 96, Göteborg, SWEDEN
Kent, T.	Gray Freshwater Biological Institute, College of Biological Sciences, University of Minnesota, P.O. Box 100, County Roads 15 and 19, Navarre, Minnesota 55392, U.S.A.
Knowles, P.F.	The Astbury Department of Biophysics, The University of Leeds, Leeds LS2 9JT, UNITED KINGDOM
Koenig, S.H.	I.B.M. Thomas J. Watson Research Center, Yorktown Heights, New York 10598, U.S.A.
Kunugi, S.	Max-Planck-Institut für Experimentelle Medizin, Abteilung Chemie, Hermann-Rein-Strasse 3, Göttingen, F.R.G.
Lepri, A.	Università di Siena, Istituto di Chimica Generale e Inorganica, Via Pian dei Mantellini 44, Siena, ITALY

Lanini, G.	Università di Firenze, Istituto di Chimica Generale e Inorganica, Via G. Capponi 7, 50121 Firenze, ITALY
Lindskog, S.	University of Umea, Department of Biochemistry, S-901 87 Umea, SWEDEN
Luchinat, C.	Università di Firenze, Istituto di Chimica Generale e Inorganica, Via G. Capponi 7, 50121 Firenze, ITALY
Maldotti, A.	Centro di Studi sulla Fotochimica e Reattività degli Stati Eccitati dei Composti di Coordinazione, CNR, Istituto Chimico, Università di Ferrara, Via L. Borsari 46, 44100 Ferrara, ITALY
Mani, F.	Università di Firenze, Istituto di Chimica Generale e Inorganica, Via J. Nardi 39, 50132 Firenze, ITALY
Mansuy, D.	Ecole Normale Supérieure, Laboratoire de Chimie, 24 Rue Lhomond, 75231 Paris, FRANCE
Marzotto, A.	Istituto di Chimica e Tecnologia dei Radioelementi, C.N.R., Corso Stati Uniti, 35100 Padova, ITALY
Maret, W.	Fachbereich 15 der Universität des Saarlandes, Fachrichtung 15.2 Biochemie, D-6600 Saarbrücken, F.R.G.
Mc Millin, D.R.	Purdue University, Department of Chemistry, West Lafayette, Indiana 47907, U.S.A.
Monnanni, R.	Università di Firenze, Istituto di Chimica Generale e Inorganica, Via G. Capponi 7, 50121 Firenze, ITALY
Montiel Montoia, R.A.	Universität des Saarlandes, Fachbereich 12.1, Werkstoffwissenschaften, D-6600 Saarbrücken, F.R.G.
Morgenstern-Badarau, I.	Université de Paris-Sud, Laboratoire de Spectrochimie des Elements de Transition, Bâtiment 420, 91405, Orsay, FRANCE
Morpurgo, G.	Università di Roma, Istituto di Chimica Generale e Inorganica, Piazzale Aldo Moro 2, 00185 Roma, ITALY

Morpurgo Ceciarelli, L.	Università di Roma, Centro di Biologia Molecolare c/o Istituto di Chimica Biologica, Piazzale Aldo Moro 2, 00185 Roma, ITALY
Nguyen-Brem, T.	Bahnhofstrasse 50, D-8051 Marzling, F.R.G.
Neuman, H.	The Weitzmann Institute of Science, Department of Biophysics, Rehovot, ISRAEL
Oester, D.A.	University of Illinois at Urbana-Champaign, School of Chemical Science, 505 St. Mathews Street, Urbana, Illinois 61801, U.S.A.
Pandeja, K.B.	University of Delhi, Department of Chemistry, Delhi, 110007, INDIA
Pinna, G.G.	Università di Sassari, Istituto di Chimica Biologica, Via Muroni 23, 07100 Sassari, ITALY
Pispisa, B.	Università di Napoli, Istituto Chimico, 80100 Napoli, ITALY
Que, L.Jr.	Cornell University, Department of Chemistry, Ithaca, New York 14850, U.S.A.
Rawer, S.	Fachbereich 15 der Universität des Saarlandes, Fachrichtung 15.2 Biochemie, D-6600 Saarbrücken, F.R.G.
Reimer, K.J.	Royal Roads Military College, Department of Chemistry, FMO Victoria, B.C. VOS 1BO, CANADA
Reinhammar, B.	Chalmers Tekniska Högskola och Göteborgs Universitet, Institutionen för Biokemi & Biofysik, S-412 96 Göteborg, SWEDEN
Robb, D.A.	University of Strathclyde, Department of Biochemistry, The Todd Centre, 31 Taylor Street, Glasgow G4 ONR, UNITED KINGDOM
Rosi, A.	Istituto Superiore di Sanità, Laboratorio delle Radiazioni, Viale Regina Elena 299, 00161 Roma, ITALY
Rotilio, G.	Università di Roma, Istituto di Chimica Biologica, Piazzale Aldo Moro 2, 00185 Roma, ITALY
Rotondo, E.	Università di Messina, Istituto di Chimica Generale e Inorganica, Via dei Verdi, 98100 Messina, ITALY

Sander, M.	Westfälische Wilhelms-Universität, Institut für Biochemie, Orleansring 23a, D-4400 Münster, F.R.G.
Scheller, K.H.	Universität Basel, Institut für Anorganische Chemie, Spitalstrasse 51, CH-4056 Basel, SWITZERLAND
Schlösser, M.	University of Bremen, Department of Chemistry, P.O. Box 330440, D-2800 Bremen, F.R.G.
Schneider-Bernlöhr, H.	Fachbereich 15 der Universität des Saarlandes, Fachrichtung 15.2 Biochemie, D-6600 Saarbrücken, F.R.G.
Scozzafava, A.	Università di Firenze, Istituto di Chimica Generale e Inorganica, Via G. Capponi 7, 50121 Firenze, ITALY
Sellin, S.	University of Stockholm, Arrhenius Laboratory, Department of Biochemistry, S-106 91 Stockholm, SWEDEN
Sigel, H.	Universität Basel, Institut für Anorganische Chemie, Spitalstrasse 51, CH-4056 Basel, SWITZERLAND
Skjeldal, L.	Universitetet i Oslo, Biokjemisk Institutt, Postboks 1041, Blindern Oslo 3, NORWAY
Steinhäuser, K.G.	Max-Planck Institut für Molekulare Genetik, Ihnestrasse 63-73, D-1000 Berlin 33, F.R.G.
Sudmeier, J.L.	University of California, Department of Chemistry, Riverside, California 92521, U.S.A.
Syvertsen, C.	Universitetet i Oslo, Biokjemisk Institutt, Postboks 1041, Blindern Oslo 3, NORWAY
Teixeira, M.G.	Centro de Quimica Estrutural, Complexo Interdisciplinar, Instituto Superior Tecnico, 1096 Lisboa-Codex, PORTUGAL
Toftlund, H.	Odense University, Department of Chemistry DK-5230 Odense, DENMARK
Weller, M.G.	Anorganische Biochemie Physiolog.-Chem. Institut der Universität Tübingen, Hoppe-Seiler-Strasse 1, D-7400, Tübingen, F.R.G.
Werner, E.	Universität Innsbruck, Institut für Anorganische Chemie und Analytische Chemie, Innrain 52a, Innsbruck, AUSTRIA

Wilson, K.J. University of Strathclyde, Department of Pure and Applied Chemistry, Thomas Graham Building, 295 Cathedral Street, Glasgow G1 1XL, UNITED KINGDOM

Xavier, A. Centro de Quimica Estrutural, Complexo Interdisciplinar, Instituto Superior Tecnico, 1096 Lisboa-Codex, PORTUGAL

Zeppezauer, M. Fachbereich 15 der Universität des Saarlandes, Fachrichtung 15.2 Biochemie, D-6600 Saarbrücken, F.R.G.

THE COORDINATION PROPERTIES OF THE ACTIVE SITE OF ZINC ENZYMES

Ivano Bertini
Istituto di Chimica Generale e Inorganica
Università di Firenze
Via J. Nardi, 39
50132 Firenze, Italy

ABSTRACT

The protein residues coordinated to zinc(II) in some zinc enzymes (carbonic anhydrase, carboxypeptidase, thermolysin, liver alcohol dehydrogenase, alkaline phosphatase, and yeast aldolase) are reviewed with the aim to propose the overall coordination number around the metal ion. This goal is attempted by analyzing the electronic spectra of the cobalt(II) substituted enzymes. ^{1}H NMR of solvent water at different external fields of the latter derivatives, as well as the ^{1}H NMR spectra of coordinated histidines, are found in some cases to provide precious structural information. The role of the zinc(II) in the catalytic mechanism is also discussed.

INTRODUCTION

The zinc enzymes are a quite popular class of enzymes which have attracted the attention of researchers from the early times of that branch of science recognized today as bioinorganic chemistry. They are metalloproteins with specific enzymatic activity which may contain a single zinc(II) ion, several zinc(II) ions or different metal ions including zinc(II). In principle the zinc ion can be functional, *i.e.* necessary for the catalytic action, or structural, *i.e.* not involved in the catalytic mechanism but necessary to keep the protein in the proper conformation (secondary, tertiary and quaternary structure). Some zinc enzymes display hydrolytic activity, *i.e.* they catalyze bond cleavage in which a reactant is water or each of its H^+ or OH^- components. Typical of this class of enzymes are the proteases, phosphatases, and β-lactamases, which hydrolyze specific bonds: peptidic bonds, phosphoric esters, and β-lactames, respectively. Carbonic anhydrase is a zinc enzyme the apparent natural function of which is the hydration of CO_2 and the dehydration of HCO_3^-; furthermore *in vitro* it catalyzes the hydrolysis of esters as well as the hydration of aldehydes. Yeast aldolase and liver alcohol dehydrogenase have metal-donor moieties related to the above enzymes. Although the latter is an oxidase, the role of the active cavity allows to make fruitful comparisons within the series of enzymes.

I. Bertini, R. S. Drago, and C. Luchinat (eds.), The Coordination Chemistry of Metalloenzymes, 1–18.

With the exception of β-lactamases for which relatively little is known, the other enzymes will be analyzed here (Table 1). This chapter deals with that part of the enzyme which includes the metal ion, the donor atoms and the residues involved in the coordination. Their coordination ability towards zinc(II) and other vicariant ions will be discussed with the aim to learn as much as possible on the structure and function of the enzymes.

Table 1. Some zinc enzymes and their catalytic role.

Zinc Enzyme	k_{cat} (s^{-1})	Catalyzed Reaction
Carbonic Anhydrase	10^6	$CO_2 + H_2O \rightleftharpoons HCO_3^- + H^+$
Carboxypeptidase A	10^3	Hydrolysis of C-terminal aminoacid residues
Thermolysin	10^2-10^3	Hydrolysis of peptidic bonds
Liver Alcohol Dehydrogenase	10	$R\text{-}CH_2OH + NAD^+ \rightleftharpoons R\text{-}CHO + NADH + H^+$
Alkaline Phosphatase	10	Hydrolysis of phosphate monoesters
Yeast Aldolase	10^2	Fructose 1,6-diphosphate $\rightleftharpoons$ dihydroxyacetonphosphate + D-glyceraldehyde-3-phosphate

Carbonic anhydrase, carboxypeptidase, and liver alcohol dehydrogenase will also be discussed at length elsewhere in the book. However, the refinement in the understanding of the behavior of these enzymes is such that it may require at this stage the different competences of various contributors.

THE RESIDUES COORDINATED AT THE ZINC ION

The X-ray structure at different degrees of refinement is available for carbonic anhydrase (CA) (1), carboxypeptidase A (CPA) (2), thermolysin (TLN) (3), and liver alcohol dehydrogenase (LADH) (4), showing the residues which are involved in the coordination at the active site and in other sites. For alkaline phosphatase (AP) the X-ray structure is refined only at 6 Å of resolution (5): the coordinating groups have been guessed through ^{113}Cd (6) and ^{13}C NMR (7). For yeast aldolase (YAL) the coordinated groups are proposed only on the basis of the analysis of the aminoacid sequence and of spectroscopic (^{1}H NMR of histidines) data (8). The atoms coordinated to the metal in these enzymes are reported in Table 2. The donor atoms of the zinc site in superoxide dismutase (SOD, see elsewhere in the book), the structure of which is known at 3 A of resolution (9), are also reported for comparison. The zinc ion in this case is of structural type.

Table 2. Metal coordinated protein residues in zinc enzymes.

Enzyme	Protein residues	Protein donors	pK_a of the coordinated water	Ref.
Alkaline Phosphatase	4 His	N_4	8 (?)	(6)
Carbonic Anhydrase	3 His	N_3	6.0-7.5	(1)
Alcohol Dehydrogenase	1 His; 2 Cys	NS_2	9.2	(4)
Carboxypeptidase A	2 His; 1 Glu	N_2O_2[a]	9 (?)	(2)
Yeast Aldolase	3 His (?)	N_3 (?)	6	(8)
Thermolysin	2 His; 1 Glu	N_2O	7 (?)	(3)
Superoxide Dismutase	3 His; 1 Asp	N_3O	not present	(9)

[a] The Glu residue behaves as bidentate

The metal ions at the catalytic sites are exposed to the solvent, so that at least a water molecule completes the coordination around the metal. The role of water, its partial or total displacement upon inhibition or during the catalytic pathway are important pieces of information for the understanding of the catalytic mechanism.

The zinc ions in the sites reported in Table 2 can successfully be replaced by other metal ions like Mn^{2+}, Cu^{2+}, Co^{2+}, in some cases VO^{2+}, etc. Among these, the cobalt(II) derivatives display activity comparable to that of the native enzymes; therefore the information obtained on the cobalt(II) proteins using the cobalt(II) ion as a spectroscopic probe can be transferred with some confidence to the native enzymes and these are the most precious data as long as the coordination polyhedron is concerned. Information obtained through the investigation of other metal derivatives will also be considered here as long as they shed light on the behavior of the native compounds.

THE ACID-BASE PROPERTIES OF COORDINATED WATER

A point of major interest in the investigation of these systems is the pK_a of the coordinated water molecule(s) because the specific rate constant of the catalyzed reaction, k_{cat}, is in general pH dependent.

Furthermore, the problem is fascinating from a chemical point of view. In a protein of M.W. 30000-80000 there are so many acidic groups with pK_a in the range 5-11 that it is very difficult to distinguish the water deprotonation from the other acidic groups (Table 3). Even the optical spectra of the cobalt(II) derivatives may be sensitive to some acidic groups nearby and in principle only slightly to the deprotonation of the coordinated water.

Table 3. Aminoacid residues with pK_a between 5 and 11.

Group	Free aminoacid	In proteins
-COOH (ω, Glu, Asp)	3-5	4-6
Imidazolium (His)	6.1	5.6-7.0
$-NH_3^+$ (α, Lys)	9.5-9.7	9.4-10.6
-OH (Tyr)	10.1	9.8-10.4
-SH (Cys)	10.3	9.4-10.8
-NH- (peptide bonds)	10.5	10.2-10.8

Table 4. Model complexes with a single coordinated water.

Compound	Donor set	pK_a of the coordinated water	Ref.
$Co(TPyMA)H_2O^{2+}$ [a]	N_4O	9.1	(10)
$Co(CR)H_2O^{2+}$ [b]	N_4O	≈8	(11)
$Co(TMC)H_2O^{2+}$ [c]	N_4O	8.4	(12)
$Co(TPT)H_2O^{2+}$ [d]	N_4O	10.8	(13)
$Co(DACODA)H_2O$ [e]	N_2O_3	9.3	(14)

[a] TPyMA=tris(3,5-dimethyl-1-pyrazolylmethyl)amine
[b] CR=Condensation product of 2,6-diacetylpyridine and dipropylenetriamine
[c] TMC=1,4,8,11-tetramethyl-1,4,8,11-tetraazacyclotetradecane
[d] TPT=N,N,N-tris(3-aminopropyl)amine
[e] DACODA=1,5-diazacyclo-octane-N,N'-diacetate

In Table 4 the simple complexes of cobalt(II) with a single water molecule coordinated are reported together with their pK_a. The complex aqua-tris (3,5-dimethyl-1-pyrazolyl)methyl amine cobalt(II) displays a pH dependence of the absorption spectra closely related to that of cobalt substituted carbonic anhydrase (10). From this table one learns that a water molecule coordinated to cobalt(II)(and hence to zinc(II)) may ionize in the pH range 9-10, or even less if the complex is tetrahedral and bipositive. In the protein such values can be substantially shifted according to the hydrophobic nature of the cavity in which the $M-OH_2$ moiety is placed.

THE ELECTRONIC SPECTRA OF COBALT(II) DERIVATIVES

The cobalt(II) derivatives of the enzymes listed in Table 1 are high spin (S=3/2) with well shaped absorption spectra in the region 5-25 $cm^{-1} \times 10^{-3}$. The position and intensity of the absorptions depend on the number and nature of the donor atoms. Therefore each spectrum should be carefully analyzed by itself in order to figure out the coordination geometry. We propose here a pragmatical criterion which cannot completely be justified on the basis of ligand field theory but has proved powerful since it has been found consistent with the structural indications obtained from a deeper spectral analysis of the systems here discussed. Furthermore the criterion provides interpretations consistent with those obtained by other techniques. The criterion is based on the maximum intensity of the absorption spectrum (15,16), as first proposed by Gray et al. (17), which may be directly related to the coordination number.

range of molar absorbance ($M^{-1}\ cm^{-1}$)	coordination number
$\varepsilon < 50$	six coordinated (pseudooctahedral)
$50 < \varepsilon < 200$	five coordinated (any geometry)
$\varepsilon > 300$	four coordinated (pseudotetrahedral)

At the borderline or in between ranges of values, equilibria between two species are proposed.

Figure 1 shows the spectra of CoBCAB (Bovine carbonic anhydrase B isoenzyme) and CoHCAB (Human, B isoenzyme) recorded at various pH values. The first consideration is due to the pH dependence clearly reflecting geometrical variations due to acid dissociations. The inset shows that at least two acid-base equilibria are capable of affecting the electronic spectra: one with larger effects presumably is the coordinated water (18), whereas the other might be a histidine hanging into the cavity (19). The pK_a of water can be estimated around 6 for CoBCAB and around 7 for CoHCAB.

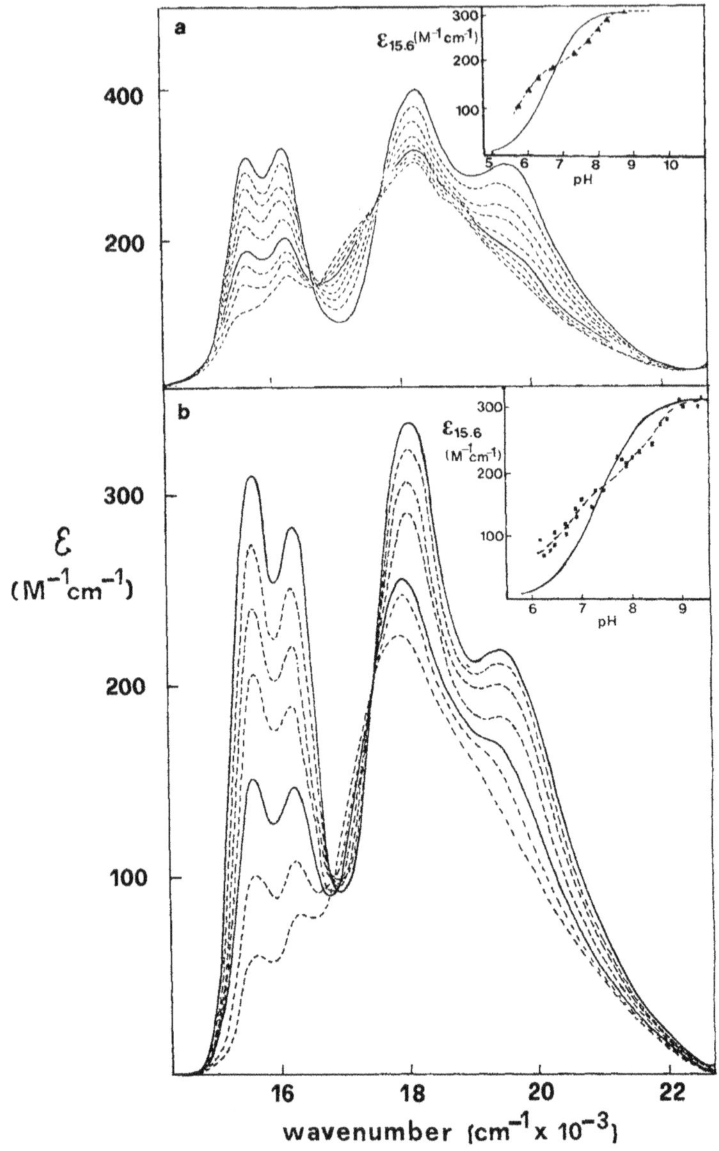

Figure 1. Electronic absorption spectra of CoBCAB (a) and CoHCAB (b) as a function of pH. a: unbuffered solutions at pH 5.8, 6.0, 6.3, 6.7, 7.3, 7.7, 7.9, 8.2, 8.8, in order of increasing $\varepsilon_{15.6}$; b: 10^{-2} M HEPES buffered solutions at pH 6.1, 6.6, 7.1, 7.8, 8.3, 8.6, 9.5, in order of increasing $\varepsilon_{15.6}$. The insets represent the intensity of the 15.6 $cm^{-1}x10^{-3}$ d-d transitions as a function of pH. The solid lines are calculated assuming a single pK_a of 6.6 (a) and of 7.35 (b), respectively (18)(from Ref. 15)

The electronic spectra of the fully acid and basic species are shown in Figure 2. According to the above criteria, the alkaline spectrum is assigned as pseudotetrahedral, the acid spectrum to an equilibrium between four- and five-coordinated species, the latter being predominant in the case of CoHCAB since the absorbance is 190 M^{-1} cm^{-1} (20,21).

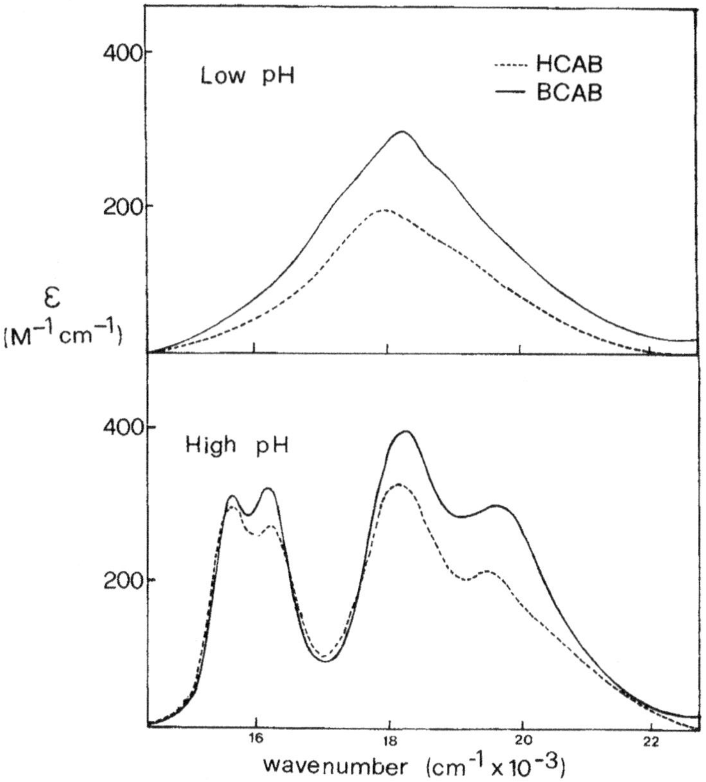

Figure 2. Electronic absorption spectra of cobalt(II) bovine carbonic anhydrase B and cobalt(II) human carbonic anhydrase B in the low and high pH limits.

The spectrum of CoCPA' (Figure 3a) is in our mind typical of five-coordination as first suggested by Gray et al (17). The implications involved in the coordinated carboxylate being bidentate (see Table 2) will be discussed later. The inhibitor Gly-L-Tyr leaves the coordination number unaltered (22) whereas N_3^- gives rise to tetracoordination (23). Also the spectrum of CoTLN first recorded by Vallee (24) is in this frame typically fivecoordinated even in presence of the inhibitor (16)

(Figure 3b). The spectra of these two enzymes are very slightly pH dependent with pK_a's of 9 and 7 respectively which correspond to the pK_a profiles of the activities.

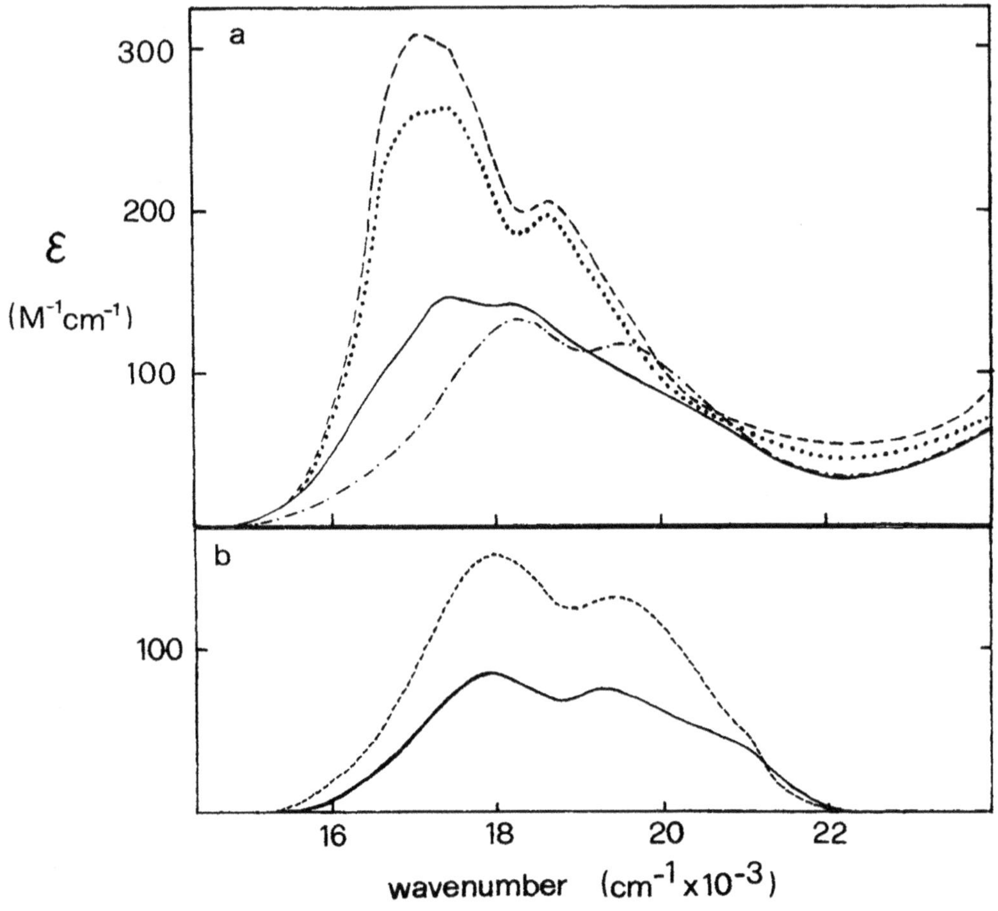

Figure 3. a: electronic absorption spectra of cobalt(II) carboxypeptidase A (——) and of its adducts with N_3^- (----), NCO^-(····), and Glycyl-L-Tyrosine (-·-·-·). b: electronic absorption spectra of cobalt(II) thermolysin (——) and of its adduct with β-phenylpropionyl-L-phenylalanine (----).

The alkaline phosphatase with a cobalt(II) ion in the catalytic site and two magnesium ions for each subunit (Co_2Mg_4AP) (25,26), and CoYAL (27), show again pH dependent spectra. The intensity of the latter (Figure 4), although proposed as pseudo-tetrahedral, is in our mind indicative of five coordination at neutral pH. At lower pH values the spectrum is reported to decrease in intensity.

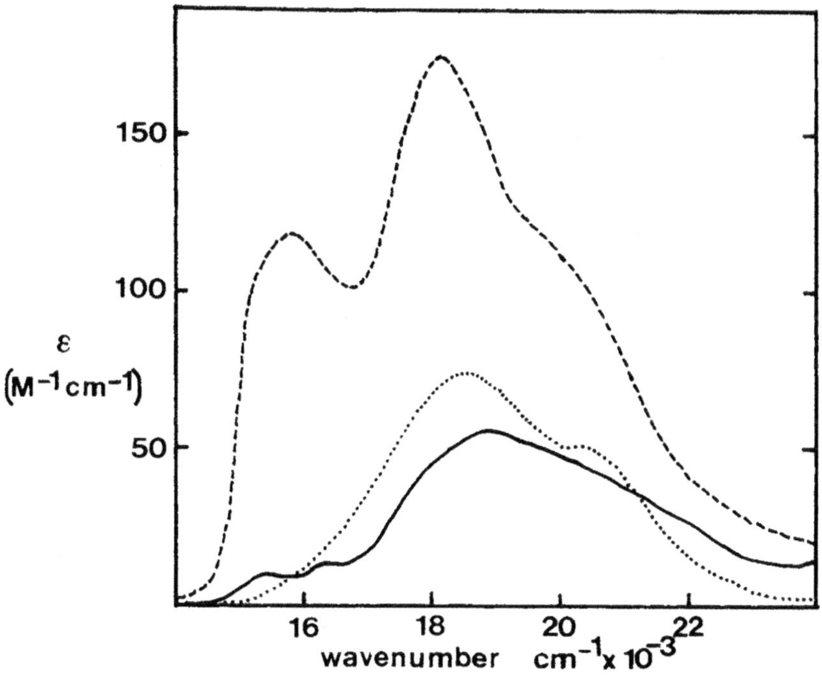

Figure 4. Electronic absorption spectra of cobalt(II)$_2$ magnesium(II)$_4$ alkaline phosphatase in the low (——) and high (----) pH limits, and of cobalt(II) yeast aldolase (····) at high pH.

The spectra of Co_2Mg_4AP display a pK_a of 7, corresponding to the pK_a of the activity profile, the high pH spectrum being indicative of five-coordination. The low pH spectrum shows small intensity which could be consistent with either a very distorted six-coordinated chromophore or a five-coordinate chromophore different from the high pH one. The spectra of CoLADH in both catalytic and structural sites (28) as well as CuCoSOD (29) are typical of tetrahedral chromophores (Figure 5).

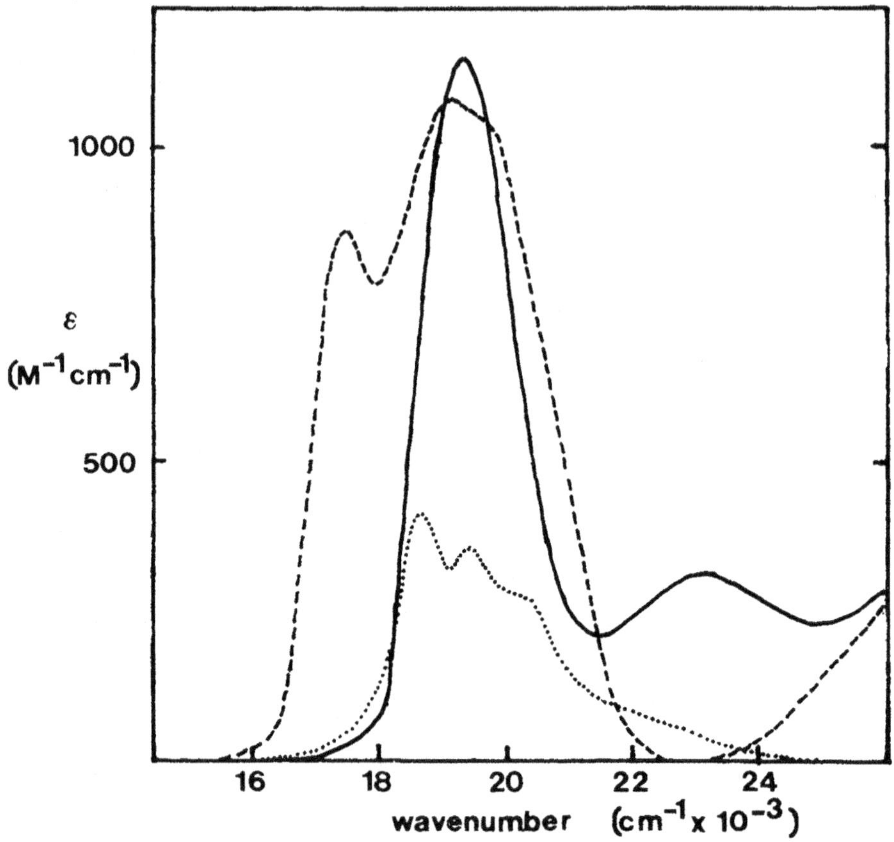

Figure 5. Electronic absorption spectra of cobalt(II) liver alcohol dehydrogenase with the cobalt(II) ion in the catalytic (——) and in the structural (---) site, and of cobalt(II)-copper(II) superoxide dismutase corrected for the absorption due to the copper(II) (···).

THE WATER ^1H NMR STUDIES

Once guessed the coordination number from the protein residues known to be coordinated at the metal, the number of water molecules or hydroxide ions should come by simple difference. However, independent data on the coordinated water have been quite precious. In principle, if the coordinated water exchanges with bulk water fast on the NMR time scale, it is possible to obtain information on the extent of water-metal ion interaction by monitoring the water nuclei. In practice such interaction is often detected when the metal ion is paramagnetic. The enhancement in solvent proton T_1^{-1} due to the coupling with the paramagnetic center may be expressed as (30)

$$T_{1p}^{-1} = K\ G\ f\ (\tau_c) \qquad (1)$$

where K is a constant containing nuclear and electronic constants and G is a geometric factor equal to $\Sigma_i \frac{n_i}{r_i^6}$, r_i being the distance of the ith proton from the metal, $f(\tau_c)$ is an expression of the type

$$\frac{7\ \tau_c}{1+\ \omega_s^2\ \tau_c^2}$$

where τ_c is the correlation time for the electron-nucleus dipolar interaction which corresponds to the electronic relaxation time for the cases here discussed.

The first point of our concern has been whether an equation of the type of equation (1) holds for the present cobalt(II) systems. Water ^1H T_1^{-1} NMR measurements have been performed at various magnetic fields between 0.01 and 300 MHz on a single system (31). Equation 1 has been found to be valid and to allow the determination of τ_c and G. The former parameter has been found to largely vary among the systems investigated; for the derivatives assigned as tetrahedral τ_c is of the order 10^{-10} - - 10^{-11} s whereas for those assigned as five-coordinated τ_c is 10^{-11} - -10^{-12} s (32). Although rather empirically τ_c has been used as a further criterion for assigning the coordination number. Even on theoretical grounds one can expect that five- and six-coordinated cobalt(II) complexes have several Kramers' doublets within 1000 cm^{-1} which allow efficient two-phonons electronic relaxation,whereas tetracoordinated complexes have only two Kramers' doublets separated by ≈10 cm^{-1}, the other excited levels being several thousand of cm^{-1} away. In this case the two phonons process is much slower than in the former case.

Also the G values are quite meaningful,although they are sums over all the protons interacting with the paramagnetic center each of them being weighted according to the sixth power of the distance. For example we have shown in model complexes that a coordinated water may provide the same G factor as a coordinated hydroxide, since the single proton of the hydroxide is closer to the paramagnetic center than the two protons of water (33). Although with many limitations, we can say that G values of $3\text{-}5\times10^{-15}pm^{-6}$ are indicative of a coordinated water or hydroxide, whereas G values smaller than 1×10^{-15} are indicative of outer - sphere interactions.

In Table 5 G and τ_c values are reported for several systems investigated. CoBCAB displays τ_c values typical of tetra-coordination at every pH and therefore the coordination is completed with a coordinated water or hydroxide; CoHCAB at low pH values shows τ_c values closer to that of five-coordinated species with a large G factor.

CoCPA and its derivative with β-phenyl-propionate show τ_c values in the range of five-coordinate species,in agreement with the electronic spectra (34). Since the X-ray structure shows that the coordinated

Table 5. G and τ_c values for some cobalt(II) derivatives of metallo-enzymes and their inhibitor adducts.

	pH = 6	pH = 9.5
CoHCAB	$\tau_c = 5x10^{-12}s$	$\tau_c = 1.1x10^{-11}s$
	$G = 5x10^{-15}pm^{-6}$	$G = 8x10^{-15}pm^{-6}$
CoBCAB	$\tau_c = 3.3x10^{-11}s$	$\tau_c = 3.2x10^{-11}s$
	$G = 3x10^{-15}pm^{-6}$	$G = 4x10^{-15}pm^{-6}$

CoBCAB + In	$\tau_c = 4\text{-}6x10^{-12}$	In = $Au(CN)_2^-$; NO_3^-;
	$G = 3\text{-}5x10^{-15}pm^{-6}$	NCS^-; CH_3COO^-
	$\tau_c = 3\text{-}4x10^{-11}s$	In = NCO^-; PTS
	$G = 1x10^{-15}pm^{-6}$	

	τ_c (s)	G (pm^{-6})
CoCPA	$3.1x10^{-12}$	$5.8x10^{-15}$
CoCPA + β-phenyl-propionate	$3.1x10^{-11}$	$2.9x10^{-15}$

glutamate residue is bidentate in the native enzyme (2), five coordination could arise from two histidines, one glutamate and one water molecule. However the two oxygens of glutamic acid are at 220 and 230 pm from the metal ion with an error of 20 pm. Therefore the carboxylate might be bound with a larger difference in the metal-oxygen distances as it may happen in inorganic models (35). Anyway we feel that the carbonyl group can be considered monodentate as far as the spectroscopic properties are concerned. Indeed, the inhibitor Gly-L-Tyr gives rise to spectra indicative of five coordinated species whereas the X-ray structure has allowed to propose that it behaves both as bidentate (50%) and monodentate (50%) (2); in the latter case the amino nitrogen interacts with another carboxylic group. On account of the large G value of CoCPA,which is reduced by a half in presence of the inhibitor β-phenyl-propionate, we propose (34) that the enzyme contains two water molecules in the coordination sphere and β-phenyl-propionate removes one of them.

^{1}H T_1^{-1} values for Co_2AP (36) and MnYAL (37) are consistent with water coordinated at the metal center. The presence of sulphur in the coordination sphere of CoLADH is thought to be responsible for not detecting coordinated water although there are enough evidences that at least one water molecule is coordinated (4).

Another piece of evidence for assigning the coordination comes from ^{1}H NMR spectra of the cobalt(II) substituted enzymes. The ^{1}H T_1^{-1} values of coordinated histidine protons are again related to τ_c and in a way also to the linewidths. Therefore five-coordinated chromophores give rise to sharp well resolved ^{1}H NMR signals whereas pseudotetrahedral chromophores give rise to unresolved spectra (39). Figure 6 shows the ^{1}H NMR spectra of CoBCAB; the spectra of CoCPA are reported elsewhere in the book.

THE IMPLICATIONS ON THE CATALYTIC MECHANISM

The general role of the metal ion in these enzymes is to act as a Lewis acid although the differences in behavior are very large. In carbonic anhydrase the zinc seems to provide a hydroxide capable of attacking carbon dioxide according to the scheme

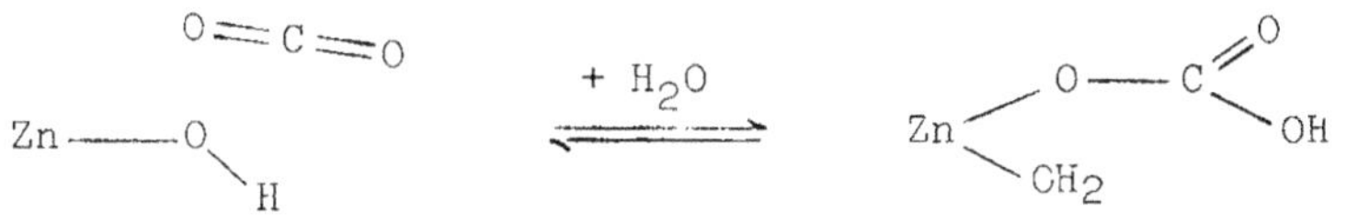

The pK_a of the coordinated water depends, among other factors, on the overall coordination number of the acidic species. The human B isoenzyme is largely five coordinated and consistently the pK_a is higher than in the case of the corresponding bovine B isoenzyme. Direct CO_2-metal interactions through oxygen has never been proved; ^{13}C studies on the copper(II) substituted derivatives have shown that CO_2 does not interact with the metal but does interact with the protein at a metal-carbon distance of ≈60-80 pm(40). Although CuCA is not active the information is still quite precious. Inhibitors work by formally substituting the coordinated OH^- in tetrahedral derivatives or binding at the site of HCO_3^- in the five-coordinated adducts.

A coordinated hydroxide has smaller nucleophilic properties than a free OH^- ion. Nevertheless one possible role of the enzyme is to provide

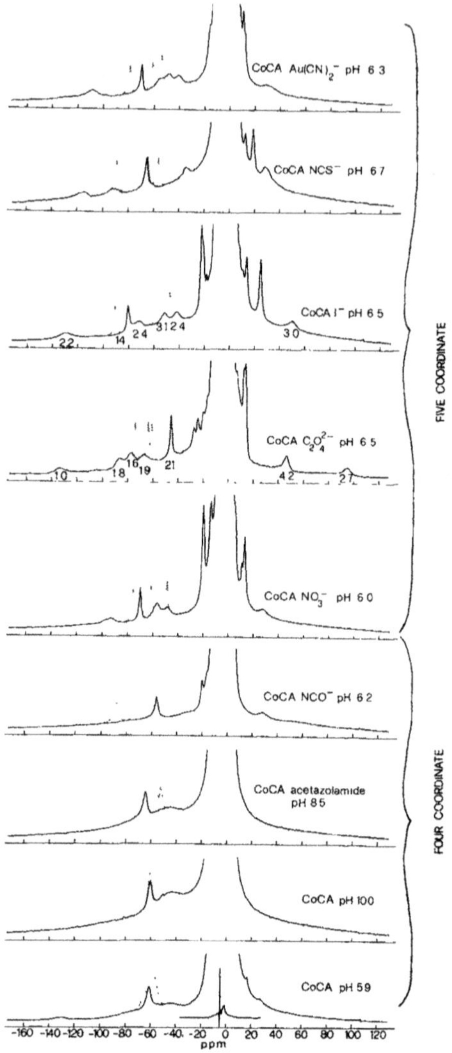

Figure 6. 60 MHz ^1H NMR spectra of cobalt(II) bovine carbonic anhydrase (CoCA), at pH 5.9 and 10.0, as well as of some of its inhibitor derivatives. Enzyme concentrations were (2-3)x10^{-3} M; inhibitor concentrations were about 10^{-2} M for NCS^-, NCO^- and acetazolamide and 10^{-1} M for the others. The dotted lines refer to the imino protons of the cobalt bound histidines which are observed in H_2O solutions. The T_1 values (ms) for some signals in the iodide and oxalate derivatives are also reported.

such an ion even at low pH values. Bicarbonate simply binds at the metal to undergo dehydration

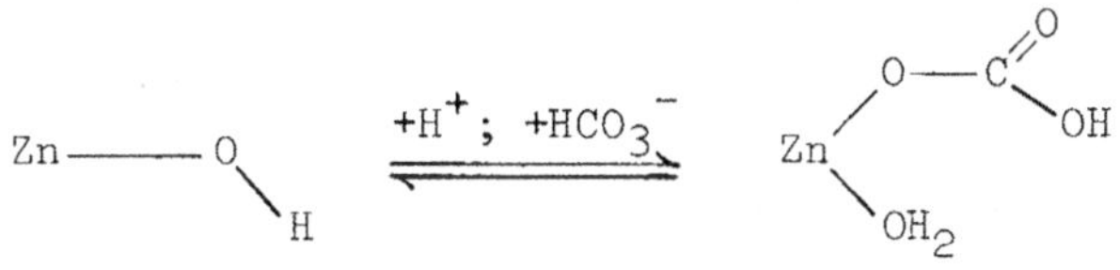

Further details on the mechanism will be given elsewhere in the book.

In CPA and TLN, substrates apparently bind the metal at the water site. Information come from X-ray studies (2,3) on substrate derivatives for those substrates which are not hydrolyzed owing to some particular properties. Assuming that the same happens when hydrolysis occurs is therefore somewhat arbitrary. We know that both nickel and cobalt CPA derivatives are active and that probably in both cases there is at least a water molecule coordinated. Indeed, the former derivative is typically six coordinated while its derivative with β-phenyl-propionate is five-coordinated(17). We propose that water is present in the coordination sphere together with the substrate

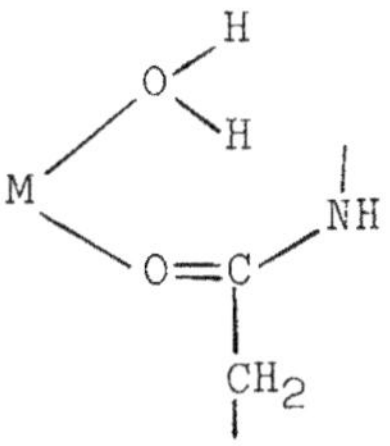

A Glu residue has been proposed to form an anhydride of the type

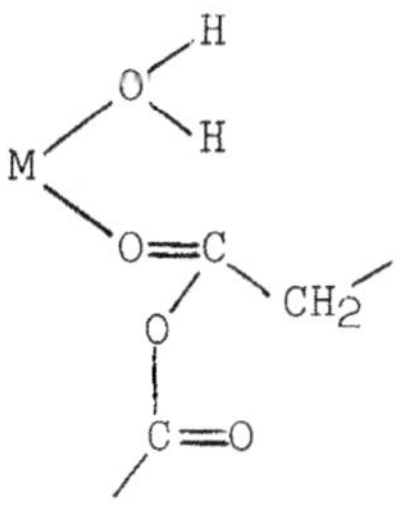

which could be hydrolyzed by the coordinated water (14).

In this frame the coordinated water may undergo deprotonation in order to allow OH^- to accomplish the nucleophilic attack. The presence of a tyrosinate at a certain stage of the reaction

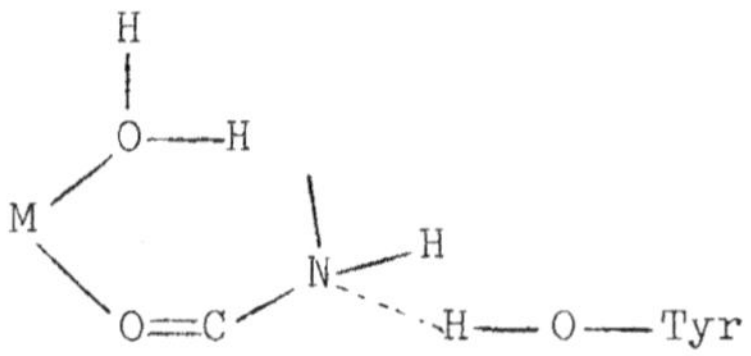

may help the water deprotonation. Therefore the role of the metal might be twofold: activation of the carbonyl carbon through substrate coordination and activation of coordinated water for the anhydride hydrolysis. The presence of water in the coordination sphere has been independently proposed by M.W. Makinen et al. for CoCPA in presence of esters as substrates (42).

The proposals do not imply that the native enzyme behaves in the same way but help in providing a general picture of what could happen along the catalytic pathway. Further comments on this point are given elsewhere in the book.

A third different behavior of zinc as a Lewis acid has been proposed for YAL. The zinc ion would increase the acidity of a coordinated histidine which on its turn would activate the carboxyl carbon (8,37). In this case the substrate would never interact with the metal at any stage.

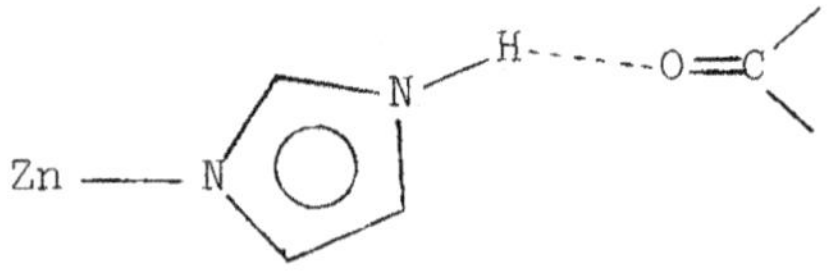

CoAP displays a pK_a around 7, in both activity and spectral properties (26); therefore the proposal of a coordinated hydroxide involved in the hydrolysis of phosphate monoesters seems reasonable although not proved (43). Phosphate then interacts with the metal ion. In LADH both the metal ion and a coordinated hydroxide play a role in oxidizing alcohols. Full details for the latter enzyme will be given elsewhere in the book.

REFERENCES

(1) Kannan, K.K.: 1980, in "Biophysics and Physiology of Carbon Dioxide", Bauer, C., Gros, G., Bartels, H., eds., Springer Verlag, Berlin Heidelberg, p. 184.
(2) Rees, D.C., Lewis, M., Honzatko, R.B., Lipscomb, W.N., and Hardman, K.D.: 1981, Proc. Natl. Acad. Sci. USA, 78, p. 3408.
(3) Colman, P.M., Jansonius, J.N., and Matthews, B.W.: 1972, J. Mol. Biol.,

16, p. 2506.
(4) Branden, C.-I., Jörnvall, H., Eklund, H., and Furugren, B.: 1975, in "The Enzymes", vol. 11A, Boyer, P.D., ed., 3rd edn., Academic Press, New York, p. 103.
(5) Sowadski, J.M., Foster, B.A., Wyckoff, H.W.: 1981, J. Mol. Biol., 150, p. 245.
(6) Otvos, J.D., and Armitage, I.M.: 1980, Biochemistry, 19, p. 4031.
(7) Otvos, J.D., and Browne, D.T.: 1980, Biochemistry, 19, p. 4011.
(8) Smith, G.M., and Mildvan, A.S.: 1981, Biochemistry, 20, p. 4340.
(9) Beem, K.M., Richardson, D.C., Rajagopalan, K.V.: 1977, Biochemistry, 16, p. 1930.
(10) Bertini, I., Canti, G., Luchinat, C., and Mani, F.: 1980, Inorg. Chim. Acta, 46, p. 291.
(11) Woolley, P.: Nature, 250, p. 677.
(12) Meier, P., Merbach, A., Burki, S., and Kaden, T.A.: 1977, J. Chem. Soc. Chem. Commun., p. 36.
(13) Dei, A., Paoletti, P., and Vacca, A.: 1968, Inorg. Chem., 7, p. 685.
(14) Billo, E.J.: 1975, Inorg. Nucl. Chem. Letters, 11, p. 491.
(15) Bertini, I., Luchinat, C., and Scozzafava, A.: 1982, Struct. Bonding, 48, p. 46.
(16) Bertini, I., and Luchinat, C.: in "Metal Ions in Biological Systems", vol. 15, H. Sigel, ed., Marcel Dekker, New York, in press.
(17) Rosenberg, R.C., Root, C.A., and Gray, H.B.: 1975, J. Am. Chem. Soc., 97, p. 21.
(18) Bertini, I., Luchinat, C., and Scozzafava, A.: 1980, Inorg. Chim. Acta, 46, p. 85.
(19) Simonsson, I., and Lindskog, S.: 1982, Eur. J. Biochem., 123, p. 29.
(20) to be published
(21) Brown, R.S., Curtis, N.J., and Huguet, J.: 1981, J. Am. Chem. Soc., 103, p. 6953.
(22) Latt, S.A., and Vallee, B.L.: 1971, Biochemistry, 10, p. 4263.
(23) see chapter VII in this book.
(24) Holmquist, B., and Vallee, B.L.: 1974, J. Biol. Chem., 249, p. 4601.
(25) Applebury, M.L., and Coleman, J.E.: 1969, J. Biol. Chem., 244, p. 709.
(26) Simpson, R.T., and Vallee, B.L.: 1968, Biochemistry, 7, p. 4343.
(27) Simpson, R.T, Kobes, R.D., Erbe, R.W., Rutter, W.J., and Valle, B.L.: 1971, Biochemistry, 10, p. 2466.
(28) Maret, W., Andersson, I., Dietrich, H., Schneider-Bernlöhr, H., Einarsson, R., and Zeppezauer, M.: 1979, Eur. J. Biochem., 98, p.501.
(29) Calabrese, L., Cocco, D., Morpurgo, L., Mondovì, B., and Rotilio, G.: 1976, Eur. J. Biochem., 64, p. 465.
(30) Bertini, I.: 1981, Comments Inorg. Chem., 1, p. 227 and refs. therein.
(31) Bertini, I., Brown, R.D., Koenig, S.H., and Luchinat, C.: submitted.
(32) Bertini, I., Canti, G., and Luchinat, C.: 1981, Inorg. Chim. Acta, 56, p. 99.
(33) Bertini, I., Canti G., Luchinat, C., and Messori, L.: Inorg. Chem. in press.
(34) Bertini, I., Canti, G., and Luchinat, C.: J. Am. Chem. Soc., in press
(35) Garner, C.D., and Mabbs, F.E.: 1976, J. Chem. Soc., Dalton Trans., p. 525; Holt, E.M., Holt, S.L., and Watson, K.J.: 1970, J. Am. Chem. Soc., 92, p. 2721.

(36) Zukin, R.S., and Hollis, D.P.: 1975, J. Biol. Chem., 250, p.835.
(37) Smith, G.M., Mildvan, A.S., and Harper, E.T.: 1980, Biochemistry, 19, p. 1248.
(38) Andersson, I., Maret, W.,Zeppezauer, M., Brown, R.D., and Koenig, S.H.: 1981, Biochemistry, 20, p. 3424.
(39) Bertini, I., Canti, G., Luchinat, C., and Mani, F.: 1981: J. Am. Chem. Soc., p. 7784.
(40) Bertini, I., Borghi, E., and Luchinat, C.: 1979, J. Am.Chem. Soc., 101, p. 7069; Bertini, I., Borghi, E., Canti, G., and Luchinat, C.: J. Inorg. Biochem., in press.
(41) Lipscomb, W.N.: 1980, Proc. Natl. Acad. Sci. USA, 77, p. 3875.
(42) Kuo, L.C., and Makinen, M.W.: 1982, J. Biol. Chem., 257, p. 24.
(43) Coleman, J.E., and Chlebowski, J.F.: 1979, in "Advances in Inorganic Biochemistry", vol. I, Eichhorn, G.L., and Marzilli, L.G., eds., Elsevier North Holland, New York, p. 2 ff..

COORDINATED SOLVENT MOLECULES IN METALLOENZYMES AND PROTEINS STUDIED USING NMRD

Seymour H. Koenig and Rodney D. Brown, III

IBM Thomas J. Watson Research Center, Yorktown Heights, New York 10598

ABSTRACT

In the early investigations of nuclear magnetic resonance in liquids, paramagnetic ions were added to regulate the magnetic relaxation rates of solvent protons and to study the ion-solvent complexes formed. The theory of relaxation that was developed as experimental techniques improved was very successful for describing the observed relaxation. However, the same theory applied to analogous studies of solutions in which these paramagnetic ions are complexed with protein gave ambiguous results. With the advent of NMRD data (i.e., relaxation rate as a function of magnetic field), it became increasingly clear that the theory adequate for aquoions was inadequate for ion-protein complexes. We trace this history, in part using data now readily attainable with greatly improved instrumentation, and show that the major reason for the inadequacy relates to the need for new models for solvent-protein interactions. The phenomena are far richer than originally anticipated.

INTRODUCTION

Bloch *et al.* (1) were the first to show that the presence of (paramagnetic) Fe^{3+}-aquoions markedly increases the spin-lattice relaxation rate of solvent water protons. Shortly thereafter, in a classic paper, "Relaxation effects in nuclear magnetic resonance absorption," Bloembergen *et al.* (2) extended the measurements of relaxation rates of solvent protons to solutions containing a variety of paramagnetic ions. They also outlined a theory of relaxation ostensibly appropriate to their conditions. It was an "outer sphere" theory, however, in which it was assumed that relaxation was due mainly to diffusion of solvent molecules in the magnetic field produced by the solute paramagnetic ions. A theory of "inner sphere" relaxation, in which relaxation occurs predominantly in hydrated complexes of the solute molecules

I. Bertini, R. S. Drago, and C. Luchinat (eds.), The Coordination Chemistry of Metalloenzymes, 19–33.

with the liganded water in reasonably rapid exchange with solution, was subsequently developed by several investigators (3,4). Characteristic of these theories is that nuclear relaxation rates should depend implicitly on temperature, through both the diffusion constant and the ligand exchange rates, and explicitly on the magnitude of the static magnetic field. This dependence on magnetic field we have called Nuclear Magnetic Relaxation Dispersion or NMRD (5); it is the NMRD of protein solutions that we address in this report.

By the end of the 1950s a variety of measurements of the dependence on temperature (6-8) and magnetic field (9,10) of the relaxation behavior of solvent protons had demonstrated that the earlier theories of relaxation, augmented to include contact (also known as scalar and hyperfine) interactions (11), were indeed quite adequate to explain the data, which were becoming increasingly refined. In their early work, Bloembergen *et al.* (2) assumed that the fluctuations in the proton-ion interaction responsible for relaxation arose entirely from the diffusional motion of solvent and solute molecules, though they recognized that relaxation of the paramagnetic spin-moments would, when sufficiently rapid, contribute observably to the correlation time of these fluctuations. The latter occurs only rarely in solution for "good relaxers" (i.e., ions with S-state configurations). When it does, there arises the possibility of a magnetic field dependence of the relaxation rate of the paramagnetic moment, which in turn produces a field dependence of the correlation time. This can introduce an ambiguity in the analysis of the data (12) that can only be resolved by an independent measurement (or estimate) of the ligand exchange rate. Though rare in solutions of aquoions, this ambiguity is often encountered in solutions of protein complexes of these same ions (13), and has been discussed in an earlier NATO School (14).

During the late 1950s and early 1960s, relaxation measurements were reported of solvent proton relaxation in solutions of Fe^{3+}-containing proteins, including myoglobin, hemoglobin and their derivatives (15-19), and conalbumin (17). Wishnia (17) pointed out at the time that the distinction between inner sphere (i.e., direct coordination of solvent to the metal ion) and outer sphere contributions to the relaxation rates could not readily be made on the basis of the data then available, most of which had been taken at a single value of magnetic field. He suggested that measurements of the magnetic field-dependence of the relaxation rates might clarify the analysis of the data. Such measurements were not to come until the end of the decade, when NMRD data for a solution of paramagnetic protein were reported for the first time (20) on native Fe^{3+}-transferrin, a blood plasma protein responsible for iron transport.

In the interim, Scheler (19) reported relaxation data for solutions of Fe^{3+}-hemoglobin and its derivatives. The remarkable result was that the replacement of a sixth-ligand H_2O of the Fe^{3+} ions of the hemes by F^-, a substitution that does not alter the paramagnetic properties of the heme groups, increased the paramagnetic contribution to the solvent proton relaxation rates by a factor of seven; i.e., the removal of solvent from the inner hydration sphere of the Fe^{3+}-ions produced an outer sphere contribution much larger than the sum of the two before the substitution. This result, since studied in greater detail by NMRD (21), should have been devastating to the growing view at the time that, the greater the number of coordinated waters and the more closely they approach the paramagnetic ions, the greater the proton relaxation rates. In particular, data were beginning to appear that showed the effect of Mn^{2+}-ions, specifically bound in tertiary and quaternary complexes of diamagnetic proteins, their substrates, and activators, typically ATP (22, 23), on solvent proton relaxation rates. Despite the caveat implicit in the "striking behavior" ("abnorme Verhalten") of the results of Scheler (19), models of these higher order complexes were being proposed in which the number of H_2O molecules coordinated to the Mn^{2+}-ions was assumed directly proportional to the relaxation rate (more strictly, the "enhancement" or the change in rate related to an appropriate standard) of solvent protons.

For some time now we have been improving the instrumentation needed for NMRD measurements: automating the apparatus; extending the accessible range of fields (now 0.01 to 55 MHz for protons); improving the temperature regulation (now $\pm$ 0.1°C); and increasing the speed and precision with which solvent proton (or deuterium) relaxation rates can be measured (now with an absolute uncertainty less than $\pm$1%, limited by sample conditions, in less than one minute). The latest results confirm our beliefs, which have been accruing since the first NMRD measurements, that the theory that had been so successful in explaining NMRD data for solutions of paramagnetic aquoions is inadequate in a variety of ways when these ions are complexed with protein. The theory is inadequate not in quantitative ways so much as in its qualitative aspects; it is not predictive, in that the variety of phenomena observed is far richer for metalloprotein solutions than for solutions of the analogous aquoions, and new interaction models are needed. Indeed, we now know that the situation for fluoromethemoglobin (which we will call the "fluoromet effect") is by no means unique; in fact, it appears quite general that replacement of an H_2O molecule liganded directly to a complexed paramagnetic ion by an anion (including a slowly exchanging OH^-) can often increase the solvent proton relaxation rates drastically. This is quite contrary to the view that the closer the approach of solvent protons, the greater their relaxation rates (once it is inferred from the temperature dependence of the relaxation data that chemical exchange is not rate-limiting).

What we know now is that there is a wealth of types of NMRD spectra, and that each protein system investigated so far requires a different model, or a different set of underlying considerations, before a successful comparison of data and theory can be made. We also know, with some certainty, that it is almost impossible to determine hydration numbers using solvent proton relaxation measurements, including the NMRD spectra. But the richness of the phenomena that are appearing more than compensates for this. What follows is a survey of data for a variety of metalloprotein systems, with emphasis on enzymes, along with a qualitative description of the biochemical inferences that can be made.

THE "FLUOROMET" MECHANISM

Figure 1 shows NMRD data (21) for two samples: one a solution of Fe^{3+}-hemoglobin, at a pH at which the solvent-donated ligand of each Fe^{3+}-ion is (known to be) water and not OH^-; and the other with this H_2O replaced by F^-. The relaxation rate of the former, the lesser rate, is small because the bound water is known to be in rather slow exchange. This contribution is predominately outer sphere, and because the proton charge of the bound water faces away from the Fe^{3+}-ions, the distance of closest approach of the solvent protons to these paramagnetic ions is maximized (21).

Addition of F^- displaces the bound H_2O, but allows hydrogen bonding of a proton of an outer sphere H_2O to the F^-, resulting in an Fe^{3+}-proton separation not much greater than that expected if the water were bound directly to the Fe^{3+} instead of the F^-. This hydrogen-bonded water is in rapid exchange; this geometry gives an NMRD spectra that is in quantitative agreement with relaxation theory (21). The interesting point to note is that this mechanism is not unique to F^-; any anion that will form hydrogen bonds should behave the same way, including OH^-. Our current opinion is that this fluoromet mechanism is important in understanding the NMRD spectra of transferrin.

Fe^{3+}-TRANSFERRIN

The first reported NMRD data (20) were for solutions of (native) Fe^{3+}-transferrin (Fe^{3+}-TFN), a plasma protein, molecular weight ~ 80,000, with two similar, noninteracting metal-binding sites per molecule (24). The NMRD spectra, to a maximum field of 6 MHz, were reported as a function of temperature and pH. The theory for outer sphere relaxation (25-27) was analyzed in some detail and shown to contribute little in this case, and the relaxation rate expected for both acid- and base-catalyzed proton exchange was derived as a function of pH. The conclusions were that, within the spirit of these mechanisms and the existing theory of inner sphere relaxation, protons on three inequivalent classes of water molecules per protein (not OH^- ions) contributed to the observed relaxation.

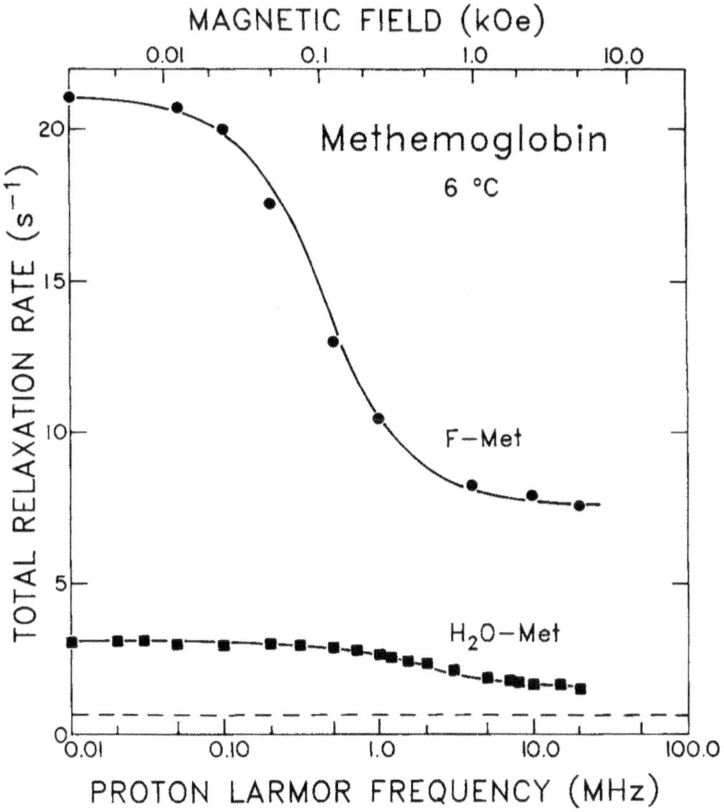

Figure 1. Total NMRD spectra for a solution of fluoro-methemoglobin (●), and a comparable sample of methemoglobin (■) at 6°C. The fluoro-methemoglobin is 1.0 mM protein in 0.05 M bis-Tris, 0.1 M NaCl, 0.1 M NaF buffer, pH 6.55. The methemoglobin sample, except for the NaF, is comparable. The relaxation contribution of the protein-free buffer is indicated by the dashed horizontal line. The solid line through the fluoro-methemoglobin data results from a least-squares comparison of the data with the usual theory of inner sphere relaxation; the results, assuming one exchanging proton, are 3.7 Å for the proton-Fe^{3+} separation (without corrections for ligand field splitting), $\tau_S = 6.0 \times 10^{-10}$ s, and an upper limit of about 5×10^{-8} s for the resident lifetime of this proton. From reference (21).

More recent NMRD data, to 10-fold higher fields, and with differential population of the two metal-binding sites (28), show that both sites contribute identically to the NMRD spectra, the form of which is more complex than the earlier theory could explain, or that present theory can handle without extensive computations (13). Qualitatively, however, one need consider no more than two inequivalent solvent binding sites, the second (as earlier) having a high activation energy and only becoming important above 30°C.

The ligands of the Fe^{3+}-ions in transferrin are believed known; there are three tyrosyl and one histidyl ligands from the protein, and one HCO_3^- from solution. The sixth, an H_2O if the simpler view of relaxation is taken, has been argued to be an OH^- on the basis of differential UV spectroscopy and proton-release data (29). Given our present understanding, this view is entirely consistent with the NMRD spectra if we invoke the fluoromet mechanism.

Cu^{2+}- AND VO^{2+}-TRANSFERRIN

NMRD spectra for Cu^{2+}- and VO^{2+}-transferrin are compared in Figure 2. The electronic configuration of Cu^{2+} has one d-hole, that of VO^{2+} one d-electron, so that their electronic properties should be (and are) similar (30, 31). Thus, naively, the NMRD spectra should also be similar, as indeed they are. However, the sixth ligand position of the vanadium is occupied by oxygen, so that there can be no inner sphere contribution to the NMRD spectra (unless the coordination sphere is expanded beyond six). Presumably the fluoromet mechanism is involved here as well. To quantitate this, however, is difficult because the theory for relaxation by Cu^{2+}, and by extension VO^{2+}, is complex (32); nonetheless, as best one can tell, the results for VO^{2+}-TFN are consistent in all aspects with the fluoromet mechanism. But what about Cu^{2+}-TFN?

By analogy with the VO^{2+}-TFN interpretation, as well as that for Fe^{3+}-TFN, one would argue that OH^- is also the sixth ligand in Cu^{2+}-TFN. However, there are chemical reasons for believing otherwise, that an apical H_2O with a fairly long Cu-O bond is the sixth ligand (28). The data are also consistent with this view. The point is that relaxation in the VO^{2+}-TFN case does not arise from direct coordination of water to the metal-ion; rather the water in the second coordination shell can have a geometry favorable for relaxation because of the possibility of hydrogen bond formation. The Cu^{2+}-TFN may be analogous, or may be predominately inner sphere. The bonding in the two cases is quite different, and is the basis for the distinction (perhaps arbitrary) between inner and outer sphere effects; the geometry of at least one proton (and perhaps two, since two waters can H-bond to the VO^{2+}) is similar for both transferrin derivatives.

Co^{2+}-CARBONIC ANHYDRASE

Co^{2+}-substituted carbonic anhydrase (Co^{2+}-CA) is the first enzyme for which NMRD data were obtained (33), the interest at the time being to see if there was a correlation between the known pH-dependence of the enzymatic activity and the NMRD spectra. A correlation was indeed found; relaxation

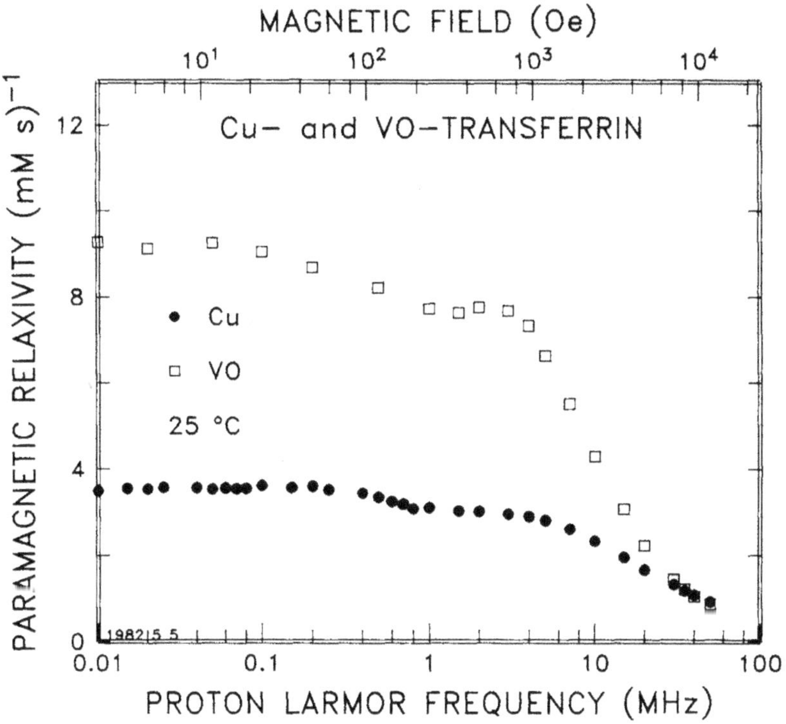

Figure 2. Paramagnetic component of the NMRD spectra for solutions of Cu^{2+}- and VO^{2+}-transferrin, (●) and (□). The protein concentrations are 0.94 and 0.91 mM , and the metal-ion concentrations 1.68 and 1.64 mM, respectively. The data are for 25°C. The buffer is 50 mM Tris-HCl, 30 mM $NaHCO_3$, at pH 8.

rates were large at high-pH where the activity for hydration of CO_2 is maximum, and low at low pH where the hydration activity vanishes. It has taken a decade to explain this correlation (34), the insight resulting from advances in understanding of the coordination chemistry of tetrahedral Co^{2+}-ions during this period (35).

A problem arose with the predictions of a popular explanation of the pH-dependent hydration activity of carbonic anhydrase; that it was linked to the ionization of a solvent-donated H_2O ligand of the tetracoordinate Co^{2+}-ion of the enzyme (Zn^{2+} in the native enzyme) to OH^-, with increasing pH. The expected pH-dependence of the solvent relaxation rate, in terms of the mechanisms known at the time, would be opposite from that observed. Water, present at low pH, would be expected to exchange rapidly, whereas it had already been argued (5) that proton exchange resulting from either acid or base catalysis is slow, too slow (in the pH range considered) to explain, for example, the high-pH NMRD data of Co^{2+}-CA. The fluoromet mechanism, only recently postulated, also does not resolve the issue. Though it is a mechanism that can explain relatively high relaxation rates in the presence of OH^- ligands (by allowing close solvent proton access to the metal-ion), the proton and, particularly, the oxygen of the OH^- ligand do not exchange with solvent via this mechanism. But the evidence is very strong that rapid relaxation in CA is associated with the rapid ligand exchange demanded by the observed enzymatic activity, particularly that relating to the redistribution of isotopes between the CO_2- HCO_3^- system and solvent (36, 37). It is unlikely, then, that the fluoromet mechanism is the dominant relaxation mechanism at high-pH. Moreover, an explanation for the low relaxation rate at low-pH is still wanting since the Co^{2+}-ions would have H_2O ligands.

As an aside, rapid exchange of a proton from the active site of carbonic anhydrase had been a puzzle arising from the enzymatic properties as well (38). The rate of turnover of CO_2 to form HCO_3^- and H^+ was observed to be more rapid than the rate at which diffusion could remove the protons produced without causing destructive pH-gradients in the vicinity of the protein. Or conversely, as the problem is generally presented, the turnover rate of protons in the dehydration of HCO_3^- is far more rapid than they can be supplied by diffusion. This problem was resolved by the realization that protons can be supplied by buffer (39, 40), which constitutes a proton reservoir with (typically) 10^4 greater proton activity than water. It was implicit in the thinking of many that the presence of buffer could resolve the relaxation issue as well. But this is not true; relaxation effects are observed as a characteristic of *solvent* protons, as an average over all of them. Relaxed protons confined to buffer ions that do not exchange rapidly with solvent would not contribute to the observed NMRD spectra. But the same arguments that showed that it was difficult to transfer protons to and from solvent (38) also apply to the transfer of protons between buffer and solvent (41). Another mechanism must be found, and indeed another mechanism has recently been proposed that is compatible with the relaxation data and the ionization of a water ligand of the metal-ions of carbonic anhydrase (34).

It has been proposed that exchange of an OH^- ligand on (tetrahedral) Co^{2+}-CA occurs by (a), the formation of a pentacoordinate intermediate by addition of an H_2O from solvent; (b), the subsequent transfer of a proton from the H_2O to the OH^-; and (c), release of the initial OH^- as an H_2O (34). Though there is no precedent for the overall processes, there is sufficient precedent for each of the steps to make this high-pH picture appear reasonable. The situation at low pH is explained by the realization that this form is really a thermal mixture of two forms, tetra- and pentacoordinate, with one and two H_2O ligands respectively. The first relaxes well, the second poorly because the electronic spin-lattice relaxation rate is increased by pentacoordination; the result is a thermal average that does not vanish as does the enzymatic hydration activity (42, 43).

The foregoing is an example of what is at issue: theory successful in explaining NMRD of solutions of paramagnetic aquoions cannot be used directly to explain NMRD spectra of solutions of these same ions complexed to protein. The problem, for the most part, is that new mechanisms, new phenomena, must be considered, almost on a protein by protein basis.

ALCOHOL DEHYDROGENASE

Alcohol dehydrogenase from horse liver (LADH), like carbonic anhydrase, is an enzyme with a tetracoordinate Zn^{2+} at the active site that can be replaced by Co^{2+} with retention of enzymatic activity (44). A second Zn^{2+}, also replaceable by Co^{2+} and which appears to have only a structural function, is ~ 20 Å from the active site. The binding of substrate, coenzymes (oxidized and reduced nicotinamide adenine dinucleotide, NAD^+ and NADH), and the enzymatic activity all have a pH-dependence that has not as yet been satisfactorily explained. Moreover, in the absence of substrate and coenzyme, it is generally agreed that solvent supplies the forth ligand of the active site metal-ion, so that solvent relaxation studies of the Co^{2+}-substituted enzyme were indicated. Early results (45) were unfortunately taken before techniques were developed for selective replacement of Zn^{2+} at the two sites, so that some of the conclusions are questionable. More recent NMRD studies (44), however, were of solutions of Co^{2+}-LADH with the substitution demonstrably at the active site. An assiduous search failed to show any paramagnetic contribution to the solvent proton water relaxation, nor to the relaxation of methyl protons in a solution of 10% CH_3OD (substrate), 90% D_2O. These results were particularly surprising, based on expectations from measurements of water and methanol relaxation (46) in ostensibly analogous Co^{2+}-CA.

The puzzle was resolved when NMRD results for (inactive) Cu^{2+}-LADH were analyzed. The electronic relaxation time of the Cu^{2+}-ions, as inferred from the NMRD spectra, is about two orders of magnitude shorter than is typical for coordinated Cu^{2+}-ions, presumably due to the two thiolate Cu^{2+}-protein bonds and the greater spin-orbit interaction expected for the covalent ligand orbitals with the sulfur atoms as contrasted with, for example, nitrogens of the histidyl ligands of carbonic anhydrase. This short relaxation time becomes the correlation time for the Cu^{2+}-proton dipolar interaction, reducing the scale of the proton relaxation rates by the same two orders of magnitude. An analogous effect for Co^{2+}-LADH would explain the absence of a paramagnetic component of its NMRD spectra, even were the shortening of the electronic relaxation rate only 10-fold.

Thus for Co^{2+}-LADH, the absence of a paramagnetic contribution to the NMRD spectra results not from absence of an interaction, but from a short correlation time for the interaction that makes its relaxation contribution small. There is every reason to believe that water has access to Co^{2+}-ions at the active sites and that slow exchange is not a limiting factor, judging by results for Cu^{2+}-LADH (44) and Mn^{2+}-LADH (47). One could not have drawn this conclusion from the NMRD data for Co^{2+}-LADH alone, and thus we have another protein that must be considered as a special case.

Mn^{2+}-CARBOXYPEPTIDASE A

Native carboxypeptidase A (CPA) contains tetracoordinate Zn^{2+}-ions at the active site, as do carbonic anhydrase and alcohol dehydrogenase, but unlike these, replacement of Zn^{2+} by Mn^{2+} in CPA does not drastically reduce or eliminate enzymatic activity. Thus relaxation studies of Mn^{2+}-CPA were reported quite early (48, 49) and, indeed, this enzyme was the first Mn^{2+}-protein complex for which extensive NMRD spectra were reported (50), including the influence of inhibitors (51). It was clear from the early work that Mn^{2+} binds weakly to CPA, was at least pentacoordinate, that binding of F^- occurred without essential alteration of the solvent proton relaxation (49), and that the latter was eliminated upon binding of the inhibitor β-phenylpropionate (48).

The alteration of the NMRD spectra upon addition of either L- or D-phenylalanine (51), both competitive inhibitors of CPA, is quite remarkable, sufficiently so that their effects were remeasured with greater precision,

Figure 3 (28). Though the temperatures and pH values differ from the early measurements, the salient features remain the same: addition of D-phe roughly halves the relaxation rate at all fields, whereas addition of L-phe lowers the rates above about 15 MHz, and raises them below. (Consider the conflicting results were two investigators to compare results, one of whom worked only at 60 MHz, the other at 14 MHz, both common fields at one time.)

Though detailed analysis will be reserved for another publication, several interesting inferences can be made directly, once it is recalled that the

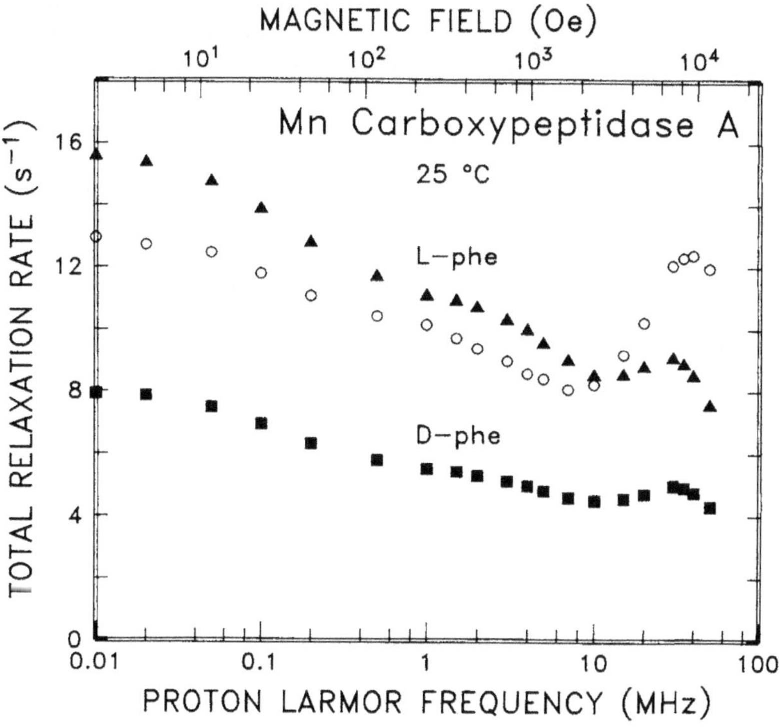

Figure 3. Total NMRD spectra for a solution of Mn^{2+}-carboxypeptidase A (○), and separate aliquots to which 50 mM L-phenylalanine (▲) and D-phenylalanine (■) were added. The data are for 25°C. The samples were 0.5 mM protein, 0.30 mM total Mn^{2+}, in 0.05 M Tris-HCl, 1 M KCl buffer at pH 6.7.

dispersion below ~ 0.6 MHz results entirely from the Mn^{2+}-ions free in solution and that the magnitude of the drop centered near 0.1 MHz provides a quantitative measure of the concentration of Mn^{2+}-aquoions (52). First, addition of 40 mM D-phe (which is ~ 10 times the dissociation constant (53)) does not alter the concentration of unbound Mn^{2+}-ions, but reduces the relaxation rate substantially, though not entirely. The effect is consistent with the view that D-phe displaces some, but not all, of the solvent-donated ligands of the bound Mn^{2+}-ions.

Second, addition of L-phe increases the apparent concentration of Mn^{2+}-aquoions almost two-fold; the NMRD contribution of the Mn^{2+}-CPA, normalized to [Mn^{2+}-CPA], before and after the addition of L-phe, is little altered at low fields, though reduced at the peak (as a little arithmetic will show). It appears as though L-phe decreases the binding of Mn^{2+}-ions, presumably because L-phe binds to the apoprotein better than to the Mn^{2+}-complex.

Once again, analysis of NMRD spectra of Mn^{2+}-CPA complexes cannot be made by taking over the theory used to explain relaxation in solutions of Mn^{2+}-aquoions. A model for the biochemical interactions must first be deduced, and the data then interpreted within this framework.

CONCLUSIONS

We have developed the instrumentation for NMRD spectroscopy to the point where the rapidity, accuracy, and ease of obtaining NMRD spectra make possible detailed investigations of the interaction of solvent with paramagnetic metalloproteins, as a function of temperature, pH, solvent composition, etc., in a relatively short time. What we are discovering, as illustrated in the foregoing, is that the theoretical description, so successful for explaining NMRD of solutions of aquoions, is quite inadequate for direct application to complexes of these ions with protein. The theory has been reexamined recently by a new approach (13, 54) designed to be more easily applied to metalloproteins than the usual theory. We do know that quantitative agreement between theory and observation is still elusive even in the most tractable case (14). But, for the present, this is not a problem; it is still premature to attempt quantitative comparisons of data and theory for protein systems. Currently, we are limited more by the variety and richness of the phenomena being uncovered. Each system appears to demand a unique approach, and the more data we take, the more we learn about both biochemistry and relaxation processes.

ACKNOWLEDGMENTS

The (unpublished) data, and associated sample preparation, relevant to Figures 2 and 3 are due to G. Canti, obtained during a stay at our laboratory sponsored by IBM Italy.

1. Bloch, F., Hansen, W. W., and Packard, P.: 1946, Phys. Rev. 70, pp. 474-485.
2. Bloembergen, N., Purcell, E. N., and Pound, R. V.: 1948, Phys. Rev. 73, pp. 679-712.
3. Solomon, I.: 1955, Phys. Rev. 99, pp. 559-565.
4. Kubo, R., and Tomita, J.: 1954, J. Phys. Soc. Jap. 9, pp. 888-919.
5. Koenig, S. H., and Schillinger, W. E.: 1969, J. Biol. Chem. 244, pp. 3283-3289.
6. Bernheim, R. A., Brown, T. H., Gutowsky, H. S., and Woessner, D. E.: 1959, J. Chem. Phys. 30, pp. 950-956.
7. Bloembergen, N., and Morgan, L. O.: 1961, J. Chem. Phys. 34, pp. 842-850.
8. Hausser, R., and Laukien, G.: 1959, Z. Physik 153, pp. 394-411.
9. Nolle, A. W., and Morgan, L. O.: 1957, J. Chem. Phys. 26, pp. 641-648.
10. Morgan, L. O., and Nolle, A. W.: 1959, J. Chem. Phys. 31, pp. 365-368.
11. Bloembergen, N.: 1957, J. Chem. Phys. 27, pp. 572-573.
12. Koenig, S. H., and Epstein, M.: 1975, J. Chem. Phys. 63, pp. 2279-2284.
13. Koenig, S. H.: 1978, J. Magn. Reson. 31, pp. 1-10.
14. Koenig, S. H., and Brown, R. D., III: 1980, *ESR and NMR of Paramagnetic Species in Biological and Related Systems*, Bertini, I., and Drago, R., Eds., Boston, D. Reidel Publishing Co., pp. 89-115.
15. Davidson, N., and Gold, R.: 1957, Biochim. Biophys. Acta 26, pp. 370-373.
16. Kon, H., and Davidson, N.: 1959, J. Mol. Biol. 1, pp. 190-191.
17. Wishnia, A.: 1960, J. Chem. Phys. 32, pp. 871-875.
18. Lumry, R., Matsumiya, H., Bovey, F. A., and Kowalsky, A.: 1961, J. Phys. Chem. 65, pp. 837-843.
19. Scheler, W.: 1963, Biochim. Biophys. Acta 66, pp. 424-433.
20. Koenig, S. H., and Schillinger, W. E.: 1969, J. Biol. Chem. 244, pp. 6520-6526.
21. Koenig, S. H., Brown, R. D., III, and Lindstrom, T. R.: 1981, Biophys. J. 34, pp. 397-408.

22. Cohn, M.: 1967, *Magnetic Resonance in Biological Systems*, Malmstrom, B. G., and Vanngard, T., Eds., Oxford, Pergamon Press, pp. 101-117.
23. Mildvan, A. S., and Cohn, M.: 1970, Advan. Enzymol. 33, pp. 1-70.
24. Aisen, P.: 1980, *Iron in Biochemistry and Medicine, II,* Jacobs, A., and Worwood, M., Eds., New York, Academic Press, pp. 87-129.
25. Pfeiffer, H.: 1961, Ann. Phys. (Leipzig) 8, pp. 1-8.
26. Pfeiffer, H.: 1962, Z. Naturforsch. 17a, pp. 279-287.
27. Pfeiffer, H.: 1963, Biochim. Biophys. Acta 66, pp. 434-439.
28. Canti, G.: 1981 (unpublished).
29. Pecoraro, V. L., Harris, W. R., Carrano, C. J., and Raymond, K. N.: 1981, Biochemistry 20, pp. 7033-7039.
30. Aisen, P., and Froncisz, W.: 1982, Biochim. Biophys. Acta 700, pp. 55-58.
31. Chasteen, N. D., White, L. K., and Campbell, R. F.: 1977, Biochemistry 16, pp. 363-368.
32. Koenig, S. H., and Brown, R. D.: 1973, Ann. N.Y. Acad. Sci. 222, pp. 752-763.
33. Fabry, M. E., Koenig, S. H., and Schillinger, W. E.: 1970, J. Biol. Chem. 245, pp. 4256-4262.
34. Koenig, S. H., Brown, R. D., III, Luchinat, C., and Bertini, I.: 1982, Biophysical J. (submitted).
35. Bertini, I., Canti, G., Luchinat, C., and Scozzafava, A.: 1978, J. Am. Chem. Soc. 100, pp. 4873-4877.
36. Silverman, D. N., and Tu, C. K.: 1976, J. Am. Chem. Soc. 98, pp. 978-984.
37. Koenig, S. H., and Brown, R. D., III: 1976, Biophysical J. 35, pp. 59-78.
38. De Voe, H., and Kistiakowsky, G. B.: 1961, J. Am. Chem. Soc. 83, pp. 274-280.
39. Khalifah, R. G.: 1973, Proc. Natl. Acad. Sci. U.S.A. 70, pp. 1986-1989.
40. Lindskog, S., and Coleman, J. E.: 1973, Proc. Natl. Acad. Sci. USA 70, pp. 2505-2508.
41. Koenig S. H., Brown, R. D., London, R. E., Needham T. E., and Matwiyoff, N. A.: 1974, Pure Appl. Chem. 40, pp. 103-113.
42. Khalifah, R. G.: 1971, J. Biol. Chem. 246, pp. 2561-2573.
43. Bertini, I., Canti, G., and Luchinat, C.: 1981, Inorganic Chimica Acta 56, pp. 99-107.
44. Andersson, I., Maret, W., Zeppezauer, M., Brown, R. D., III, and Koenig, S. H.: 1981, Biochemistry 20, pp. 3424-3432, and references therein.

45. Sloan, D., Young J. M., and Mildvan, A. S.: 1975, Biochemistry 14, pp. 1998-2008.
46. Jacob, G. S., Brown, R. D., III, and Koenig, S. H.: 1980, Biochemistry 19, pp. 3754-3765.
47. Andersson, I., Maret, W., Zeppezauer, M., Brown, R. D., III, and Koenig, S. H.: 1981, Biochemistry 20, pp. 3433-3438.
48. Navon, G., Shulman, R. G., Wyluda, B. J., and Yamane, T.: 1968, Proc. Natl. Acad. Sci. U.S.A. 60, pp. 86-91.
49. Navon, G., Shulman, R. G., Wyluda, B. J., and Yamane, T.: 1970, J. Mol. Biol. 51, pp. 15-30.
50. Koenig, S. J., Brown, R. D., and Studebaker, J.: 1971, Cold Spring Harbor Symp. Quant. Biol. 36, pp. 551-559.
51. Quiocho, F. A., Bethge, P. H., Lipscomb, W. N., Studebaker, J. F., Brown, R. D., and Koenig, S. H.: 1971, Cold Spring Harbor Symp. Quant. Biol. 36, pp. 561-567.
52. *cf.* Brown, R. D., III, Brewer, C. F., and Koenig, S. H.: 1977, Biochemistry 16, pp. 3883-3896 for description of the technique.
53. Studebaker, J., Brown, R. D., and Koenig, S. H.: 1970, Fourth International Conference on Magnetic Resonance in Biological Systems, Oxford (abstract).
54. Koenig, S. H.: 1982, J. Magn. Reson. 47, 441-453.

^{113}Cd NUCLEAR MAGNETIC RESONANCE STUDIES OF ZINC METALLOPROTEINS

J.L. Sudmeier and D.B. Green

Department of Chemistry
University of California
Riverside, California 92521 (U.S.A.)

1. INTRODUCTION

Since the first commercial multinuclear Fourier transform nmr spectrometers became available in about 1973, the use of multinuclear nmr for studying metalloproteins has now reached a degree of maturity. This chapter is an overview of ^{113}Cd nmr studies of zinc metalloproteins, in which ^{113}Cd ions substituted for zinc are observed directly by nmr. Practical requirements and limitations in regard to sample, nmr spectrometer, and relaxation mechanisms are discussed. Types of information which can be obtained include the types of ligand atoms, numbers of metal binding sites of various types, numbers and mode of binding of substrate molecules, and microscopic thermodynamic and kinetic processes occurring at or near the metal ions. Examples from the chemical literature are presented, with a distinct bias towards our own work.

The idea of replacing the zinc ions with isotopically enriched ^{113}Cd, and monitoring events in the active site of zinc metalloproteins captured the imagination of a number of researchers in the early 1970's. The large chemical shift range and favorable spin quantum number, $I = \frac{1}{2}$, along with similar chemical reactivity to zinc gave ^{113}Cd nmr special promise. Serious efforts to observe ^{113}Cd-metalloprotein nmr signals began in our labs and perhaps other labs as early as 1974.

The first successful spectra were reported in 1976 by Armitage et al. (1) involving alkaline phosphatase and various carbonic anhydrases. Reports from our group (2), Ellis' (3), Forsen's (10) and others ensued rapidly. In retrospect, the major difficulties in obtaining the first spectra probably arose as much from the condition of the samples as from the capabilities or use of the nmr spectrometers.

I. Bertini, R. S. Drago, and C. Luchinat (eds.), The Coordination Chemistry of Metalloenzymes, 35–47.

Table I. ^{113}Cd Chemical Shifts of Metalloproteins

Compound	Shift (ppm, approx)	Probable Ligands O	O^-	OH^-	N	N^-	S^-	Reference
1. dil. $CdSO_4$	0	6						
2. carbonic anhydrases	215-230	1			3			4
" "	275			1	3			"
" "	350-400				3	1		5
" "	375				3		1	this work
3. insulin	165	3			3			6
	200	2		1	3			"
	-36	2	4					"
4. concanavalin A-(S1)	40		5		1			7
-(S2)	-130		6					"
5. superoxide dismutase	310-330		1		3			"
6. liver alcohol dehydrogenase								
catalytic site	450-530	1			1		2	8
structural site	750						4	"
7. metallothionein	600-700						4	9
8. parvalbumin, troponin C, calmodulin	-90 to -120		6					10

2. CHEMICAL SHIFTS

Table I is a representative sample of the ^{113}Cd chemical shifts detected to date in metalloproteins. A good deal of excellent work on alkaline phosphatase (see ref. 11 and loc. cit.) was omitted because the exact distribution of nitrogen and oxygen ligands giving rise to the two major peaks (40-80 ppm and 140-170 ppm) is still uncertain. The important contributions of Forsen and coworkers (10) are mentioned only briefly, since their work deals with Ca^{2+} and Mg^{2+} rather than Zn^{2+}-binding proteins, which is our main emphasis here.

We have attempted to classify the proteins according to the identity and ionic charge of the ligand donor atoms, insofar as they are known. The symbol "O" refers to neutral oxygen from water, alcohol OH groups, etc., and "O^-" refers to negatively charged oxygen from carboxylate groups, tyrosinate groups, etc. Hydroxyl groups are listed separately because each hydroxyl produces ∿40 ppm downfield shifts instead of the upfield shifts produced by the "O^-" groups referred to above. The symbol "N" refers to imidazole nitrogen and "N^-" to ionized benzenesulfonamide nitrogen. "S^-" refers to ionized cysteine sulfhydryl or bisulfide sulfur. When the exact distribution of various ligand atoms is unknown or unspecified, we have entered the total numbers of ligands between the relevant columns.

What emerges is a fairly satisfying correlation of ^{113}Cd chemical shifts with varying distributions of ligand atoms. Empirical correlations on this limited data base are probably best avoided, particularly when the metal coordination numbers vary from 4 to 6, and there is little theoretical grounds for substituent additivity. Nevertheless, we cannot resist the observation that when $^{113}Cd^{2+}$ is surrounded by all S^- groups, the chemical shift is 750 ppm vs. dil. $CdSO_4$. Similar values for N^-, N, and O^- are about 500, 300, and -200 ppm respectively. When various combinations of ligands are present, weighted averaging reproduces the major trends in Table I, although superoxide dismutase doesn't fit very well. In our opinion the chemical shift trends are good enough to have predictive value, and there is one example in the literature (6) which is discussed below.

3. SENSITIVITY

Even though a good deal has been published since 1976, obtaining ^{113}Cd nmr spectra of metalloproteins is still not a trivial or routine procedure, because of the low sensitivity inherent in the method. Months have been spent searching for these weak signals in vain. Here we will discuss some of the practical considerations in obtaining a signal.

3.1. Samples

Several hundred milligrams of highly purified protein is required. A protein concentration of 100 mg/ml (10%) is desirable. The cost of the required few milligrams of highly enriched (∿96%) ^{113}Cd isotope is

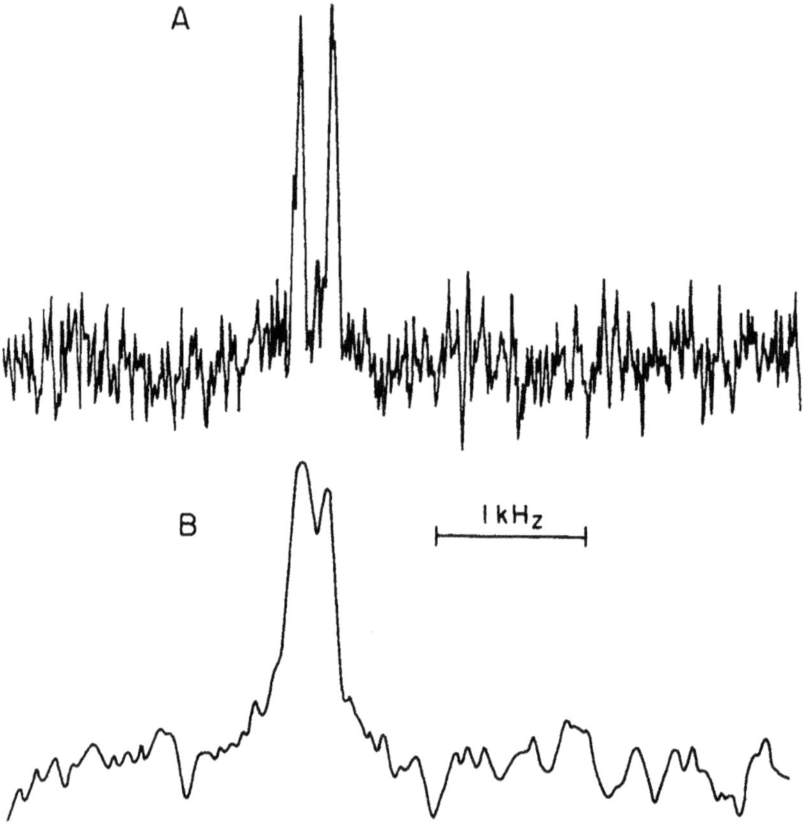

Figure 1. ^{113}Cd nmr spectrum of ∿4 mM bovine carbonic anhydrase plus 1 equivalent ∿95% ^{15}N-enriched benzenesulfonamide, pH ∿ 8. A) Bruker WH90 (2.1T magnet), 15 mm sample containing ∿5 ml sample. Acquisition time 0.27s, flip angle 30°, line broadening 10 Hz. Total acquisition ∿13 hours. B) Nicolet 300 Widebore (7.1T magnet), 200 mm sideways probe containing ∿7 ml sample. Acquisition time 1 sec, flip angle 40°, line broadening 70 Hz. Total acquisition time 1 hr.

usually negligible, perhaps $10 per spectrum. Chemical exchange broadening may arise from unsuspected contaminants such as adventicious bicarbonate (4). The presence of isozymes must similarly be avoided. In general, the presence of a strong inhibitor will aid in the detection of an initial signal.

3.2. Nmr spectrometers

A Fourier transform instrument with every available sensitivity advantage, including excellent probes designed to accommodate large samples (e.g., 4-5 ml), good filters, "traps", quarter-wave cables, and preamplifiers, quadrature phase detection, and good lineshape is required. A signal-to-noise of 80:1 on 1.0M $CdSO_4$ (single pulse) is acceptable, but better than 300:1 is currently attainable on some commercial instruments. Freedom from systematic noise, i.e., "glitches", spurious frequencies, intermodulation frequencies, harmonics, cable "crosstalk", etc. is vitally important. Ability to tune and match the probe with sample installed in the magnet is vital. If a superconducting magnet is used, use of the sideways probe design can give a sensitivity advantage of between 2 and 3 (12), (or time savings of 4 to 9). Linewidths less than $\sim$20 Hz have seldom been observed, so sample spinning is usually unnecessary.

For a single pulse, sensitivity increases as the ratio of magnetic field strength to the 7/4 power (12). But what magnetic field in real experiments give the best performance depends upon each sample and upon how performance is defined. Figure 1 shows the ^{113}Cd nmr spectra of ^{113}Cd-bovine carbonic anhydrase at two different magnetic field strengths and illustrates some of the advantages and disadvantages of the two fields. The enzyme is inhibited by ^{15}N-enriched benzenesulfonamide, which accounts for the spin coupling of 190 Hz due to direct ^{113}Cd-^{15}N bond formation (5). Spectrum 1A was obtained in a conventional iron magnet at 2.1T, and the latter in a sideways probe in a 7.1T widebore superconducting magnet. An increase in field from 2.1T to 7.1T translates to a S/N increase by a factor of 8, providing sideways probes are used at 7.1T. For about the same total amount of sample, spectrum 1A required 13 hours and 1B required only 1 hour. Different sensitivity enhancement parameters were employed, and in neither case were the acquisition conditions well optimized. However, we can say that sensitivity increased about an order of magnitude at the higher field strength. The linewidth in 1B is $\sim$120 Hz, about ten times greater than in 1A. Such degradation in linewidth may be unacceptable, despite the large sensitivity increase, which leads us next to a consideration of relaxation mechanisms.

3.3. Relaxation mechanisms

In metalloproteins such as superoxide dismutase and alkaline phosphatase it has been found that ^{113}Cd relaxation consists of comparable amounts of dipolar relaxation from nearby protons and chemical shift anisotropy (CSA) (7,11). Calculations using the Bloembergen, Purcell, and Pound (BPP) theory (13) for isotropic reorientation with lifetime τ_c indicate negligible scalar relaxation due to ligand ^{14}N atoms for protein molecules, principally because ^{14}N lifetimes are too short.

Figures 2 and 3 show the results of similar calculations using BPP theory for T_1, linewidth, and nuclear Overhauser enhancement factor (NOEF) for ^{113}Cd. The protons for the dipolar calculations are the five imidazole protons at an assumed distance of 3.6Å and two water protons at 3.1Å that one would find in a ^{113}Cd-carbonic anhydrase having a single coordinated water molecule. The CSA relaxation is calculated using $\Delta\sigma$ values in the chemical shift tensor (for an assumed axial symmetry) of 100 and 400 ppm.

If dipolar relaxation dominates, then sensitivity at the higher field strength will surely be lost for carbonic anhydrase, assuming $\tau_c = 10^{-8}$ sec due to longer T_1 values (Figure 2, upper). On the other hand, if $\Delta\sigma > 300$ ppm, then CSA will dominate, and T will be smaller at the higher field strength (though not greatly field dependent for $\tau_c = 10^{-8}$ and longer). Sensitivity will, therefore, be enhanced at the expense of greater linewidth. In many cases, the use of proton decoupling has led to the disappearance of the ^{113}Cd signal, even for quite large proteins (>80,000 daltons), indicating a breakdown in the assumption of a rigid isotropically tumbling molecule (11).

It seems clear for the sample in Figure 1 that CSA is the dominant mechanism, exhibiting the characteristic dependence of linewidth on H_o field strength to the second power. With four nitrogen ligands, although one is an ionized sulfonamide $-NH^{\ominus}$, $\Delta\sigma$ may be as large as 1000 ppm. Chemical exchange broadening also depends on H_o field strength to the second power (14), although is not likely to be a factor in the present case, where a tightly bound inhibitor is present.

The evidence is that CSA has played a major role in the relaxation mechanisms of virtually all ^{113}Cd nmr studies of metalloproteins to date. All things considered, we feel that low-field iron magnet nmr systems should be retained for multinuclear studies of nuclei having large chemical shift ranges, including the heavy metals and ^{15}N.

4. INFORMATION CONTENT

What types of information can one expect to gain from the method?

4.1. Types of ligand atoms

As evidenced by Table I and associated discussion, within certain limits, chemical shifts can be used to draw conclusions regarding the identity of ligand atoms.

4.2. Number of ligands

Through isotopic labelling of ligand donor atoms and observation of any spin-spin coupling to the ^{113}Cd, it is possible to count the number

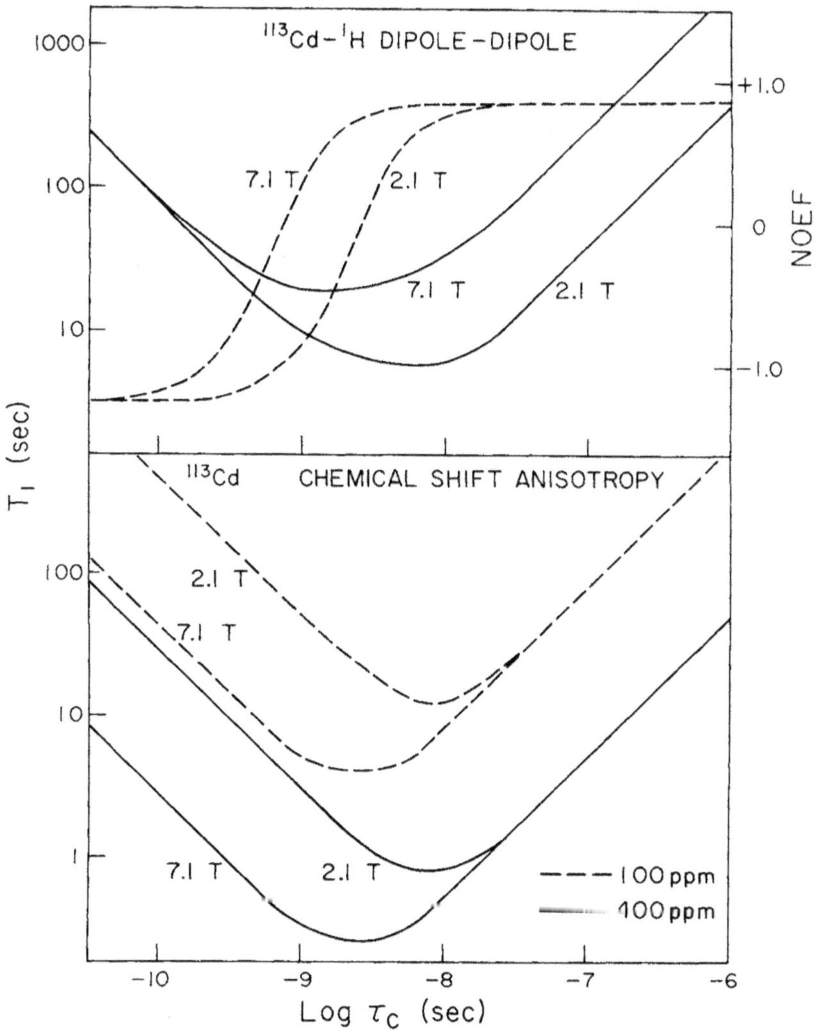

Figure 2. Calculated ^{113}Cd T_1 values and nuclear Overhauser enhancement factors (NOEF) vs. molecular correlation time, τ_c, for rigid isotropic reorientation. Upper: ^{113}Cd dipolar relaxation by 5 imidazole protons at 3.6Å and 2 protons at 3.1Å, simulating carbonic anhydrase with a single coordinated water. The dashed line accompanies the NOEF scale on the right. Lower: ^{113}Cd relaxation by chemical shift anisotropy for various values of $\Delta\sigma$, assuming axially symmetric g tensors.

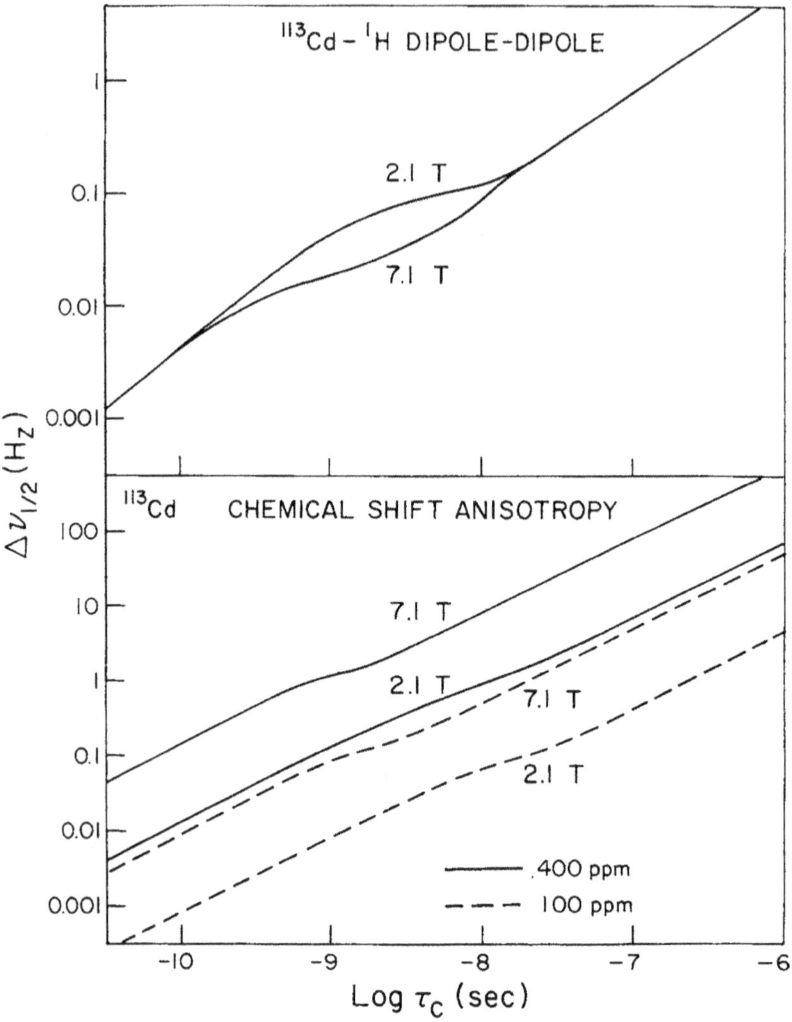

Figure 3. Calculated ^{113}Cd linewidths vs. molecular correlation time, τ_c, for rigid isotropic reorientation. Upper: ^{113}Cd dipolar relaxation by 5 imidazole protons at 3.6Å and 2 protons at 3.1Å, simulating carbonic anhydrase with a single coordinated water. Lower: ^{113}Cd relaxation by chemical shift anisotropy for various values of $\Delta\sigma$, assuming axially symmetric g tensors.

of ligands bound to the metal. For example, in ^{113}Cd carbonic anhydrases, it was shown that only a single $^{13}CN^-$ binds to the metal, based on the observed splitting pattern (a doublet), even when large excess of $^{13}CN^-$ was present (4). Using protein bioenrichment with ^{15}N-labeled histidine, it should similarly be possible to determine the number of imidazole ligands in the metal coordination sphere.

4.3. Mode of ligand binding

From the detection of spin coupling to isotopically enriched ligands such as ^{15}N-labeled benzenesulfonamides (Figure 1 and reference 5), it has been possible to discover the way in which these carbonic anhydrase inhibitors bind to the metal. This removed previous ambiguities in the literature regarding inner-sphere vs. outer-sphere binding.

4.4. Number of strong metal binding sites

^{113}Cd nmr led to the discovery of previously unknown Ca^{2+} binding sites in the insulin hexamer. A group of six carboxylate groups in the central cavity can actually bind a total of three Ca^{2+} ions as shown by follow-up studies using europium fluorescence (15) (Figure 4).

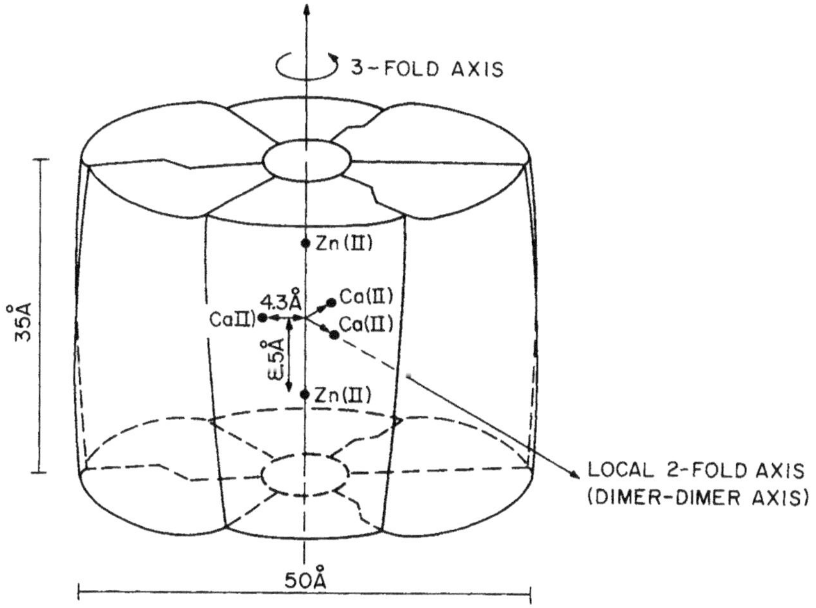

Figure 4. Three-dimensional structural representation of the proposed $Zn_2Ca_3Ins_6$ insulin hexamer.

4.5. Accessibility of bound metals to bulk solution

In the insulin hexamer, the ^{113}Cd resonance from the site deeply buried in the central cavity did not shift with pH or halide concentration (6). The ^{113}Cd resonance from the metal in the Zn site did titrate with pH, however, and shifted upon addition of chloride and bipyridyl. Similar observations allowed assignment of the two ^{113}Cd resonances in troponin C (10). Accessibility of the central binding site to entering or leaving ^{113}Cd was also probed by "capping" the molecule with non-labile Co(III) (6).

4.6. Intramolecular metal-metal distances

In ^{113}Cd substituted SOD, addition of Cu^{2+} paramagnetically broadened the ^{113}Cd resonance, showing the close proximity of the sites (7). Reduction to diamagnetic Cu^{+1} restored the resonance. The use of the paramagnetic Gd^{3+} in Ca^{2+}-binding proteins has been used similarly (10). The existence of metal clusters has been established by the establishment of ^{113}Cd-^{113}Cd spin coupling in metallothioneins (16).

4.7. Thermodynamic measurements

^{113}Cd nmr has been used to measure the microscopic pK_a values associated with pH titration of carbonic anhydrases (4), values ranging from 9.2 to 9.7 (Figure 5). This was accomplished only after scrupulous exclusion of bicarbonate from the samples. The dissociation constants for bicarbonate were also determined. Relative binding strengths of various metals have been estimated by displacement of ^{113}Cd ions with competing metals, including Zn^{2+}, Ca^{2+}, and Mg^{2+} (6,10,11).

4.8. Kinetic measurements

Significant variations in linewidth have been observed in ^{113}Cd-carbonic anhydrases as a function of pH and bicarbonate concentrations (4). So far, no rigorous attempt has been made to perform lineshape analysis involving three-site exchange among H_2O, OH^-, and HCO_3^-, but certainly the potential exists. Several authors, however, have used ^{113}Cd nmr to estimate limits of chemical lifetimes, for example in calmodulin, where the two tightly bound ^{113}Cd ions are visible and the two weakly bound ions are invisible due to lifetime broadening (10).

In Figure 6, we show the ^{113}Cd spectrum of human carbonic anhydrase C (HCAC) with a single bisulfide ion bound to the metal. The doublet (J = 48 Hz) is ascribed to the two-bond coupling of ^{113}Cd to the proton. Such spin coupling is not observed for OH^- binding, probably because of small OH^- proton lifetime. In contrast, the proton on HS^- must be long-lived compared to 3, allowing a lower limit of about 0.2 sec to be established for this proton lifetime.

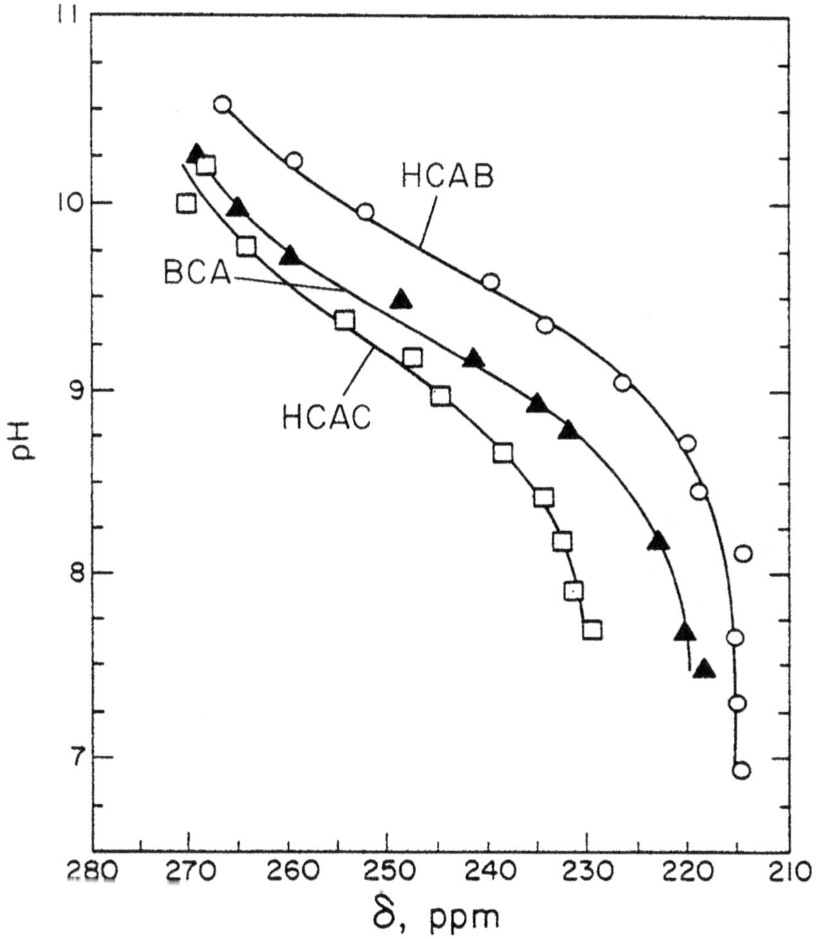

Figure 5. ^{113}Cd chemical shift of ^{113}Cd-carbonic anhydrases *vs*. pH. Enzyme samples were exhaustively dialyzed *vs*. CO_2-free water and pH was increased by addition of CO_2-free NaOH solution. Sample concentrations: human carbonic anhydrase B (HCAB), 7.0 m*M*; human carbonic anhydrase C (HCAC), 3.4 m*M*; and bovine carbonic anhydrase (BCA), 4.5 m*M* (from reference 4 with permission).

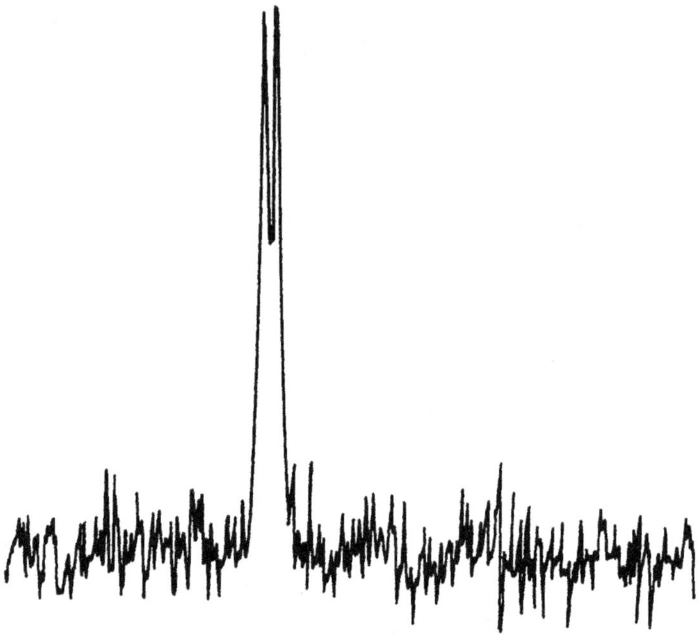

Figure 6. ^{113}Cd nmr spectrum of ∿4 mM human carbonic anhydrase C containing 1 equivalent of HS^-, pH = 9.2. The chemical shift is 373.8 ppm and the spin coupling is 48 Hz. Magnetic field strenght is 2.1 T. All other parameters as in Figure 1A.

5. CONCLUDING REMARKS

The main disadvantage of ^{113}Cd as a probe of Zn^{2+} metalloproteins is that Cd^{2+} is not Zn^{2+}. Being about 30% larger, Cd^{2+} often binds about an order of magnitude weaker than Zn^{2+}. Ca^{2+} binding proteins do not suffer this disadvantage. Because of their nearly identical sizes, Ca^{2+} binding sites generally accept Cd^{2+} with equal affinity. Binding of inhibitors tends to be several orders of magnitude stronger for Cd^{2+} than Zn^{2+}, and pK_a values of bound water molecules tend to be several units higher. This can be an advantage if one wants to deliberately perturb the metal-associated pK_a values, or increase the stability of inhibited species or intermediates.

Most of the other types of information, however, should extrapolate very well to zinc proteins, using reasonable caution and chemical sense.

ACKNOWLEDGEMENTS

Most of this work was supported by U.S. Public Health Service Grant GM25877. We are grateful to J.L. Evelhoch, N.B.-H. Jonsson, L.A.E. Tibell, and R.E. Hurd for experimental assistance, and to R.H. Palmieri for editorial assistance.

REFERENCES

(1) Armitage, I.M., Pajer, R.T., Schoot-Uiterkamp, A.J.M., Chlebowski, and Coleman, J.E.: 1976, J. Amer. Chem. Soc. 98, pp. 5710-5711.
(2) Sudmeier, J.L. and Bell, S.J.: 1977, J. Amer. Chem. Soc. 99, pp. 4499-5000.
(3) Bailey, D.B., Ellis, P.D. and Cardin, A.D.: 1978, J. Amer. Chem. Soc. 100, pp. 236-237.
(4) Jonsson, N.B.-H., Tibell, L.A.E., Evelhoch, J.L., Bell, S.J., and Sudmeier, J.L.: 1980, Proc. Natl. Acad. Sci. USA 77, pp. 3269-3272.
(5) Evelhoch, J.L., Bocian, D.F. and Sudmeier, J.L.: 1981, Biochemistry 20, pp. 4951-4954.
(6) Sudmeier, J.L., Bell, S.J., Storm, M.C., and Dunn, M.F.: 1981, Science 212, pp. 560-562.
(7) Bailey, D.B., Ellis, P.D., and Fee, J.A.: 1980, Biochemistry 19, pp. 591-596.
(8) Bobsein, B.R. and Myers, R.J.: 1981, J. Biol. Chem. 256, pp. 5313-5316.
(9) Suzuki, K.T. and Maitani, T.: 1978, Experientia Spec. 34, pp. 1449-1450.
(10) Drakenberg, T. and Lindman, B.: 1978, FEBS Letters 92, pp. 346-350; Forsen, S., Thulin, E. and Lilka, H.: 1979, ibid., 104, pp. 123-126; Forsen, S., Thulin, E., Drakenberg, T., Krebs, J. and Seamon, K.: 1980, ibid., 117, pp. 189-194.
(11) Otvos, J.D. and Armitage, I.M.: 1980, Biochemistry 19, pp. 4031-4043.
(12) Hoult, D.I. and Richards, R.E.: 1976, J. Magn. Reson. 24, pp. 71-85.
(13) Bloembergen, N., Purcell, E.M. and Pound, R.V.: 1948, Phys, Rev. 73, p. 679.
(14) Sudmeier, J.L., Evelhoch, J.L. and Jonsson, N.B.-H.: 1980, J. Magn. Reson. 40, pp. 377-390.
(15) Alameda, G.K., Birge, R.R., Evelhoch, J.L.: submitted to Biochemistry.
(16) Otvos, J.D. and Armitage, I.M.: 1979, J. Amer. Chem. Soc. 101, pp. 7734-7736.

CARBONIC ANHYDRASE: STRUCTURE, KINETICS, AND MECHANISM

Sven Lindskog, Sirag A. Ibrahim, Bengt-Harald Jonsson, and Ingvar Simonsson

Department of Biochemistry, University of Umeå, Umeå, Sweden

The carbonic anhydrase-catalyzed hydration of carbon dioxide must involve the following elementary reactions. 1. The binding of CO_2. 2. The binding of H_2O. 3. Breaking of an O-H bond in H_2O. 4. Formation of an O-C bond. 5. Dissociation of HCO_3^-. 6. Dissociation of H^+. Kinetic evidence pertaining to the rates and relative order in time of these chemical events is discussed. A kinetic reaction scheme is presented, and this is related to the structure of the active site of the enzyme as known from X-ray diffraction studies.

INTRODUCTION

The zinc-containing metalloenzyme carbonic anhydrase catalyzes the simple, reversible reaction:

$$CO_2 + H_2O \rightleftharpoons HCO_3^- + H^+ \quad (1)$$

The enzyme can also catalyze other reactions, such as the hydration of aldehydes and the hydrolysis of certain esters, for example, 4-nitrophenyl acetate.

Three genetically and immunologically distinct isoenzyme forms of carbonic anhydrase are known to occur in mammals (1). These forms have different kinetic properties but they have homologous structures. They have probably arisen through duplications of an ancestral carbonic anhydrase gene and subsequent divergence.

Isoenzyme II (or C) occurs in the red blood cells of all investigated mammals and in several other tissues. It is one of the most efficient of

I. Bertini, R. S. Drago, and C. Luchinat (eds.), The Coordination Chemistry of Metalloenzymes, 49–64.

all known enzymes catalyzing the hydration of CO_2 with a maximal turnover rate of $1 \times 10^6\ s^{-1}$ at 25 °C.

Isoenzyme I (or B) occurs in the red cells of most mammals and in parts of the alimentary canal. It is less efficient than isoenzyme II with a maximal CO_2 hydration turnover rate of about $1 \times 10^5\ s^{-1}$ at 25 °C.

Isoenzyme III was recently discovered in skeletal muscle. It is even less efficient than isoenzyme I with a CO_2 hydration turnover rate of about $1 \times 10^4\ s^{-1}$ at 25 °C.

The enzymic catalysis of CO_2 hydration must comprise the elementary steps listed in Table 1. Some of these events may occur as discrete reactions steps, while some may be concerted in a single reaction step.

The purpose of this chapter is to summarize present evidence concerning the time course of these events and how they are related to the structure of the active site of carbonic anhydrase. We shall emphasize the highly efficient type II isoenzymes but we shall also discuss the kinetics of the type I isoenzymes. The mechanism of type III isoenzymes will not be considered since too little is yet known about their kinetic properties.

Before we go into the intricacies of carbonic anhydrase kinetics, let us first describe the scene of the catalytic events, the enzymic active site as revealed by X-ray diffraction studies.

Table 1. Elementary reactions in carbonic anhydrase-catalyzed CO_2 hydration

1. Binding of CO_2
2. Binding of H_2O
3. Breaking of O-H bond in H_2O
4. Formation of O-C bond: CO_2-HCO_3^- transformation
5. Dissociation of HCO_3^-
6. Dissociation of H^+

STRUCTURE

The carbonic anhydrase molecule has an ellipsoidal overall shape with the approximate dimensions 4.1x4.1x4.7 nm. The active site is located in a cavity having a roughly conical shape. The cavity is about 1.5 nm wide at the entrance and about 1.6 nm deep reaching almost to the center of the molecule. The zinc ion is near the apex of the cone and liganded to three imidazole groups. The metal ion is bound to the N_τ atoms from His 94 and His 96 and the N_π atom from His 119. A fourth ligand, probably derived from the solvent, is also observed. The structure of human carbonic anhydrase II has recently been refined by Dr. Alwyn Jones

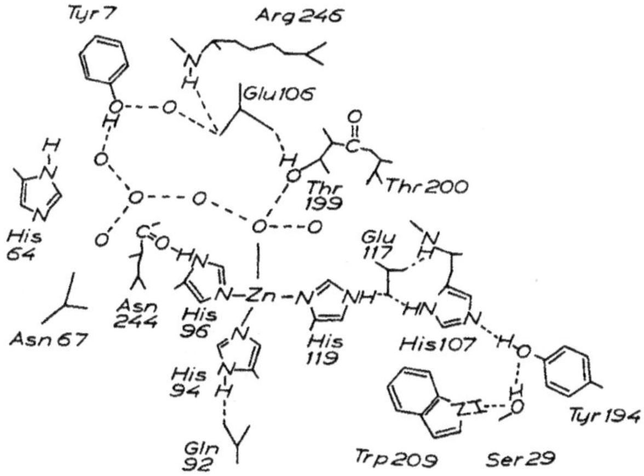

Fig. 1. Schematic drawing of hydrogen-bond patterns in the active site of human carbonic anhydrase II. The positions of seven densities assigned to ordered water molecules are indicated by circles.

at the Wallenberg Laboratory in Uppsala. He reports that the ligand geometry in the refined structure is close to tetrahedral with only small deviations from regular symmetry.

Each one of the four zinc ligands is hydrogen bonded to a protein group. The hydrogen-bond patterns deduced from the refined structure are shown schematically in Fig. 1.

The hydrogen bonding of the three imidazole ligands is the same as previously reported (2). One of the ligands, His 119, interacts with a buried glutamic acid residue, Glu 117, potentially carrying a negative charge. However, it seems possible that Glu 117 accepts three hydrogens in bonds to His 119, His 107 and the peptide NH group of His 107.

The zinc-coordinated solvent molecule is bonded to Thr 199 which interacts with another glutamic acid residue, Glu 106. The refined structure reveals that Glu 106 is bonded to the peptide NH group of Arg 246 and, most importantly, that Glu 106 is not quite shielded from the solvent but bonded to a solvent molecule which also interacts with Tyr 7.

A number of additional densities are observed indicating the presence of an ordered water structure. This structure seems to connect Tyr 7 and the zinc-bound solvent molecule. His 64 may also be associated with the ordered water structure.

If a line through Thr 199, Zn, and Gln 92 in Fig. 1 is imagined, the surface of the active site cavity to the left of this line is dominated by hydrophilic amino acid side chains, such as those of His 64 and Asn 67, while the right hand side has a hydrophobic character. Most of the ordered water molecules are in the hydrophilic part of the active site but one is found in a hydrophobic environment interacting with the zinc-bound solvent molecule.

Of the 17 amino acid residues shown in Fig. 1, 14 are invariant in all known sequences of carbonic anhydrases I, II, and III (I). The variable ones are residues 64, 67, and 200. Residue 200 is Thr or Asn in isoenzymes II and Thr in isoenzymes III, but His in isoenzymes I. In fact, the presence of His 200 in type I isoenzymes is the only unique feature that clearly distinguishes their active sites from those of the type II isoenzymes. All sequenced type I and type II isoenzymes have His at position 64. The type II isoenzymes all have Asn at position 67, and most type I isoenzymes have His 67. However, horse carbonic anhydrase I has Gln 67 (3). Distinctive features of type III isoenzymes are Lys 64 and Arg 67 in the hydrophilic part of the active site, and another basic residue, Arg 91, in the outer portion of the hydrophobic part.

KINETICS

Transfer of H^+ Between Active Site and Reaction Medium

The kinetic and inhibitor-binding properties of carbonic anhydrases I and II depend on the ionization state of a group in the active site. The pK_a of this activity-linked group is near 7 under usual experimental conditions.

For each turnover cycle of CO_2 hydration, one H^+ ion must be transferred from the active site to the reaction medium. If the donor of this H^+ ion is the activity-linked group and the acceptor is H_2O or OH^-, then this proton transfer step would limit the turnover rate to $10^3 - 10^4\ s^{-1}$ at pH 7 (4). However, the observed CO_2 turnover rates at pH 7 are $6 \times 10^4\ s^{-1}$ and $5 \times 10^5\ s^{-1}$ for the human isoenzymes I and II, respectively. Therefore, proton transfer must proceed by some alternative mechanism. It was proposed by several authors (5-7) that H^+ might be transferred rapidly between the activity-linked group and buffer molecules.

$$EH^+ + B \rightleftharpoons E + BH^+ \qquad (2)$$

We have shown experimentally that the CO_2 turnover rate depends on the buffer concentration (8). The observed kinetic pattern is analogous to that of a "ping-pong" mechanism with CO_2 and buffer as substrates. For human isoenzyme II, apparent rate constants for H^+ transfer from EH^+ to B were estimated to be about $2 \times 10^8\ M^{-1}s^{-1}$ with 1,2-dimethylimidazole or N-methylimidazole as buffers. We have recently found that the CO_2 turnover rate catalyzed by the human isoenzyme I also depends on the buffer concentration (S.A. Ibrahim and S. Lindskog, unpublished results). We estimate that the corresponding rate constants for isoenzyme I are about 3×10^7 $M^{-1}s^{-1}$. Thus, it appears that this buffer-dependent step is slower in isoenzyme I than in isoenzyme II. However, at a high buffer concentration, say 50 mM, this step would not limit the turnover rate of isoenzyme I nor that of isoenzyme II. Hence, the important kinetic difference between the two isoenzymes must reside in some other step or steps.

At very low buffer concentrations, the CO_2 turnover rate catalyzed by human isoenzyme I tends to become buffer independent. This buffer-independent rate might be due to the reaction

$$EH^+ + H_2O \rightleftharpoons E + H_3O^+ \qquad (3)$$

If so, the results suggest that H^+ transfer from EH^+ to H_2O has a rate constant of about 10^4 s^{-1}.

Intramolecular H^+ Transfer

Some steady-state kinetic parameters obtained for human isoenzymes I and II at high pH and high buffer concentration are given in Table 2. Under these conditions the buffer-dependent step is not rate limiting.

As shown in Table 2, the values of k_{cat} as well as K_m for CO_2 have isotope effects (value in H_2O/value in D_2O) exceeding 1, whereas the isotope effect in k_{cat}/K_m is near 1. For isoenzyme II, the isotope effect is about 4 which suggests that a H^+ transfer step limits the turnover rate (9). Venkatasubban and Silverman (10) have shown that the isotope effect depends exponentially on the atom fraction of deuterium in H_2O/D_2O mixtures. They suggest that the rate-limiting step involves the transfer of 2 or more protons.

Table 2. Isotope effects in steady-state kinetic parameters for CO_2 hydration catalyzed by human isoenzymes I and II. Buffer: 50 mM 1,2-dimethylimidazole-H_2SO_4, pH 8.9, with Na_2SO_4 to give an ionic strength of 0.2 M. Temperature: 25 °C. pH values in D_2O were estimated by addition of 0.4 to the meter reading.

	Isoenzyme I			Isoenzyme II		
	H_2O	D_2O	Ratio	H_2O	D_2O	Ratio
k_{cat} ($10^{-5} \times s^{-1}$)	1.2	0.7	1.7	10	2.5	4.0
K_m (mM)	4.2	2.5	1.7	8.3	2.2	3.8
k_{cat}/K_m ($10^{-7} \times M^{-1}s^{-1}$)	2.9	2.8	1.0	12	11	1.1

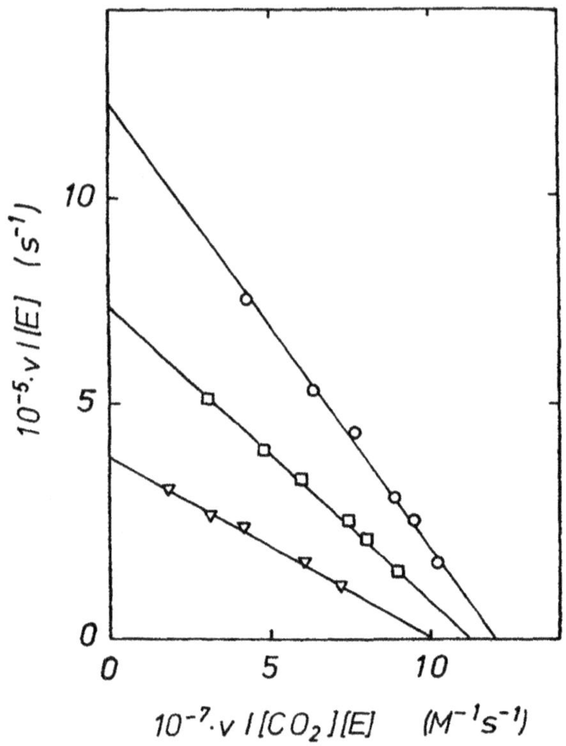

Fig 2. Inhibition by SCN^- of CO_2 hydration catalyzed by human carbonic anhydrase II in 50 mM 1,2-dimethylimidazole-H_2SO_4 buffer at pH 8.7 and 25 °C. Inhibitor concentrations were 0, 0.8, and 3.2 mM.

We have proposed that this rate-limiting step is an intramolecular H^+ transfer between the activity-linked group and another ionizing group in the active site, a "proton-transfer" group (9). Evidence that this step is not concerted with CO_2-HCO_3^- interchange is presented in the following section. Thus, we propose that H^+ transfer from the activity-linked, catalytic group to the solution is a complex reaction involving at least two steps, for example

$$EH^+ \rightleftharpoons {}^+HE \quad (4)$$
$${}^+HE + B \rightleftharpoons E + BH^+ \quad (5)$$

This means that the catalytic group appears to be unable rapidly to deliver a H^+ ion directly to the buffer, but the buffer-dependent step rather involves the "proton-transfer" group.

The data discussed so far do not unequivocally establish the order between the intramolecular step and the buffer-dependent step. However, we have some recent results suggesting that the reaction sequence of eqs. 4 and 5 is the correct one. These results concern the inhibition of CO_2 hydration by a monovalent anion at high pH. It has long been known that the binding of an anionic inhibitor requires the protonated form of the activity-linked group. At high pH and low CO_2 concentration, the unprotonated species, E, will be the dominating enzyme form in the steady state. Thus, the anionic inhibitor should have a small effect on k_{cat}/K_m. However, at high CO_2 and buffer concentrations, the protonated form EH^+ will accumulate in the steady state if eqs. 4 and 5 operate. Therefore, the anionic inhibitor should affect k_{cat}. This means that an uncompetitive inhibition pattern is predicted. The results presented in Fig. 2 show that an uncompetitive pattern is, indeed, observed. Pocker and Deits (11), working with bovine isoenzyme II, have also obtained uncompetitive anion inhibition patterns at high pH, but their interpretation differs from ours.

The most straightforward candidates for the catalytic group and the proton-transfer group in isoenzyme II are zinc-bound H_2O and His 64, respectively. If this assignment is correct, the intramolecular H^+ transfer step would correspond to the H_2O splitting reaction postulated in Table 1. Thus,

$$\text{His-E-Zn}^{2+}\text{-OH}_2 \rightleftharpoons {}^{+}\text{H-His-E-Zn}^{2+}\text{-OH}^- \qquad (6)$$

In human isoenzyme I, the observed isotope effects in k_{cat} and K_m are only about 1.7 (Table 2). If we assume that these isotope effects come from an analogous, intramolecular H^+ transfer step with an intrinsic isotope effect of 4, then we estimate a rate constant for H^+ transfer from catalytic group to proton-transfer group of about $8 \times 10^5\ s^{-1}$ taking also into account a possible isotope effect of similar magnitude in the buffer-dependent step. In isoenzyme II, the corresponding rate constant would be about $1 \times 10^6\ s^{-1}$. Hence, intramolecular H^+ transfer might proceed with approximately equal rates in the two isoenzymes and, therefore, the major kinetic difference should reside in some other steps.

The human isoenzyme I has three titrable active site histidines, His 64, His 67, and His 200 with pK_a values estimated from ^1H NMR measurements as 4.7, 6.0, and 6.1, respectively (12). Of these, His 200 is closest to the metal ion and seems to be the best candidate for the proton-transfer group in isoenzyme I, but His 67 may also play a role. The low pK_a of His 64 suggests that this residue would not be a sufficiently good H^+ acceptor to contribute significantly.

The Binding and Interconversion of CO_2 and HCO_3^-

Rates of CO_2-HCO_3^- exchange at chemical equilibrium can be estimated from the carbonic anhydrase-induced line broadening of ^{13}C NMR resonances from labelled substrates (13,14). The catalyzed exchange rates, v_{exch},

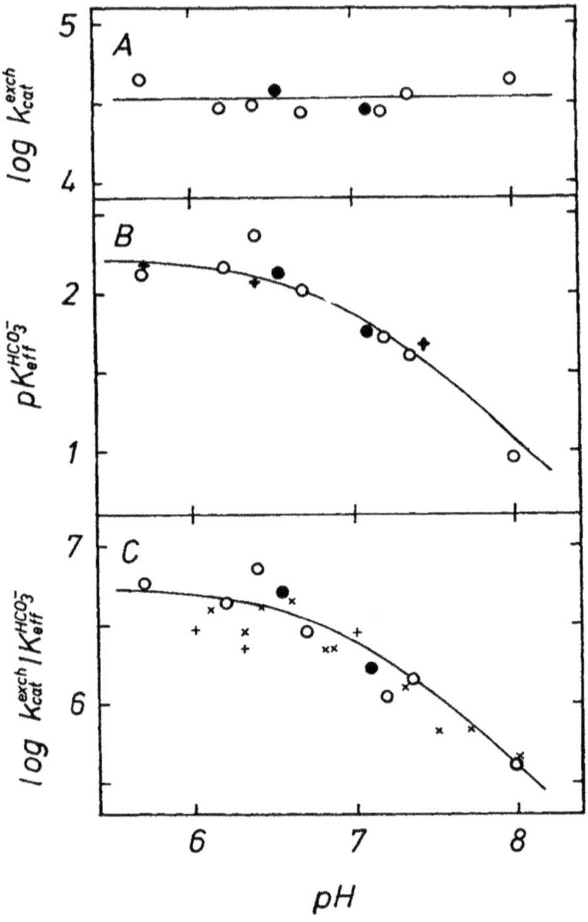

Fig. 3. pH dependence of CO_2-HCO_3^- exchange kinetic parameters for human isoenzyme I.
(A), log k_{cat}^{exch} (s^{-1}); (B), pK_{eff} (M) with respect to HCO_3^-. Crosses represent values obtained from HCO_3^- inhibition of 4-nitrophenyl acetate hydrolysis; (C) log(k_{cat}^{exch}/K_{eff}) ($M^{-1}s^{-1}$); + represent values of k_{cat}/K_m for HCO_3^- dehydration obtained in steady-state experiments; x are such values calculated from CO_2 hydration data using the Haldane relation. Open circles refer to 90 % H_2O and 10 % D_2O and filled circles to 99.5 % D_2O.

were found to display Michaelis-Menten behaviour.

$$v_{exch}/[E_{tot}] = k_{cat}^{exch} \times [S]/(K_{eff} + [S]) \qquad (7)$$

For human isoenzymes I and II, the values of k_{cat}^{exch} are nearly independent of pH whereas K_{eff} is strongly pH dependent. Data for human isoenzyme I are shown in Fig. 3.

With both isoenzymes, the exchange is equally rapid in the absence and in the presence of added buffers. Furthermore, equal exchange rates were obtained in H_2O and in D_2O. These observations are strong evidence that the buffer-dependent step and the intramolecular H^+ transfer step are separate from the steps involved in the CO_2-HCO_3^- exchange pathway. The exchange kinetic results fit with the simple reaction scheme,

$$E + CO_2 \underset{k_{-1}}{\overset{k_1}{\rightleftharpoons}} E\text{-}CO_2 \underset{k_{-2}}{\overset{k_2}{\rightleftharpoons}} EH^+\text{-}HCO_3^- \underset{k_{-3}}{\overset{k_3}{\rightleftharpoons}} EH^+ + HCO_3^- \qquad (8)$$

In case of human isoenzyme II, the mean value of k_{cat}^{exch} in the pH range 6.2 to 7.5 is $1.8 \times 10^6\ s^{-1}$. This value is larger than the maximal turnover rate constants for CO_2 hydration and HCO_3^- dehydration. Therefore, steps on the CO_2-HCO_3^- exchange pathway do not limit the turnover rates at substrate saturation.

The maximal exchange rate constant, k_{cat}^{exch}, depends on all the first order rate constants of eq. 8, k_{-1}, k_2, k_{-2}, and k_3. In case of human isoenzyme II, it is not possible to decide if any one of these rate constants is much smaller than the others and, hence, to determine the value of k_{cat}^{exch}.

In case of human isoenzyme I, the mean value of k_{cat}^{exch} in the pH range 5.7 to 8.0 is $3.5 \times 10^4\ s^{-1}$. This value is about 50 times smaller than that obtained for isoenzyme II. Clearly, the important kinetic difference between isoenzymes I and II resides in the CO_2-HCO_3^- exchange pathway. For isoenzyme I it is possible to speculate about the magnitudes of the rate constants in eq. 8. Thus, the maximal value of k_{cat} for CO_2 hydration is $1.2 \times 10^5\ s^{-1}$ (Table 2). Therefore, rate constants k_2 and k_3 must exceed this value. On the other hand, the maximal value of k_{cat} for HCO_3^- dehydration that we have measured (pH 6) is $3.6 \times 10^4 s^{-1}$. Hence, it seems likely that k_{cat}^{exch} is essentially determined by k_{-1} and/or k_{-2}. If we make the assumption that the small, neutral CO_2 molecule exchanges rapidly with the active site, we arrive at k_{-2} as the best candidate for the step that limits the value of k_{cat}^{exch} in isoenzyme I. The simplest explanation of the observed isotope effect of 1 in k_{cat}^{exch} would then be that this reaction step does not involve H^+ transfer but the removal of an OH^- moiety from HCO_3^-.

Human isoenzyme I appears to bind HCO_3^- about 10 times more strongly than isoenzyme II. It is tempting to speculate that His 200 is involved in promoting this stronger substrate binding. The rates of substrate binding are not known. However, rate constants for the binding of a number of anionic inhibitors have been estimated (1). These rate constants are usually in the range 10^8 - 10^9 $M^{-1}s^{-1}$. These values are not very far from those expected for diffusion-controlled reactions. Presumably, CO_2 and HCO_3^- bind to carbonic anhydrase with rate constants of similar magnitudes.

The inhibition of CO_2-HCO_3^- exchange by the monovalent anion, Cl^-, is shown in Fig. 4. A competitive pattern is observed, suggesting that no significant ternary enzyme-substrate-inhibitor complexes are formed under the conditions used. This finding strengthens our hypothesis that the uncompetitive anion inhibition pattern obtained at high pH in the steady state (Fig. 2) is not due to the formation of such ternary complexes but rather depends on the accumulation of EH^+ in the steady state.

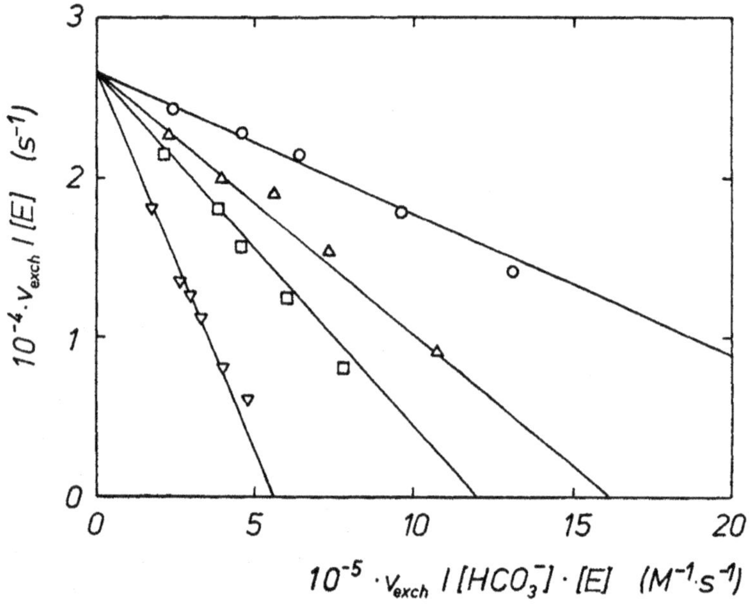

Fig. 4. Inhibition by Cl^- of CO_2-HCO_3^- exchange catalyzed by human carbonic anhydrase I. The solvent was 90 % H_2O and 10 % D_2O and no buffer was added. Temperature, 25 °C; pH 6.8. Inhibitor concentrations were 0, 4, 7, and 20 mM.

The Dissociation of H_2O

Water is a substrate or product of many enzymic reactions. Yet, it is usually not possible to obtain information on the rates of binding and dissociation of this substrate. Due to its extraordinary catalytic efficiency, carbonic anhydrase offers unique possibilities to extract such information. Our knowledge of the exchange rate of substrate H_2O in carbonic anhydrase comes from the work of Silverman and coworkers. They measure two types of ^{18}O exchange at chemical equilibrium; the exchange of ^{18}O between CO_2 and water and the exchange of ^{18}O between ^{12}C- and ^{13}C-containing species of CO_2. The occurrence of the latter exchange reaction is strong evidence that ^{18}O derived from one substrate molecule stays in the active site long enough to be incorporated into another substrate molecule. From these two exchange rates, the rates of two chemical exchange processes can be calculated (15). One of these corresponds to the CO_2-HCO_3^- exchange discussed in the previous section. The other rate, R_{H_2O}, is the rate of dissociation from the active site of H_2O carrying oxygen derived from HCO_3^-.

To be kinetically competent, the rate constant for the H_2O dissociation step must be at least as large as the maximal turnover rate constant, k_{cat}, for HCO_3^- as substrate. In case of human isoenzyme II, a maximal value of k_{cat} for HCO_3^- of 6×10^5 s^{-1} has been measured (9). In the absence of added buffers and at a high pH (pH 8), Tu et al. (16) observe a value of k_{H_2O} ($R_{H_2O}/[E_{tot}]$), of about 1×10^5 s^{-1}. Under these conditions, the oxygen derived from HCO_3^- will essentially be carried by enzyme having the basic forms of the catalytic group and the proton-transfer group. Obviously, product H_2O cannot dissociate from this enzyme species at a sufficient rate. However, at lower pH and in the presence of buffers, k_{H_2O} is larger. With 50 mM buffer at pH 7, k_{H_2O} is 1×10^6 s^{-1} (15) which is sufficient.

The simplest explanation of these findings is that the buffer-dependent step and the intramolecular H^+ transfer step discussed in previous sections are required for the rapid release of product H_2O from the active site. This is most easily understood if one assumes that the basic form of the catalytic group is zinc-bound OH^- exchanging rather slowly with bulk H_2O. Thus,

$$\text{His-E-Zn}^{2+}\text{-}^{18}OH^- \xrightleftharpoons[B]{BH^+} {}^{+}\text{H-His-E-Zn}^{2+}\text{-}^{18}OH^- \longleftrightarrow$$

$$\text{His-E-Zn}^{2+}\text{-}^{18}OH_2 \xrightleftharpoons[H_2{}^{18}O]{H_2{}^{16}O} \text{His-E-Zn}^{2+}\text{-}^{16}OH_2 \qquad (9)$$

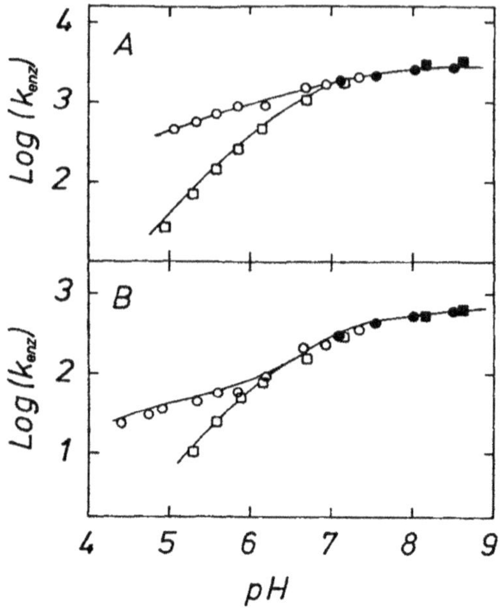

Fig. 5. pH dependence of 4-nitrophenyl acetate hydrolase activity of human carbonic anhydrases I and II. (A), isoenzyme II; (B), isoenzyme I. Circles represent results obtained without sulfate, squares are with 50 mM Na_2SO_4. Open symbols are results obtained in 50 mM MES-NaOH buffers and filled symbols 50 mM HEPES-NaOH buffers.

The Titration Behaviour of Active Site Groups

As discussed in the previous sections, we interpret the kinetic behaviour of carbonic anhydrases I and II in terms of a mechanism involving two ionizing groups in the active site, a catalytic group and a proton-transfer group. In human and bovine isoenzyme II, both of these groups have apparent pK_a values near 7 when the ionic strength is kept constant at 0.2 M with SO_4^{2-} as the only anion present. Sulfate was employed for this purpose because divalent anions were considered non-inhibitory. However, recent work by Bertini et al. (17) and by Koenig et al. (18) suggests that sulfate is not as "innocent" as previously thought.

We have investigated the effects of sulfate on the pH-rate profile of the 4-nitrophenyl acetate hydrolase activity and on the ^1H NMR resonance belonging to the C-2 hydrogen of His 64 in bovine isoenzyme II in MES-NaOH and HEPES-NaOH buffers (19). These buffers appeared to be noninhibitory at the concentrations used. The results showed that, in the absence of sulfate, both the catalytic activity and His 64 have complex titration curves. Sulfate inhibits at low pH, but the binding of sulfate appears to depend very strongly on ionic strength.

The results were interpreted in terms of an electrostatic interaction between the catalytic group and His 64. The sulfate inhibition could be explained by assuming that strong binding of SO_4^{2-} requires that both the catalytic group and His 64 are protonated. Thus, the following scheme seems to apply, where X and Y represent the catalytic group and His 64, respectively.

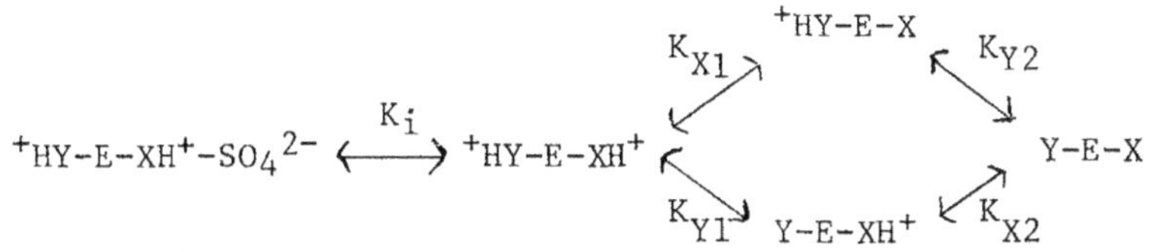

(10)

The effects of sulfate on the pH-rate profiles of human isoenzymes I and II were also investigated. The results, shown in Fig. 5, are in accordance with eq. 10. Estimates of the microscopic pK_a values defined in eq. 10 are given in Table 3. The K_i values for SO_4^{2-} are estimated for an ionic strength of 0.2 M. The most significant difference in the values of Table 2 is the lower pK_a for the group Y in human isoenzyme I. In this isoenzyme, Y presumably corresponds essentially to His 200.

Table 3. Estimated microscopic pK_a values and SO_4^{2-} inhibition constants for various forms of carbonic anhydrase. The K_i values apply at an ionic strength of 0.2 M.

Enzyme	pK_{X1}	pK_{Y1}	pK_{X2}	pK_{Y2}	K_i (mM)
Bovine II	5.1	5.1	6.8	6.8	2
Bovine Co^{2+}-II	5.7	5.8	7.0	7.1	5
Human II	5.4	5.0	7.0	6.6	3
Human I	5.5	4.4	7.1	6.0	2

MECHANISM

The features of the kinetic mechanism of carbonic anhydrase that have been discussed in previous sections are summarized in the scheme shown in Fig. 6. This scheme is in agreement with the observed pH dependences of the steady-state and equilibrium kinetic parameters.

Considering the structure of the active site as it is known from X-ray diffraction studies (Fig. 1), it seems that the most likely

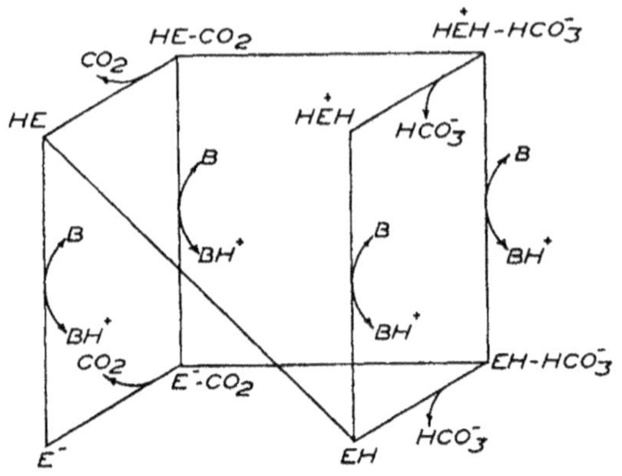

Fig. 6. Scheme of the kinetic mechanism of carbonic anhydrase. Protonated "proton-transfer" group is indicated by H^+ to the left of E, and protonated "catalytic" group by H^+ to the right of E. Vertical lines represent buffer-mediated H^+ transfer. The diagonal line represents the intramolecular H^+ transfer step.

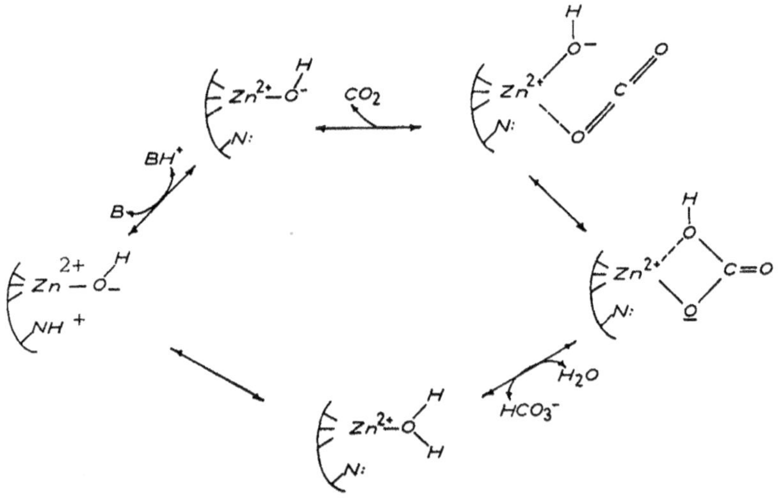

Fig. 7. The "zinc-hydroxide" mechanism. The pathway that may dominate during CO_2 hydration at high pH is illustrated.

candidates for the proton-transfer group are His 64 in type II isoenzymes and His 200 in type I isoenzymes as proposed repeatedly in this chapter. The chemical nature of the catalytic group is still being actively debated. The crystal structure seems to allow a choice between the zinc-bound solvent molecule and Glu 106, but there is no direct evidence available that conclusively discriminates between these alternatives. However, we feel that the kinetic and inhibitor-binding properties are most simply explained on the basis of a "zinc-hydroxide" mechanism as illustrated in Fig. 7. If the inhibition of the enzyme by sulfate is considered, this model requires that the pK_a of zinc-bound water can be as low as 5 under certain conditions. It is up to bio-inorganic chemists to find out if this is chemically possible or not.

ACKNOWLEDGEMENTS

We thank Dr. Alwyn Jones of the Wallenberg Laboratory at the University of Uppsala for permission to describe unpublished results from the refinement of the crystal structure of human carbonic anhydrase II. One of us (SAI) gratefully acknowledges the receipt of a fellowship from the International Seminar in Chemistry, University of Uppsala. Research on carbonic anhydrase in the authors' laboratory is financially supported by a grant from the Swedish Natural Science Research Council (K 2911). Finally, we thank Ms Agneta Ehman and Mr Sixten Johansson for excellent help in preparing the manuscript.

REFERENCES

(1) Lindskog, S.: 1982, in Advances in Inorganic Biochemistry (Eichhorn, G.L. and Marzilli, L.G., eds.), Vol. 4, Elsevier North-Holland, New York, p. 115

(2) Kannan, K.K., Notstrand, B., Fridborg, K., Lövgren, S., Ohlsson, A., and Petef, M.: 1975, Proc. Natl. Acad. Sci. U.S.A. 72, pp. 51-55.

(3) Jabusch, J.R., Bray, R.P., and Deutsch, H.F.: 1980, J. Biol. Chem. 255, pp. 9196-9204.

(4) Eigen, M., and Hammes, G.G.: 1963, Adv. Enzymol. 25, pp. 1-38.

(5) Khalifah, R.G.: 1973, Proc. Natl. Acad. Sci. U.S.A. 70, pp. 1986-1989.

(6) Prince, R.H., and Woolley, P.R.: 1973, Bioorg. Chem. 2, pp. 337-344.

(7) Lindskog, S., and Coleman, J.E.: 1973, Proc. Natl. Acad. Sci. U.S.A. 70, pp. 2505-2508.

(8) Jonsson, B.-H., Steiner, H., and Lindskog, S.: 1976, FEBS Lett. 64, pp. 310-314.

(9) Steiner, H., Jonsson, B.-H., and Lindskog, S.: 1975, Eur. J. Biochem. 59, pp. 253-259.

(10) Venkatasubban, K.S., and Silverman, D.N.: 1980, Biochemistry 19, pp. 4984-4989.

(11) Pocker, Y., and Deits, T.L.: 1981, J. Am. Chem. Soc. 103, pp. 3949-3951.

(12) Campbell, I.D., Lindskog, S., and White, A.I.: 1974, J. Mol. Biol. 90, pp. 469-489.

(13) Koenig, S.H., Brown, R.D., London, R.E., Needham, T.E., and Matwiyoff, N.A.: 1974, Pure Appl. Chem. 40, pp. 103-113.

(14) Simonsson, I., Jonsson, B.-H., and Lindskog, S.: 1979, Eur. J. Biochem. 93, pp. 409-417.
(15) Silverman, D.N., Tu, C.K., Lindskog, S., and Wynns, G.C.: 1979, J. Am. Chem. Soc. 101, pp. 6734-6740.
(16) Tu, C.K., Wynns, G.C., and Silverman, D.N.: 1981, J. Biol. Chem. 256, pp. 9466-9470.
(17) Bertini, I., Canti, G., Luchinat, C., and Scozzafava, A.: 1977, Biochem. Biophys. Res. Commun. 78, pp. 158-160.
(18) Koenig, S.H., Brown, R.D., and Jacob, G.S.: 1980, in Biophysics and Physiology of Carbon Dioxide (Bauer, C., Gros, G., and Bartels, H., eds.), Springer-Verlag, Berlin, Heidelberg, New York, pp. 238-243.
(19) Simonsson, I., and Lindskog, S.: 1982, Eur. J. Biochem. 123, pp. 29-36.

METAL ION PROMOTED HYDROLYSIS OF NUCLEOSIDE 5'-TRIPHOSPHATES

Helmut Sigel

Institute of Inorganic Chemistry, University of Basel, Spitalstrasse 51, CH-4056 Basel, Switzerland

INTRODUCTION

The transfers of phosphoryl and nucleotidyl groups are among the most fundamental processes in biochemistry [1-4]. As all the enzymes catalyzing these reactions require metal ions for activity it is not surprising that the metal ion promoted dephosphorylation of nucleoside 5'-triphosphates (NTP) has long been recognized [5-8]. Indeed, the easiest way to obtain more insight into this type of reactions is to study the transfer of a phosphoryl group to a water molecule, e.g. of an organic triphosphate [8], and by accelerating this process through the addition of metal ions one may hope to reveal the ways by which metal ions promote such reactions [9]. A condition for being able to do so is, however, a detailed knowledge of the coordination chemistry of nucleotides; fortunately, the situation has in this respect considerably improved recently [10-12].

To learn something about the reactive intermediates in the dephosphorylation processes of nucleoside 5'-triphosphates we have taken the approach not to study so much a single system in great detail but rather to compare the reactivity of several systems with each other; these are listed in Figure 1. This approach was taken because a different reactivity must originate from the structural differences of the nucleic base moieties, as all the rest of these nucleoside 5'-triphosphates is identical.

The results described in this article are all based on experiments carried out in the absence of buffers, because these inhibit the metal ion accelerated dephosphorylation [13]. The pH of the solutions was adjusted with relatively concentrated NaOH or $HClO_4$ using a glass stick; the change in volume was thus negligible. The concentration of liberated phosphate was determined with molybdate reagent taken in

I. Bertini, R. S. Drago, and C. Luchinat (eds.), The Coordination Chemistry of Metalloenzymes, 65–78.

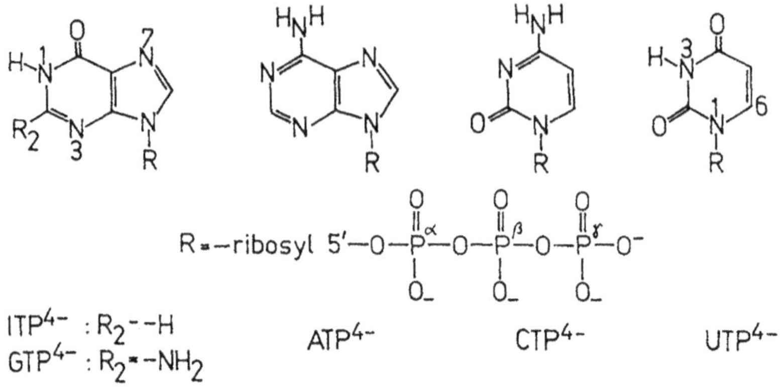

Figure 1: Structures of the nucleoside 5'-triphosphates (NTP^{4-}) considered in this study.

suitable intervals; the procedure of Hirata and Appleman [14], as altered by Schneider and Brintzinger [7], was used. [NTP] at the time, t, is given by $[NTP]_t = [NTP]_o - [PO_4^{3-}]_t$, where $[NTP]_o$ is the initial concentration of NTP, and $[NTP]_t$ and $[PO_4^{3-}]_t$ are at the time, t. The free PO_4^{3-} initially present [13] was taken into account in the calculations; the amounts of diphosphate and AMP formed are negligible [6].

To be able to compare our experimental results also with earlier studies [7] in which first-order rate constants had been used for quantifying the rate, we determined also first-order constants, k (s^{-1}), from the slope of the straight line portion of a log $[NTP]_t$-time plot. The corresponding pH_{av} was obtained by averaging the pH values measured for those samples that gave points on the straight-line portion (see [13,15,16]).

For mechanistic considerations where an exact relationship between rate and pH is essential and to eliminate uncertainties that might arise from the variation of pH or from side reactions, the following procedure [17] was used. The initial rate of dephosphorylation, $v_o = d[PO_4^{3-}]/dt$ (Ms^{-1}), was determined from the slope of the tangent of the $[PO_4^{3-}]_t$-time curve at the time, t = 0. The corresponding initial pH of the reaction solution, i.e., pH_o, was determined analogously [15,16]. To obtain v_o of a given system at a particular pH_o, two experiments were carried out in this pH range (one slightly above and the other below the desired value) and these results were then interpolated to the desired pH_o (see [9]).

RESULTS AND DISCUSSION

The results for two typical representatives, namely for the pyrimidine-nucleotide UTP [18] and the purine-nucleotide ATP [15], are summarized in Figure 2 and the following points are immediately evident:

(i) The dephosphorylation rate of the metal ion-free nucleotide systems is identical, hence the base moieties have no influence on the reaction (vide infra and Figure 7).

(ii) In the presence of Cu^{2+} (1:1), however, the dephosphorylation of ATP is much more accelerated than the one with UTP; a comparison at the 'peaks' shows a rate difference of a factor of about 50. This result proves that here the base moieties influence the rates.

(iii) In the mixed ligand systems consisting of NTP, Cu^{2+} and 2,2'-bipyridyl (Bpy), there is again no difference in the dephosphorylation rate which is very low. In fact at pH <7 the rate is even smaller than in the free ATP and UTP systems; in other words, in the ternary $Cu(Bpy)(NTP)^{2-}$ complexes the triphosphate is *protected* toward hydrolysis [9]. Considering the transport of these sensitive nucleotides in biological systems, this observation seems interesting.

What are the reasons for these fascinating results given in (ii) which have analogously also been observed with Zn^{2+} [9,13]? It is wellknown that in the monomeric $Cu(ATP)^{2-}$ and $Zn(ATP)^{2-}$ complexes the metal is not only coordinated to the phosphate chain, it also interacts with N-7 of the purine residue [12]. In contrast to this there is no metal ion-base interaction in $Cu(ATP)^{2-}$ or $Zn(UTP)^{2-}$. Hence, one may conclude that the metal ion-purine interaction is among the reasons responsible for the enhanced rate [9]. This conclusion is indeed confirmed by the observations summarized in (iii); by the formation of the mixed ligand complexes $M(Bpy)(ATP)^{2-}$ the base moiety is released from the coordination sphere of the metal ion [19-21]. Moreover, for several M^{2+} systems with ATP^{4-} it has in addition been proven that in aqueous solution a significant part of the $M(Bpy)(ATP)^{2-}$ complexes contains an intramolecular stack between the purine moiety of ATP and the aromatic system of 2,2'-bipyridyl [20,21]; these intramolecular stacks were also confirmed in the solid state by crystal structure analysis [22]. Similar stacks, though less stable, occur also in $M(Bpy)(UTP)^{2-}$ complexes [21,23]. Hence, as neither in $M(Bpy)(ATP)^{2-}$ nor in $M(Bpy)(UTP)^{2-}$ any metal ion base interaction exists, it is not surprising anymore that both nucleoside 5'-triphosphate systems show the same dephosphorylation properties. The reduced activity of $M(Bpy)(NTP)^{2-}$ in comparison with $M(ATP)^{2-}$ and $M(UTP)^{2-}$ originates in the inhibited formation of the reactive intermediate.

In Figure 3 are shown some results of another pyrimidine-nucleotide, namely of CTP. It is evident that the dephosphorylation of CTP increases with the increasing concentration of $Cu(CTP)(OH)^{3-}$ [9,16]. A comparison of the upper part of Figure 3 with the Cu^{2+}-UTP system in Figure 2 reveals that the Cu^{2+} promoted dephosphorylation is within experimental error identical up to pH 8 in both pyrimidine systems. Indeed, it was concluded also for Cu^{2+}-UTP that the reactive intermediate must be related to $Cu(UTP)(OH)^{3-}$. The decrease of the reactivity in Cu^{2+}-UTP at pH >8 (Figure 2) is attributed [18] to the formation of $Cu(UTP-H)(OH)^{3-}$; a species in which N-3 is ionized (see Figure 1) [19].

There is an old suggestion that the reactive intermediate of the metal ion promoted hydrolysis of an organic triphosphate should contain two metal ions [8]. Indeed, if increasing concentrations of Cu^{2+} are added to a Cu^{2+}-CTP [9] or a Cu^{2+}-UTP system (see upper part of Figure 4) [18] the reaction increases further. This has the consequence that 2:1 systems of Cu^{2+} and CTP or UTP, if measured in dependence on pH, increase their reactivity at pH 5 rather suddenly and also in quite a pronounced way [9,18]. As exactly the same observation is also made in the Cu^{2+}-methyltriphosphate systems [7], this means that the pyrimidine residue is *not* involved in this reactivity increase. It must rather mean that two metal ions are coordinated to the phosphate chain.

That the reactive intermediate is a 2:1 Cu^{2+}-UTP complex is confirmed by Job's series [25] which have been carried out at pH 5.2 (see lower part of Figure 4) and pH 7.8 [18]. Exactly the same results have been obtained for the Cu^{2+}-CTP system. It may be noted in this connection that the ability of nucleoside triphosphates to coordinate two metal ions is wellknown [9,26,27].

The linear dependence between log v_o and log [NTP] for the 1:1 and 2:1 systems of Cu^{2+} and UTP or CTP shows that the reactive intermediate is a monomeric complex [9,18] which is in accordance with the low self-association tendency of pyrimidine-nucleotides [12]. As Cu^{2+} in 10^{-3} M solutions begins to form hydroxo-complexes at pH about 5 it is

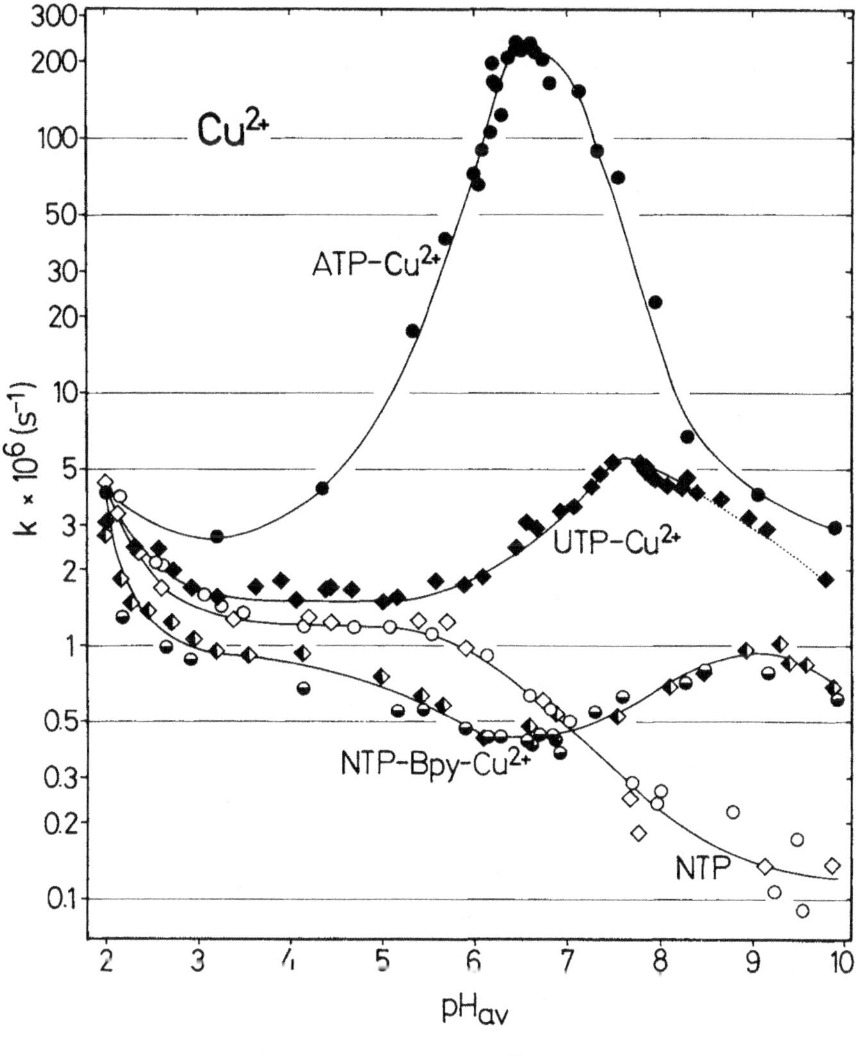

Figure 2. Comparison of the Cu^{2+} promoted dephosphorylation of ATP (●, [15]) and UTP (◆, [18]) (always in the ratio 1:1) in dependence on pH, characterized as the first order rate constants k (s^{-1}). In addition is given: ATP (○) or UTP (◇) alone, and ATP (◒) or UTP (◆) in the presence of Cu^{2+} and 2,2'-bipyridyl (1:1:1). The concentration of all reagents was 10^{-3} M; I = 0.1, $NaClO_4$; 50°C. The dotted line portion indicates uncertainty due to precipitation.

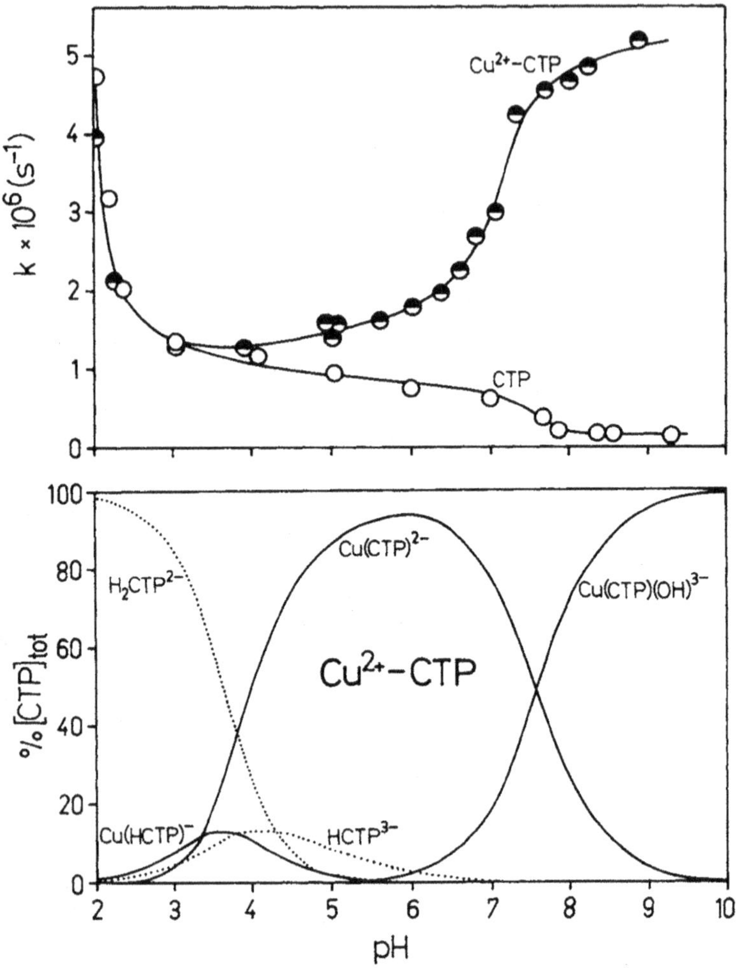

Figure 3. *Upper part:* First-order rate constant, k (s^{-1}), for the dephosphorylation of CTP, alone (○) and in the presence of Cu^{2+} (1:1, ◓), in dependence on pH [16]. $[CTP]_{tot} = 10^{-3}$ M; I = 0.1, $NaClO_4$; 50°C. *Lower part:* Effect of pH on the concentration of the species present in an aqueous solution of $[CTP]_{tot} = [Cu^{2+}]_{tot} = 10^{-3}$ M at I = 0.1 and 25°C [16]. Results are given as the percentage of total CTP (or Cu^{2+}) present. The dotted lines indicate the free CTP species and the solid lines the CTP complexes (see also [24]).

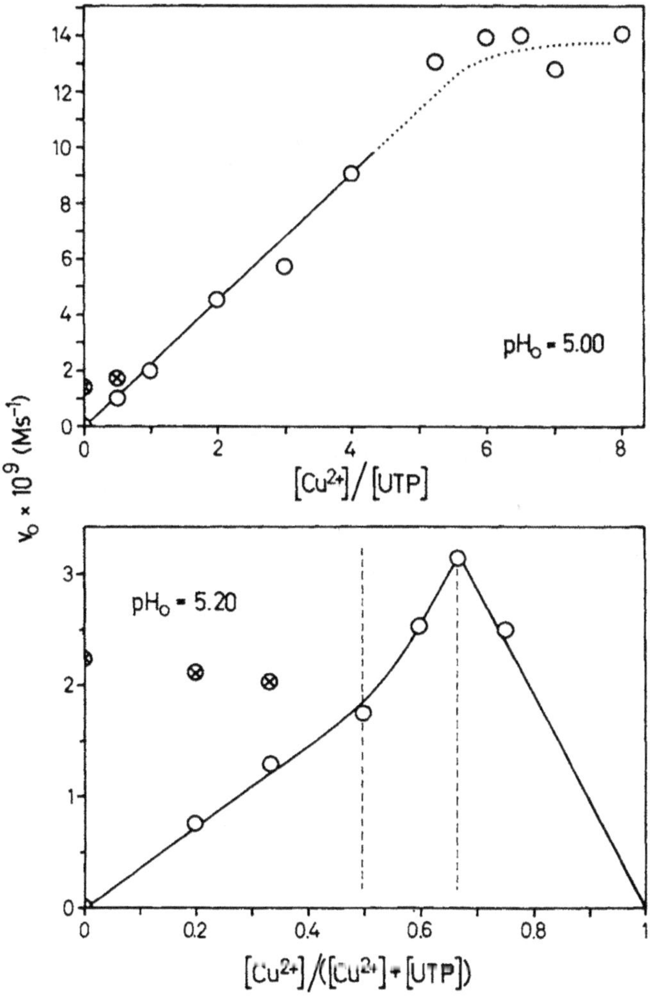

Figure 4. *Upper part:* Dependence of the initial rate v_o of the Cu^{2+} promoted dephosphorylation of UTP (10^{-3} M) on the ratio $[Cu^{2+}]/[UTP]$. The dotted line portion indicates uncertainty due to precipitation. *Lower part:* Job's series of the Cu^{2+}-UTP system at $[Cu^{2+}]_{tot} + [UTP]_{tot}$ = constant = 2×10^{-3} M. The vertical dashed lines give the positions of the ratios Cu^{2+} : UTP = 1:1 or 2:1. I = 0.1, $NaClO_4$; 50°C. The measured points (⊗) were corrected for the dephosphorylation of uncomplexed UTP [18].

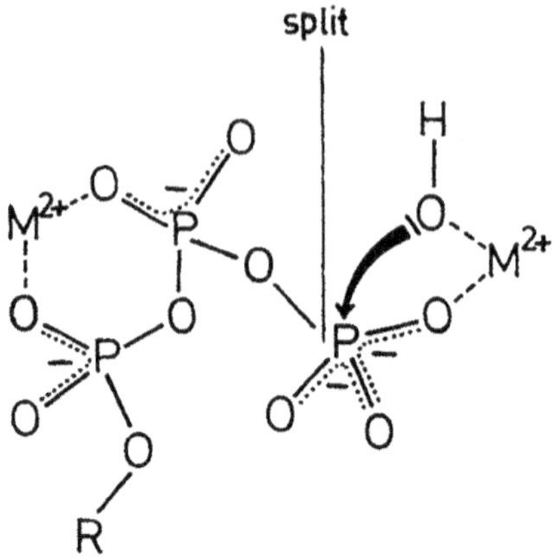

Figure 5. Tentative and simplified structure of the reactive complex formed during the metal ion promoted dephosphorylation of organic triphosphates undergoing only a metal ion-phosphate coordination, like pyrimidine-nucleoside 5'-triphosphates or methyltriphosphate

concluded that the reactive species is a complex of the type $Cu_2(R\text{-}TP)(OH)^-$, where R-TP represents an organic triphosphate, e.g. methyltriphosphate, CTP^{4-} or UTP^{4-}, which undergoes only a metal ion-phosphate (but no base) interaction [18]. This situation is tentatively illustrated in Figure 5.

For the reactivity observed in 1:1 systems of metal ions and pyrimidine-nucleoside 5'-triphosphates at pH >7 (see e.g. Figure 2) the situation as depicted in Figure 5 offers two explanations.

(i) The position of the equilibrium

$$2\ M(NTP)(OH)^{3-} \rightleftharpoons M_2(NTP)(OH)^- + NTP^{4-} + OH^-$$

or of $M(NTP)(OH)^{3-} + M(NTP)^{2-} \rightleftharpoons M_2(NTP)(OH)^- + NTP^{4-}$

may be such that some few percentages of $M_2(NTP)(OH)^-$ are formed. As the reactivity of the 2:1 complexes is much larger than of the 1:1 species, the formation of some few percentages would be enough to explain the observed reactivity.

(ii) The (α),β,γ chelated metal ion of $M(NTP)^{2-}$ releases partly the (α),β group(s) in an intramolecular equilibrium by the formation of the hydroxo-complex $M(NTP)(OH)^{3-}$; such a species with a $(M\text{-}OH)^+$ unit coordinated to the γ group, the α,β groups being free, should also show some reactivity.

It must now be emphasized that the situation with purine-nucleoside 5'-triphosphates is very different [9]. The relation between the initial rate v_o and the concentration of M^{2+}-ATP in 1:1 and 2:1 systems is reproduced in Figure 6; it is evident that here a square dependence exists what means that the reaction proceeds via dimeric intermediates. Indeed, the self-stacking tendency of purine-nucleotides is wellknown and the facilitation of this stacking process by metal ions has been characterized [12]. There is evidence that a metal ion coordinated to the phosphate chain of one ATP coordinates also to N-7 of another ATP, both ATP species being stacked; in other words, the stacking tendency is promoted by the formation of a metal ion bridge.

That the self-stacking and metal ion coordinating ability of the purine moiety is crucial for the Cu^{2+} promoted dephosphorylation of such nucleoside 5'-triphosphates is confirmed by the results shown in Figure 7. ATP, GTP and ITP are in their dephosphorylation differently accelerated by Cu^{2+}, and indeed, the self-association tendency of these nucleotides is different, as well as the metal ion binding properties of their bases [12]. It is interesting to note that the corresponding mixed ligand systems containing 2,2'-bipyridyl behave all alike, due to the formation of $Cu(Bpy)(NTP)^{2-}$ [9]. As discussed before, in these ternary complexes a metal ion-purine interaction is prevented.

The structure deduced [9] for the reactive intermediate of the metal ion promoted dephosphorylation of purine-nucleoside 5'-triphosphates has the following features. The purine moieties are stacked and a shift of the M^{2+} along the phosphate chain of one NTP into an α,β coordination is induced by binding of this metal ion to N-7 of the other NTP thus leading to a γ-phosphate group (see Figure 1) ready for nucleophilic attack by OH^- or H_2O; this γ group is further labilized by coordination of an additional M^{2+} [or $M(OH)^+$]. This reactive intermediate is related to the one shown in Figure 5 but the metal ion shift into the α,β position is facilitated by the dimer formation and the metal ion-base coordination.

One could view this reactive intermediate also by saying that one ATP is needed to bring the other into the reactive state. This view is further supported by the results given in Figure 8: adenosine, phosphate and tryptophanate inhibit the reaction to different extents, while AMP facilitates the dephosphorylation [28]. It appears that AMP, having the N-7 and one phosphate group, is able to take over the

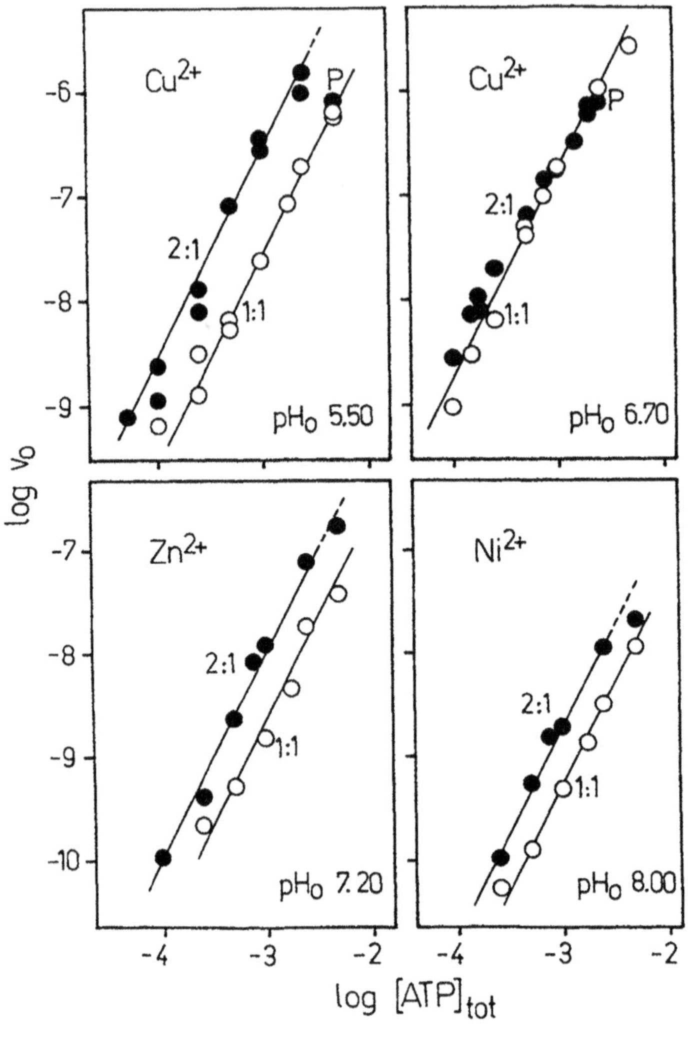

Figure 6. Relationship between the dephosphorylation rate of ATP and the total concentrations of M^{2+} and ATP. Dependence of v_o (Ms^{-1}) on $[M^{2+}]_{tot} = 2[ATP]_{tot}$ (●) or $[ATP]_{tot} = [M^{2+}]_{tot}$ (○) for Cu^{2+}, Zn^{2+}, and Ni^{2+} at different values of pH_o (I = 0.1, $NaClO_4$; 50°C). In the experiments labeled P, precipitation was observed; the dashed lines indicate also uncertainty due to precipitation. Reproduced by permission of the American Chemical Society from reference 9.

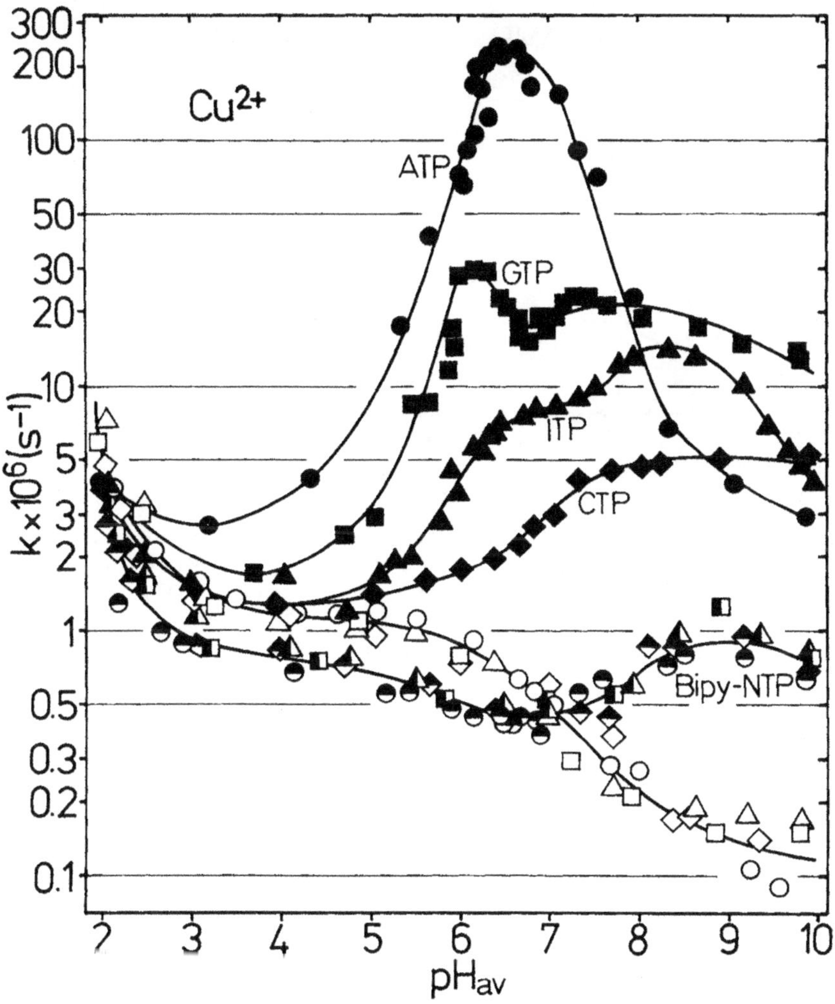

Figure 7. Comparison of the Cu^{2+} promoted dephosphorylation of ATP, GTP, ITP, and CTP in dependence on pH, characterized as the first-order rate constant k (s^{-1}): ATP (○), GTP (□), ITP (△), CTP (◇) alone; ATP (●), GTP (■), ITP (▲), or CTP (◆) in the presence of Cu^{2+} (1:1); and ATP (half-filled ○), GTP (half-filled □), ITP (half-filled △), or CTP (half-filled ◇) in the presence of Cu^{2+} and 2,2'-bipyridyl (1:1:1). The concentration of all reagents was 10^{-3} M; I = 0.1, $NaClO_4$; 50°C. Reproduced by permission of the American Chemical Society from reference 9.

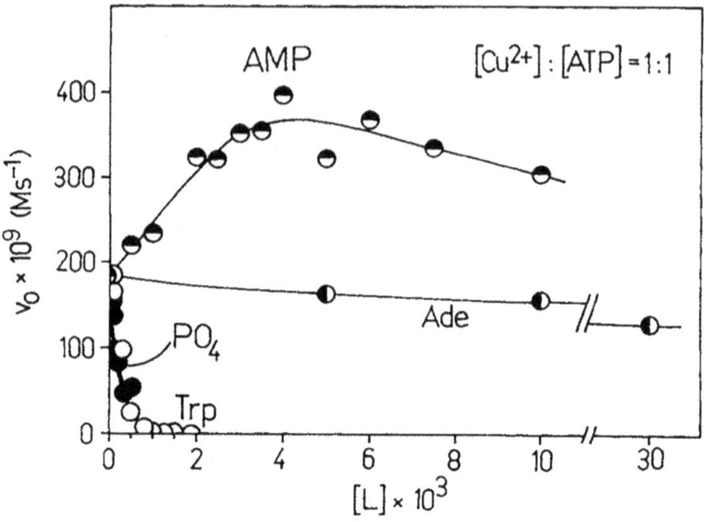

Figure 8. Influence of adenosine (Ade), phosphate, tryptophan (Trp), or adenosine 5'-monophosphate (AMP) on the rate of dephosphorylation of the Cu^{2+}-ATP system. Dependence of v_o on the concentration of L (= Ade, PO_4, Trp, or AMP) at pH_o = 6.70. $[Cu^{2+}]_{tot}$ = $[ATP]_{tot}$ = 10^{-3} M; I = 0.1, $NaClO_4$; 50°C [28].

role of the "structuring" ATP, thus creating mixed AMP/ATP stacks and forcing more ATP into the reactive form. As one might expect, this result contrasts with the effect that AMP has on the UTP-Cu^{2+} 1:1 system; in this case the reaction is inhibited by the formation of a ternary $Cu(UTP)(AMP)^{4-}$ complex [18].

As many of the enzyme-nucleotide systems operating in nature contain two or more metal ions [3,29], reactive intermediates similar to those discussed here could well play a role also there. For example, the coordination M(α,β)-M(γ) should facilitate a break between the β and the γ group thus allowing a transfer either of a nucleoside diphosphate or of a phosphate group. Other coordination types can as well be envisaged, e.g., M(α)-M(β,γ) promoting a nucleoside monophosphate or diphosphate transfer.

It appears that the crucial part in creating these different reactive intermediates has to be played by the nucleic base moiety. For example, if the purine system stacks with an indole residue of an enzyme the triphosphate chain could in this way be positioned toward a metal ion such that the activation of the desired bond results. Stacking interactions between the indole residue of tryptophanate and ATP in ternary complexes are known [21,30,31] as are the corresponding hydrophobic interactions with the isopropyl residue of leucinate [32]; hydrogen bonding [33] could also play a role in positioning the nucleotide and the metal ion toward each other. Regarding the role of the metal ions it is evident that these could (at least partly) be replaced by hydrogen bonds with the oxygens of the phosphate chain or by ionic interactions, e.g., with an arginine residue.

This work is supported by the Swiss National Science Foundation.

REFERENCES

(1) B. S. Cooperman, *Met. Ions Biol. Syst.* 5, 79-125 (1976).
(2) A. S. Mildvan, *Adv. Enzymol. Relat. Areas Mol. Biol.* 49, 103-26 (1979).
(3) G. L. Eichhorn, *Met. Ions Biol. Syst.* 10, 1-21 (1980).
(4) A. S. Mildvan and L. A. Loeb, *Adv. Inorg. Biochemistry* 3, 103-23 (1981).
(5) C. Liébecq and M. Jacquemotte-Louis, *Bull. Soc. Chim. Biol.* 40, 67-85, 759-65 (1958); C. Liébecq, *ibid.* 41, 1181-88 (1959).
(6) M. Tetas and J. M. Lowenstein, *Biochemistry* 2, 350-7 (1963).
(7) P. W. Schneider and H. Brintzinger, *Helv. Chim. Acta* 47, 1717-33 (1964).
(8) D. L. Miller and F. H. Westheimer, *J. Am. Chem. Soc.* 88, 1514-17 (1966).
(9) H. Sigel and P. E. Amsler, *J. Am. Chem. Soc.* 98, 7390-400 (1976).
(10) R. W. Gellert and R. Bau, *Met. Ions Biol. Syst.* 8, 1-55 (1979).
(11) R. B. Martin and Y. H. Mariam, *Met. Ions Biol. Syst.* 8, 57-124 (1979).
(12) K. H. Scheller, F. Hofstetter, P. R. Mitchell, B. Prijs, and H. Sigel, *J. Am. Chem. Soc.* 103, 247-60 (1981).
(13) P. E. Amsler and H. Sigel, *Eur. J. Biochem.* 63, 569-81 (1976).
(14) A. A. Hirata and D. Appleman, *Analyt. Chem.* 31, 2097-99 (1959).
(15) D. H. Buisson and H. Sigel, *Biochim. Biophys. Acta* 343, 45-63 (1974).

(16) H. Sigel, D. H. Buisson, and B. Prijs, *Bioinorg. Chem.* 5, 1-20 (1975).
(17) A. A. Frost and R. G. Pearson, "Kinetics and Mechanism", J. Wiley, New York, 1953.
(18) H. Sigel and F. Hofstetter, submitted for publication.
(19) H. Sigel, *J. Am. Chem. Soc.* 97, 3209-14 (1975).
(20) P. Chaudhuri and H. Sigel, *J. Am. Chem. Soc.* 99, 3142-50 (1977).
(21) P. R. Mitchell, B. Prijs, and H. Sigel, *Helv. Chim. Acta* 62, 1723-35 (1979).
(22) (a) P. Orioli, R. Cini, D. Donati, and S. Mangani, *J. Am. Chem. Soc.* 103, 4446-52 (1981). (b) W. S. Sheldrick, *Angew. Chem.* 93, 473-4 (1981); *Angew. Chem. Int. Ed. Engl.* 20, 460 (1981).
(23) Y. Fukuda, P. R. Mitchell, and H. Sigel, *Helv. Chim. Acta* 61, 638-47 (1978).
(24) H. Sigel, *J. Inorg. Nucl. Chem.* 39, 1903-11 (1977).
(25) P. Job, *C. R. Hebd. Seances Acad. Sci.* 196, 181-3 (1933).
(26) T. A. Glassman, C. Cooper, G. P. P. Kuntz, and T. J. Swift, *FEBS Letters* 39, 73-4 (1974).
(27) C. Miller Frey and J. Stuehr, *Met. Ions Biol. Syst.* 1, 51-116 (1974).
(28) H. Sigel, F. Hofstetter, and V. Scheller-Krattiger, results to be published.
(29) B. S. Cooperman, A. Panackal, B. Springs, and D. J. Hamm, *Biochemistry* 20, 6051-60 (1981).
(30) H. Sigel and C. F. Naumann, *J. Am. Chem. Soc.* 98, 730-9 (1976).
(31) H. Sigel in "Coordination Chemistry - 20"; ed. D. Banerjea; published by IUPAC through Pergamon Press: Oxford and New York, 1980; pp 27-45.
(32) H. Sigel, B. E. Fischer, and E. Farkas, submitted for publication.
(33) L. G. Marzilli and T. J. Kistenmacher, *Acc. Chem. Res.* 10, 146-52 (1977).

MODELS OF METALLOENZYMES: CARBOXYPEPTIDASE A

John T. Groves* and Reginald M. Dias
Department of Chemistry
The University of Michigan
Ann Arbor, Michigan 48109

ABSTRACT

Mechanistic and structural studies on bovine pancreatic carboxypeptidase A are reviewed. Pertinent model studies of metal catalyzed acyl transfer reactions are also reviewed. Results for the hydrolysis of a metal-complexing lactam (1), in which the metal is held perpendicular to the plane of the amide are described. Rate enhancements for Cu^{2+}-1 are found to be greater than 10^6 with respect to the hydrolysis of the same ligand with no metal present. Similarly, a rate enhancement of 10^4-10^5 has been observed for Zn^{2+}-1. A mechanism involving nucleophilic attack of a metal-bound hydroxide is proposed for this process.

INTRODUCTION

Carboxypeptidase A (CPA) is a hydrolytic metalloenzyme that cleaves peptides from the free carboxyl end of the molecule (2). Despite decades of study, there are still significant ambiguities regarding the mode of action of this enzyme. Particularly, there is no molecular rationale for the fast rate of amide hydrolysis by CPA. Further, model compound studies have not provided simple precedents for this catalytic activity. It is the purpose of this chapter to review the present situation regarding CPA and to describe data supporting the importance of geometrical factors in metal catalyzed amide hydrolysis.

From the pH-enzymic activity data for carboxypeptidase A, it has been shown that the enzyme is active in the pH range 6.9-9.0 with the maximum activity occurring around neutrality. The natural enzyme is known to contain Zn^{2+} at the active site (3,4) with one g atom of the metal ion per mole of the enzyme. Apoenzyme produced by removal of the metal ion by competitive complexation with chelating agents like 1,10-phenanthroline is completely inactive (5), but the enzyme activity is restored on adding the required amount of metal ion.

I. Bertini, R. S. Drago, and C. Luchinat (eds.), The Coordination Chemistry of Metalloenzymes, 79–92.

CPA has been shown to possess both peptidase and esterase activity. Other transition metal ions like Co^{2+}, Ni^{2+}, Mn^{2+} and Fe^{2+} are known to be good substitutes (6,7) for Zn^{2+} but with varying degrees of enzymic activity.

Bovine pancreatic CPA has been extensively studied. The amino acid sequence of the enzyme was determined by Neurath and his associates (8). Besides the metal ion, Glu-270 and Tyr-248 were also identified as catalytic groups at the active site. From selective blocking experiments it has been shown that Glu-270 and Tyr-248 are vital in the enzymic peptidase activity (9). But as an esterase the Tyr-248 does not play any significant role. Studies with natural substrates and synthetic substrates like benzoyl-glu-(D,L)phenylalanines it is apparent that the enzyme is specific for substrates with terminal carboxyl groups in the L-configuration. The preference for either aromatic or a branched aliphatic chain has led to the conclusion that the active site is in the vicinity of a hydrophobic pocket or an apolar region that encapsules the substituent on the terminal amino acid group.

The X-ray study on crystalline CPA was carried out by Lipscomb (10,11) which has revealed structural information on the enzyme to a resolution of 2 Å. On the basis of combined X-ray diffraction and chemical sequencing results, the Zn^{2+} coordinating ligands were shown to be His-69, Glu-72 and His-196 and a water molecule. The X-ray structural work, however, has only been possible on a complex of the enzyme with a pseudosubstrate, which is not rapidly cleaved by the enzyme. The structural investigation reveals that the carbonyl oxygen of the amide bond displaces the coordinated water molecule. A generalized picture of this complex is shown in Figure 1. While the crystalline material is

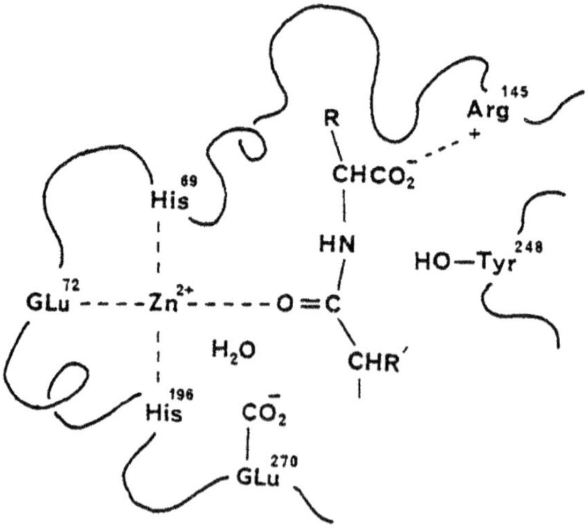

Figure 1. Mode of Binding Between CPA and a pseudo-Substrate Adapted from Ref. 11.

enzymatically active, the X-ray diffraction study of the complex with a poor substrate yields results for a static or a nonproductive complex. Thus, the structural information on enzyme-pseudo substrate complex, though extremely useful, may not totally represent the binding of the enzyme and the typical reactive substrates.

Investigations on the hydrolysis catalyzed by CPA on its natural and artificial substrates, esters and amides have resulted in some conflicting opinions regarding the mechanism of the hydrolytic reactions. Using mixed organic solvents and sub-zero temperatures, it has been possible to considerably slow down the hydrolysis of an ester O-(trans-p-chlorocinnamoyl)-L-β-phenylacetate catalyzed by CPA. From kinetics of this hydrolysis carried out by spectrophotometric methods, Makinen and Kaiser (12) have provided evidence for the existence of a covalently bound acyl-enzyme intermediate. From a combination of kinetic evidence and the structural as well as the chemical aspects of the enzyme, the acyl-enzyme intermediate is believed to have formed by a nucleophilic attack of the Glu-270 on the substrate carbonyl.

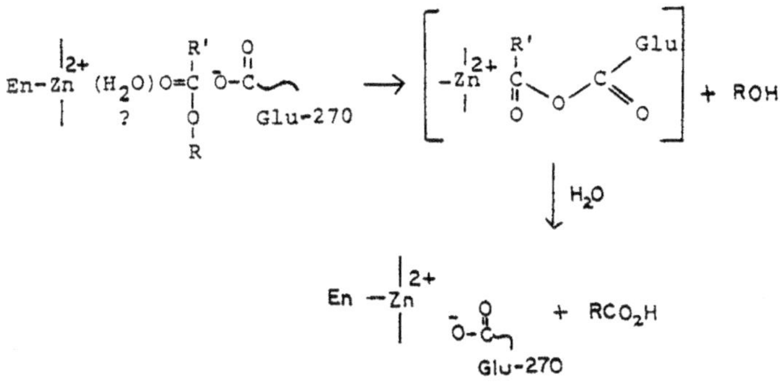

These conclusions, however, are in conflict with the results of O-18 exchange experiments on an amide substrate like N-benzoylglycine (13). CPA catalyzes the O-18 exchange in N-benzoylglycine only in the presence of an added amino acid like phenylalanine. As the principle of microscopic reversibility would dictate, a hydrolytic enzyme should also catalyze the synthesis of peptides. Thus, the observed O-18 exchange

$$\phi\text{-}\overset{O}{\overset{\|}{C}}\text{-}\underset{H}{N}\text{-}CH_2\text{-}\overset{O^*}{\overset{\|}{C}}\text{-}OH \;+\; \underset{H_2}{N}\text{-}\overset{\phi}{\overset{|}{C}}H\text{-}CO_2H \;\xrightleftharpoons{\;Enz,\ H_2O^{18}\;}\; \phi\text{-}\overset{O}{\overset{\|}{C}}\text{-}\underset{H}{N}\text{-}CH_2\text{-}\overset{O}{\overset{\|}{C}}\text{-}\underset{H}{N}\text{-}\overset{\phi}{\overset{|}{C}}H\text{-}CO_2H$$

in the presence of an added amino acid can be explained as that taking place during the hydrolysis of the dipeptide synthesized by CPA. By the same principle, the existence of an acyl-enzyme intermediate in the hydrolytic step would require O-18 exchange to take place without the added amino acid, which is not observed. From these results, Breslow has concluded that the hydrolysis proceeds through a general base participation by the Glu-270.

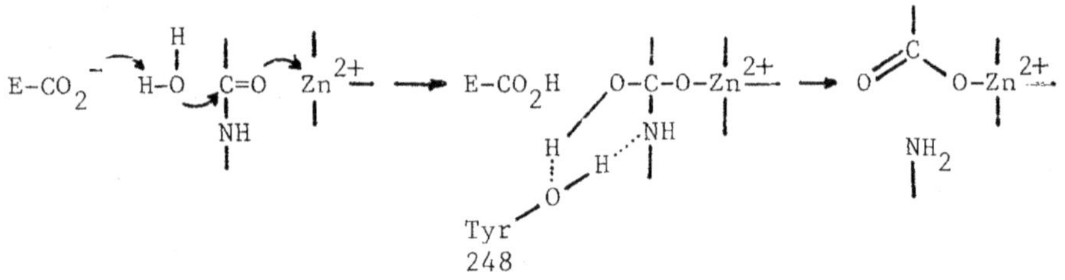

A third mechanism which should be considered is that the role of zinc ion is to coordinate a hydroxide ion to facilitate nucleophilic attack at the amide acyl carbon.

METAL ION CATALYZED HYDROLYSIS OF AMIDES AND ESTERS

Rate enhancements as a consequence of metal catalysis in the hydrolysis of carboxylic acid and phosphoric acid esters have been extensively investigated. Hydrolysis of amino acid esters catalyzed by divalent transition metal ions was first reported by Kroll (14) and by Bender (15,16). Metal ions like Cu^{2+}, Co^{2+} and Mn^{2+} were found to promote the hydrolysis of ethylglycinate in the pH range 7.3 to 7.9 and 25°C, under which conditions the free ester is ordinarily stable. It was observed that maximal rate of hydrolysis occurred in solutions constituted with a metal/ester ratio of unity and that the rate of hydrolysis under identical conditions increased with increasing tendency of the metal ions to coordinate with the amines. For phenylalanine ethyl ester the rate enhancement due to Cu^{2+} was 10^6 at pH 7.3.

α-Amino amides and dipeptides are also susceptible to catalyzed hydrolysis by divalent metal ions (17). However, the rate enhancements brought about are very much lower than those observed for the corresponding esters, and is due to the greater basicity or the poor leaving

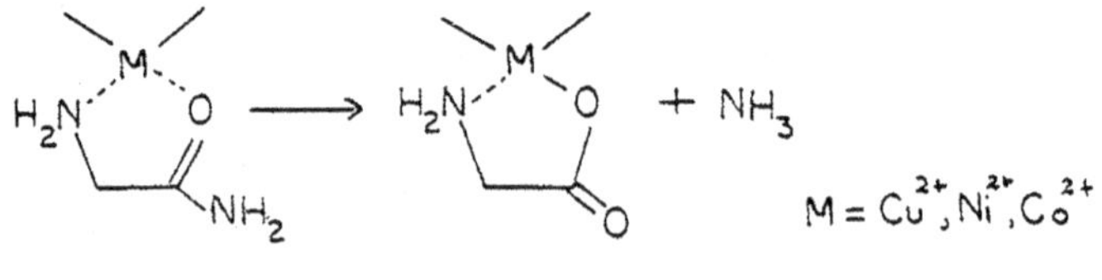

group nature of the amino function. In the hydrolysis of glycineamide the effectiveness of the metal ions were found to be in the order $Cu^{2+} > Co^{2+} > Ni^{2+}$. Further, it was observed that Zn^{2+} was a much poorer catalyst and in absolute magnitude of catalytic ability it was no better than H^+.

An extremely effective metal ion catalyzed reaction, conceptually similar to the process of hydrolysis, has been investigated by Breslow and coworkers (18,19). At neutral pH and 25°C, the half life for the cupric ion catalyzed hydration of 1,10-phenanthroline-2-nitrile is less

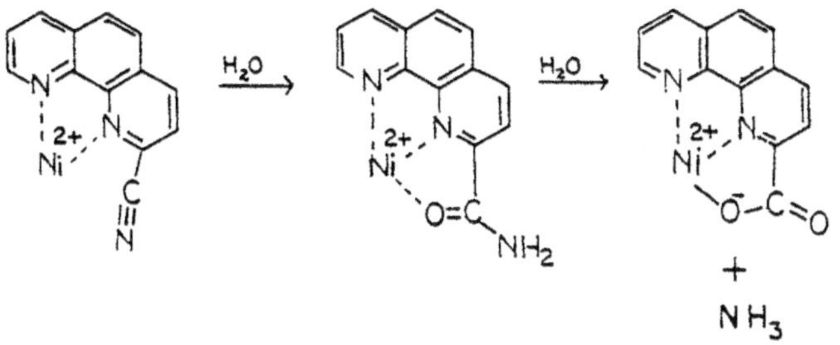

than 10 sec. Under saturating conditions of complexation, the Ni^{2+} catalyzed hydration of 1,10-phenanthroline-2-nitrile shows a linear dependence of rate constant on the hydroxide ion concentration in the pH range 5.0-7.2. The experimental results are expressed by the rate law $k_{obs} = k_1[^{-}OH]$ where $k_1 = 1.6 \times 10^6\ M^{-1}\ min^{-1}$. The second order rate constant for the Ni^{2+} catalyzed process is about 10^7 that of the alkaline hydration in the absence of the metal. A more positive $\Delta S^{\neq}$ over the uncatalyzed reaction seems to account for the rate enhancement brought about by Ni^{2+}. The metal ion was also found to accelerate the hydrolysis of the amide to the corresponding 1,10-phenanthroline-2-carboxylic acid, but with a rate enhancement of only four hundred fold over the base catalyzed reaction.

As a model for metalloenzymes, the hydrolysis of glycineamides and dipeptides catalyzed by cis-$[Co(en)_2]^{+3}$ has been investigated by Buckingham and Sargeson (20,21). Catalysis of amide hydrolysis by Co^{3+} was shown to occur by two mechanisms: (1) by coordination of Co^{3+} to the acyl oxygen and (2) by nucleophilic attack of a cobalt-bound hydroxyl group.

In evaluating the catalytic ability of metal-bound hydroxide as a catalyst in hydrolytic reaction, it is necessary to estimate the nucleophilicity of metal bound nucleophile in comparison to the free hydroxide ion. Ability of the metal-bound hydroxide to add rapidly to carbonyl substrates in bimolecular reactions has been demonstrated (22,23).

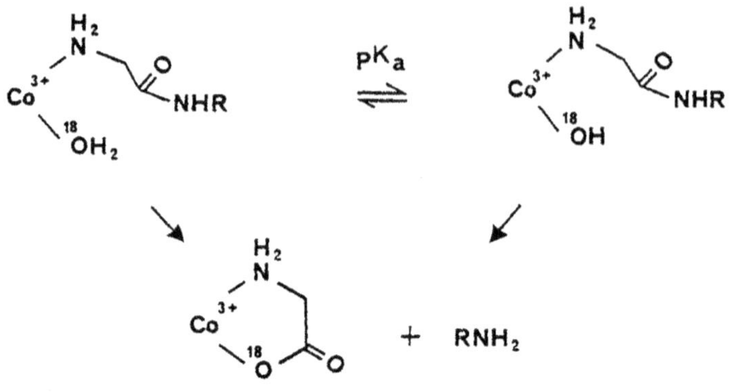

Using different non-labile complexes of transition metal ions of varying pK_a values of the metal-aquo species, a general investigation on the hydrolysis of propionic anhydride has been reported (24).

The results of this study reveal that the hydrolysis of anhydride is catalyzed by M-OH and not $M-OH_2$. It was also observed that the catalysis involves the direct attack by M-OH on the carbonyl center. The catalytic ability of the M-OH was found to depend on the pK_a of the metal-bound water. This dependence, however, was found to be rather small with an increase in k_{M-OH} by less than a factor of 10^2 when the pK_a increased by 6.

The participation by metal ions in the intramolecular nucleophilic attack by a carbonyl group on the adjacent carbonyl center has been investigated by Fife and Squillacote (25,26). In this study the hydrolysis of N-(2-phenanthrolyl)phthalamic acid and related compounds were investigated.

The hydrolysis of this compound was found to be inhibited in the presence of divalent metal ions like Cu^{2+}, Ni^{2+}, Co^{2+} and Zn^{2+}. Space filling models of the compound as well as spectrophotometric evidence suggest that the metal ions are chelated by the aromatic nitrogens. Thus, it was concluded that, though the initial nucleophilic attack by carboxyl group was assisted by the metal ion, the consequent tetrahedral intermediate would be nonproductive due to its increased stability.

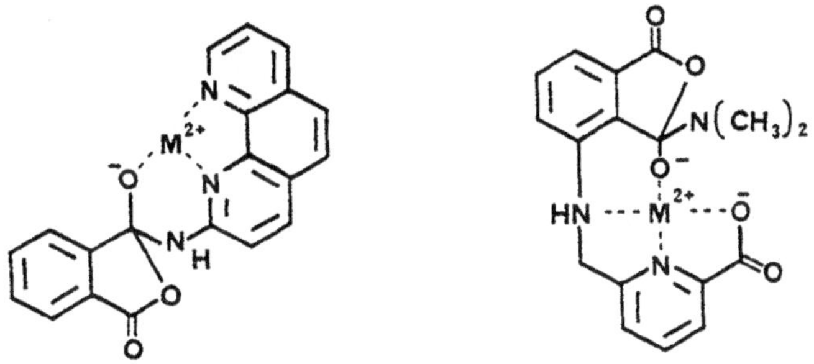

RESULTS AND DISCUSSION

The metal ion catalyzed hydrolyses of amides described so far have not given particular attention to the relative orientation of the metal ion with respect to the plane of the carbonyl group. The metal ion can interact with the carbonyl function from two distinct positions in space as seen in relation to the plane described by the amide bond. These two specific orientations are shown below. This geometric consideration

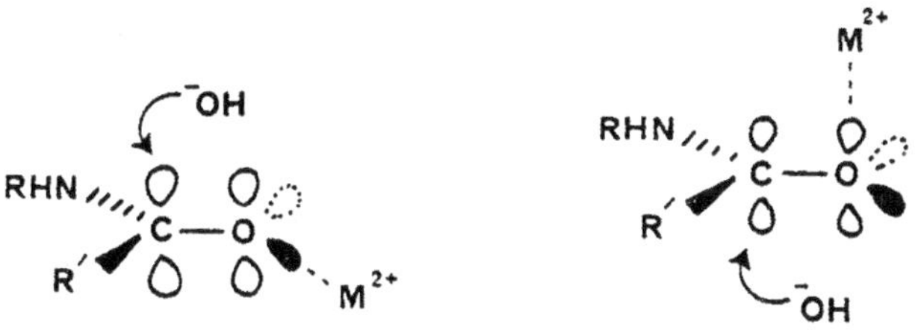

could be an important aspect of mechanism involving nucleophilic catalysis by metal-hydroxo species. Accordingly, we chose to investigate metal ion catalyzed hydrolysis of metal-bind amides in which the metal ion was held above the plane of the amide group.

We have described the synthesis of a metal coordinating lactam (1) which achieves this perpendicular arrangement (27). Formation constants for various metal complexes of 1 are shown in Table I. Titration of Cu^{2+}-1 revealed an acidic proton with a pK_a of 7.2 (28).

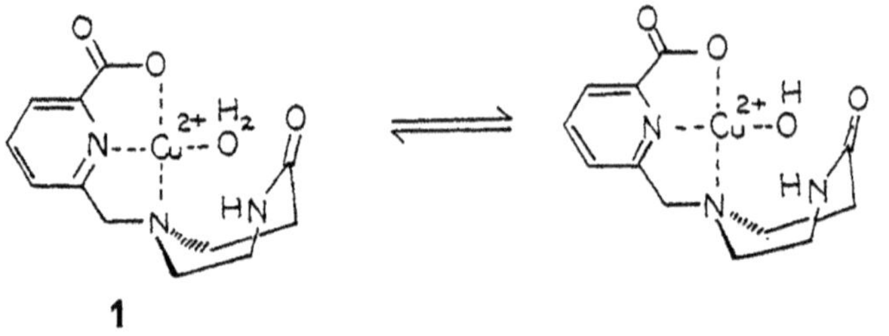

Table I. Hydrolysis Rate Constants at pH 7.0, 50°C.

metal	K_f	$k_{obs}\ S^{-1}$	k_{rel}
no metal (1)		3.7×10^{-12} [a]	1.0
1-Cu^{2+}	1.0×10^{8}	7.6×10^{-6}	2.1×10^{6}
1-Zn^{2+}	1.4×10^{5}	0.4×10^{-6}	1.1×10^{5}
1-Ni^{2+}	4.7×10^{4}	0.1×10^{-6} [b]	
1-Co^{2+}	1.2×10^{4}	0.1×10^{-6} [b]	

[a] Calculated from second-order rate constant.

[b] Estimated value, less than 10% reaction over 192 hr.

Cu^{2+} MEDIATED AMIDE HYDROLYSIS.

The Cu^{2+} mediated hydrolysis of 1 was examined at 50°C in the pH region 4.8-8.3. A plot of these data is shown in Figure 2. In the higher pH region, the rate constant has a small dependence on pH. As the pH is lowered, k_{obs} decreases rapidly. But at the lower end, the data show curvature towards another pH independent region. Reactions below pH 5.0 are extremely slow and thus no extensive investigations were possible.

At pH 7.6, the rate constant for the Cu^{2+} mediated hydrolysis of 1 is $1.38 \times 10^{-5}\ S^{-1}$. This rate constant is 0.9×10^{6} times greater than that calculated for the base catalyzed hydrolysis of 1 at the same pH. Rate enhancements of this magnitude catalyzed by divalent metal ions in the hydrolysis of aliphatic amides have not been observed in other systems and indicate the importance of metal-amide geometry in determining the degree of rate enhancement.

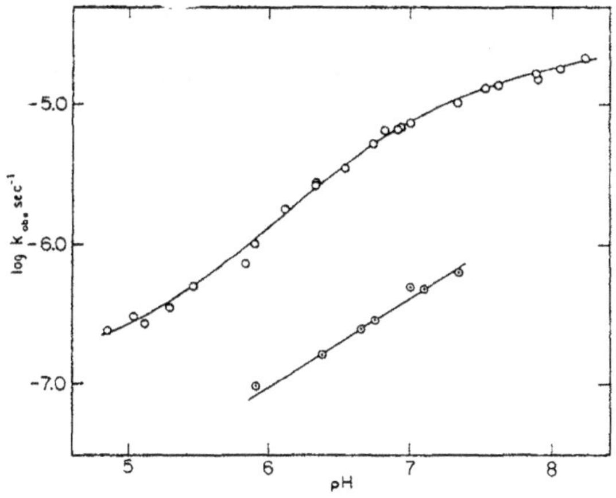

Figure 2. Plot of log k_{obs} vs. pH for the hydrolysis of 1-Cu^{2+} (O) and 1-Zn^{2+} (⊙) in water at 50°. Solid line for 1-Cu^{2+} indicates calculated values according to equation 1.

Sigmoidal pH rate constant profiles are observed in the intramolecular nucleophilic reactions of urethanes. The sigmoidal pH-rate constant profile for the copper catalyzed reaction of 1 can be understood in terms of the metal-aquo and metal-hydroxo equilibrium. The data presented in Figure 2 give a good fit to equation (1),

$$k_{obs} = k_1\left[\frac{a_H}{K_a+a_H}\right] + \left(k_2 + k_{OH}[^-OH]\right)\left[\frac{K_a}{K_a+a_H}\right] \qquad (1)$$

where k_1 (1.41×10^{-7} S^{-1}) is the rate constant for the Cu^{2+}-OH_2 catalyzed reaction, k_2 (1.72×10^{-5} S^{-1}) is the rate constant for the Cu^{2+}-OH catalyzed reaction and k_{OH} (0.3 M^{-1} S^{-1}) is the second-order rate constant for the base-catalyzed reaction of Cu^{2+}-OH species (27). The quantities in the brackets express the mole fraction of the appropriate species. In order to fit the data with equation (1), it was necessary to choose a value of $K_a = 10^{-7.13}$ very close to the titrimetric value. The actual fit of data to equation (1) is illustrated by the solid line in Figure 1.

The observed pH-rate constant profile can be rationalized to be resulting from amide hydrolysis catalyzed by Cu^{2+}-OH_2 species (or Cu^{2+}-OH and a proton, a kinetic equivalent) in the low pH region and that catalyzed by Cu^{2+}-OH species (or a kinetic equivalent) in the higher pH region. Further, a base-catalyzed term, k_{OH}, associated with the Cu^{2+}-OH species is apparent towards the higher pH side. The reactivities of the metal-aquo and metal-hydroxo complexes differ by a factor of 10^2.

In the alkaline region, titration experiments show the existence of a metal-hydroxo complex. The formation of the tetrahedral intermediate can come about from two routes: nucleophilic addition of metal-bound hydroxide (path a) and electrophilic activation of the carbonyl group toward external hydroxide attack (path b) (27). It is not possible to

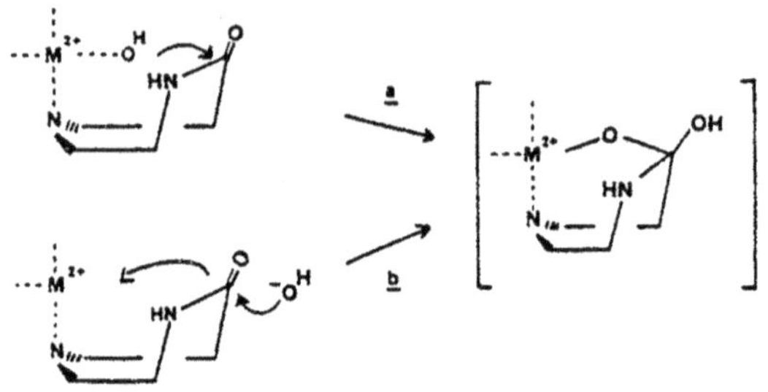

choose unambiguously between these kinetically equivalent mechanisms. Space-filling models indicate that the metal-bound hydroxide is within the van der Waals radius of the acyl carbon. The proximity of this nucleophile as well as its ideal location for a perpendicular attack on the amide plane should be highly in favor of an internal attack leading to the tetrahedral intermediate. The overall pH dependence is very similar in absolute rate as well as shape to the cyclization of 2-aminomethylbenzamide which can only proceed through a nucleophilic mechanism.

A plausible mechanism for the nucleophilic addition of metal-bound hydroxide in shown in Scheme I. This scheme accounts for the terms k_2[M-OH] and k_{OH}[M-OH][OH] in equation (1). In this scheme the breakdown of the tetrahedral intermediates to the products are considered to be rate determining. The terms k_2 and k_{OH} are associated with the intermediates T and T_-.

A comparison of the rate constants for the metal catalyzed and uncatalyzed hydrolyses of 1 is given in Table I. At pH 7.0, the catalytic abilities of the four metal ions are in the order Cu^{2+} > Zn^{2+} > Ni^{2+}, Co^{2+}. This is a reversal of trend between Zn^{2+} and Ni^{2+} when compared with some of the hydrolysis reactions catalyzed by these metal ions. The rate enhancement of 1.1×10^5 brought about by Zn^{2+} in the hydrolysis of 1 is quite significant since zinc has been found to be a poor catalyst for such acyl transfer reactions.

CONCLUSIONS.

The results of this investigation show that copper, and more interestingly, zinc are highly effective in promoting amide bond cleavage at physiological pH when the metal ion is placed above the amide π bond. By contrast, situations in which the metal ion is forced to lie in the

Scheme I

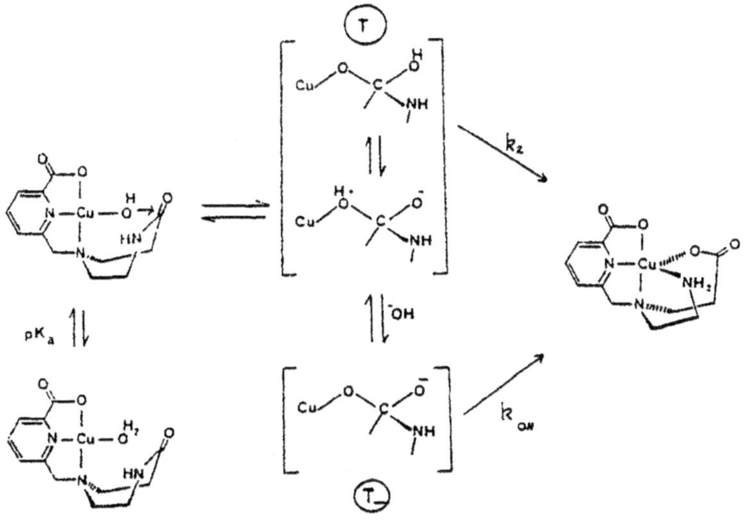

plane of the amide bond have shown limited acceleration or even inhibitory effects (25). The two important factors that may be influential in the overall catalytic process are (1) the chemical reactivity of the catalyst and (2) geometric relationship of the catalyst to the amide carbonyl.

The metal ion reduces the effective negative charge of the coordinated hydroxide. This makes the metal-bound hydroxide a weaker nucleophile than a free hydroxide ion. However, due to the substantial lowering of pK_a of coordinated water, the metal ion can act as a buffer for hydroxide around neutral pH. Thus, in the metal-hydroxo mechanism, the reduced reactivity of the metal-bound hydroxide can be considered to be more than compensated by the increased effective local concentration of the hydroxide.

The importance of orbital control in the breakdown of the tetrahedral intermediate in reactions representing amide and ester hydrolysis has been demonstrated (29). There are certain close similarities between the metal-bound tetrahedral intermediate in 1 and the model compounds used in the investigation of stereoelectronic control during hydrolysis. The chelating nature of the metal-bound tetrahedral intermediate mandates certain conformational restrictions. The plane passing through "M-O-C-O"

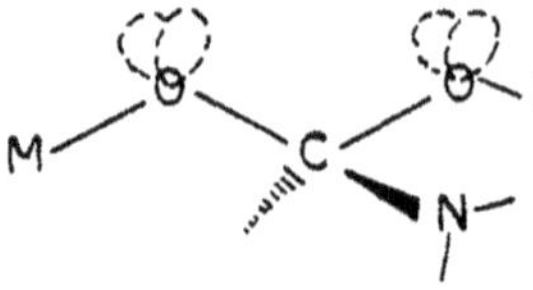

bisects the angle between the lone pair of electrons on the oxygen atoms. The two lone pair orbitals behind this plane are antiperiplanar to the C-N bond. In the base-catalyzed hydrolysis, the tetrahedral intermediate involved should be considered much more flexible with free rotations about C-O bonds. Thus, in the metal-bound intermediate anchimeric assistance from antiperiplanar lone pairs should favor explusion of nitrogen. An ideal spatial arrangement of metal-bound hydroxide for a nucleophilic attack at the carbonyl carbon as well as anchimeric assistance for expulsion of nitrogen from the intermediate are two important geometric features of this model. Thus, subtle geometric factors that lead to spectacular rate enhancements must be carefully considered in relating metal catalysis in model compounds to the role of metal ions in the enzymes.

REFERENCES

(1) Abstracted from the Ph.D. thesis of RMD, The University of Michigan.

(2) (a) Ambler, R. P.: 1967, Methods Enzymol. 11, p. 155; (b) Ambler, R. P.: 1967, ibid. 11, p. 445.

(3) Smith, E. L. and Hanson, H. T.: 1949, J. Biol. Chem. 179, p. 802.

(4) Vallee, B. L. and Neurath, H.: 1954, J. Am. Chem. Soc. 76, p. 5006.

(5) (a) Vallee, B. L. and Riordan, J. F.: 1968, Brookhaven Symp. Biol. 21, p. 91; (b) Smith, E. L.: 1949, Proc. Nat. Acad. Sci. U.S.A. 35, p. 80.

(6) (a) Coleman, J. E. and Vallee, B. L.: 1960, J. Biol. Chem. 235, p. 390; (b) Coleman, J. E. and Vallee, B. L.: 1961, ibid. 236, p. 2244.

(7) Vallee, B. L., Rupley, J. A., Coombs, T. L. and Neurath, H.: 1958, J. Am. Chem. Soc. 80, p. 4750.

(8) Neurath, H., Bradshaw, R. A., Ericsson, L. H., Babin, D. R., Petra, P. H. and Walsh, K. A.: 1969, Brookhaven Symp. Biol. 21, p. 1.

(9) Sokolvsky, M. and Vallee, B. L.: 1967, Biochemistry 6, p. 700.

(10) (a) Reeke, G. N., Hartsuck, J. A., Ludwig, M. L., Quiocho, F. A., Steitz, T. A. and Lipscomb, W. A.: 1967, Proc. Natl. Acad. Sci. U.S.A. 58, p. 2220; (b) Lipscomb, W. A., Hartsuck, J. A.; Reeke, G. N., Quiocho, F. A., Bethge, P. H., Ludwig, M. L., Steitz, T. A., Muirhead, H. and Coppolla, J. C.: 1968, Brookhaven Symp. Biol. 21, p. 24.

(11) Lipscomb, W. A.: 1970, Acc. Chem. Res. 3, p. 81.

(12) Makinen, M. W., Yamamura, K. and Kaiser, E. T.: 1976, Proc. Natl. Acad. Sci. U.S.A. 73, p. 3882.

(13) Breslow, R. and Wernick, D. L.: 1977, Proc. Natl. Acad. Sci. U.S.A. 74, p. 1303.

(14) Kroll, H.: 1955, J. Am. Chem. Soc. 74, p. 2036.

(15) Bender, M. L. and Turnquest, B. W.: 1955, J. Am. Chem. Soc. 77, p. 4271.

(16) Bender, M. L. and Turnquest, B. W.: 1957, J. Am. Chem. Soc. 79, p. 1889.

(17) Meriwether, L. and Westheimer, F. H.: 1956, J. Am. Chem. Soc. 78, p. 5119.

(18) Breslow, R., Fairweather, R. B. and Keana, J.: 1967, J. Am. Chem. Soc. 89, p. 2135.

(19) Fairweather, R. B.: Ph.D. Thesis, Columbia University, 1967.

(20) (a) Buckingham, D. E., Davis, C. E., Foster, D. M. and Sargeson, A. M.: 1970, J. Am. Chem. Soc. 92, p. 5571; (b) Buckingham, D. A., Foster, D. M. and Sargeson, A. M.: 1970, ibid. 92, p. 6151.

(21) Buckingham, D. A., Keene, F. R. and Sargeson, A. M.: 1974, J. Am. Chem. Soc. 96, p. 4981.

(22) (a) Chaffee, E.; Dasgupta, T. P. and Harris, G. M.: 1973, J. Am. Chem. Soc. 95, p. 4169; (b) Palmer, D. A. and Harris, G. M.: 1974, Inorg. Chem. 13, p. 965.

(23) Buckingham, D. A., Harrowfield, J. MacB. and Sargeson, A. M.: 1973, J. Am. Chem. Soc. 95, p. 7281.

(24) Buckingham, D. A. and Engelhardt, L. M.: 1975, J. Am. Chem. Soc. 97, p. 5916.

(25) Fife, T. H. and Squillacote, V. L.: 1977, J. Am. Chem. Soc. 99, p. 3762.

(26) Fife, T. H. and Squillacote, V. L.: 1978, J. Am. Chem. Soc. 100, p. 4787.

(27) Groves, J. T. and Dias, R. M.: 1979, J. Am. Chem. Soc. 101, p. 1033.

(28) Chambers, R. R., Ph.D. Thesis, The University of Michigan, 1981.

(29) Deslongchamps, P.: 1975, Tetrahedron 31, p. 2463.

SPECTROSCOPIC STUDIES ON COBALT(II) CARBOXYPEPTIDASE A

Ivano Bertini, Giovanni Lanini, Claudio Luchinat, and
Roberto Monnanni
Istituti di Chimica Generale e Inorganica,
Università di Firenze,
Via G. Capponi, 7
50121 Firenze, Italy

Abstract: On the ground of electronic spectra and ^1H NMR data it is proposed that a coordinated water molecule in the cobalt(II) derivative of carboxypeptidase A is subject to deprotonation with a pK_a of about 9. A coordinated hydroxide might be formed at lower pH in the enzyme-substrate adduct as a transient species of catalytic relevance.

The role of the zinc(II) ion in the catalytic mechanism of carboxypeptidase A is still unclear, despite the brilliant structural studies performed on the native enzyme and its inhibitor or pseudo-substrate derivatives. As discussed elsewhere in the book (1,2), it is usually accepted that the substrate binds the metal through its peptide carbonyl oxygen, but the nucleophile attacking the carbonyl carbon has not yet been identified. Among the possible candidates there is a metal coordinated hydroxide (3); this proposal would require the availability of two open coordination sites at the metal. Although the X-ray data on the native enzyme suggest the presence of a single water molecule bound to the zinc(II) ion (4), a recent structure of the enzyme adduct with the inhibitor Gly-L-Tyr (5) has shown the ligand to be in equilibrium between a monodentate and a bidentate arrangement; in the latter, the metal donors are the carbonyl oxygen and the amine nitrogen of the Gly residue. From earlier experiments on the manganese(II) derivative it also appears that the metal can simultaneously bind a water molecule and a fluoride ion (6).

In order to shed further light on the possibility for the active site metal ion to achieve five-coordination we have initiated a thorough spectroscopic study on the cobalt(II) derivative of carboxypeptidase A from bovine pancreas. The cobalt(II) derivative (CoCPA) is the closest analogue to the native enzyme (7). Furthermore, cobalt(II) is particularly suitable for spectroscopic investigations: as discussed in chapter I, a spectroscopic criterion has been proposed to discriminate among the various coordination numbers, based on the intensity of the electronic spectra and on ^1H NMR data.

Following the above criterion, the electronic spectra of CoCPA are assigned as five coordinated (Figure 1); the assignment is confirmed

I. Bertini, R. S. Drago, and C. Luchinat (eds.), The Coordination Chemistry of Metalloenzymes, 93–97.

by the spectrum of the Gly-L-Tyr derivative, which behaves at least partially as bidentate, and by those of the N_3^- and NCO^- derivatives which, on the other hand, are typical of pseudotetrahedral geometry (see Figure 3 A in chapter I).

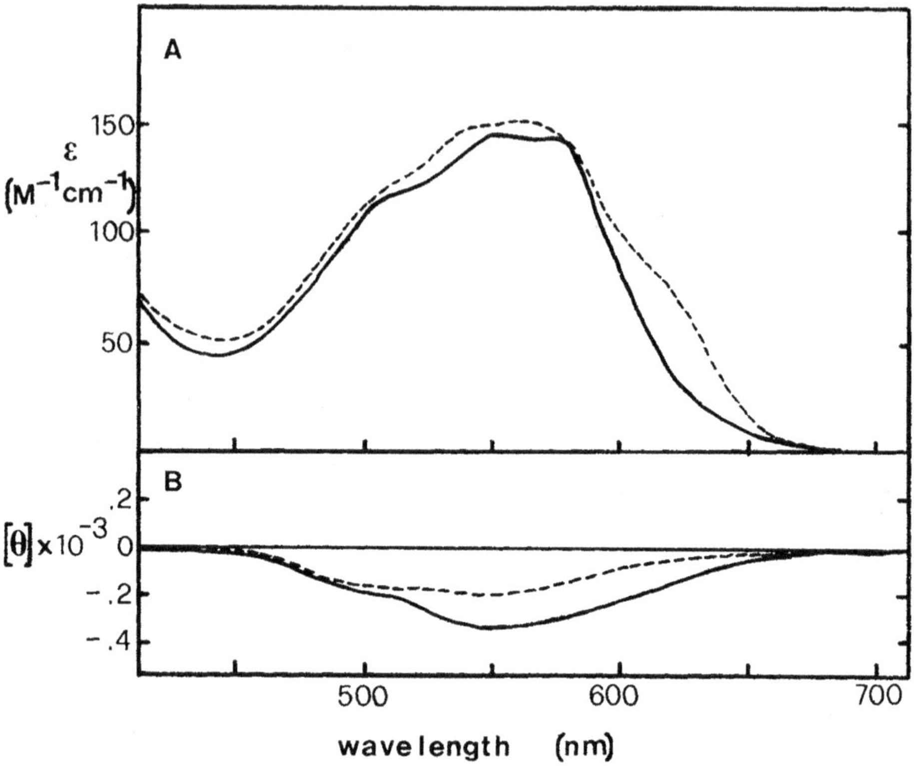

Figure 1. Electronic (A) and CD (B) spectra of cobalt(II) carboxypeptidase A at pH 7.0 (——) and 9.5 (---).

Finally, water proton relaxation measurements on the pure cobalt(II) derivative and the β-phenylpropionate adduct have shown that water is present in the coordination sphere in both cases, and the electronic relaxation times are again indicative of five- rather than four-coordination (8).

Once the five-coordinated geometry is established, the possibility exists for the coordinated carbonyl group of the substrate to be attacked by the adjacent coordinated water: the latter could ionize to OH^- during the process. Therefore, it would be important to establish whether the coordinated water is capable of ionizing in a suitable pH region. It is well established that there is at least one enzyme ionization which takes place in the active site, the pK_a from activity measurements being 8.8 (7). Such ionization is also reflected in a slight change of the electronic and CD spectra (Figure 1). However, the latter variation is somewhat too small to be taken as an evidence of deprotonation occurring in the first coordination sphere.

A much better indication comes from the ^{1}H NMR spectra (8). Owing to the presence of the paramagnetic metal ion the signals of the coordinated histidines undergo large isotropic shifts, allowing their observation well outside the usual proton chemical shift range (8,9). Both histidines in carboxypeptidase A are coordinated through the N1 nitrogen; therefore two of the four protons of the imidazole rings (namely the 4-H protons) will be in meta position with respect to the coordinated nitrogen and will not undergo severe paramagnetic broadening (9) (Figure 2). The pH dependence of the isotropic shift of such signals is sizeable and comparable to that observed in cobalt(II) carbonic anhydrase (9), where the water deprotonation is now well established (see elsewhere in the book) (1,10,11).

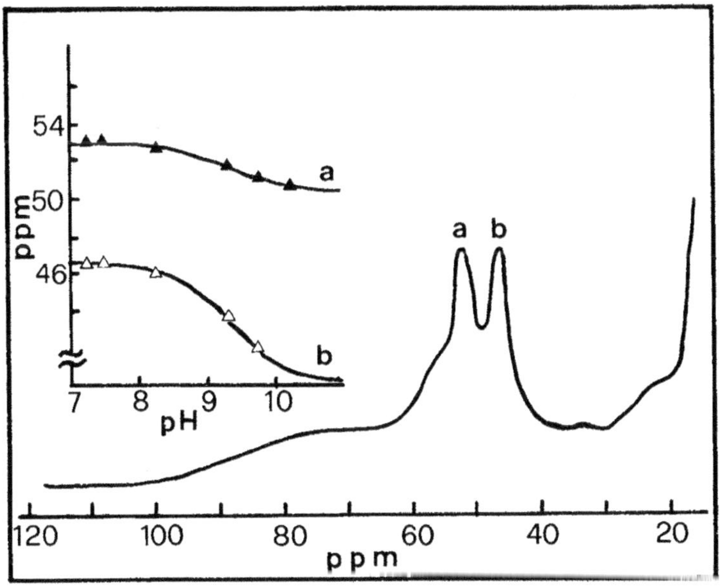

Figure 2. 60 MHz ^{1}H NMR spectrum of cobalt(II) carboxypeptidase A in D_2O solution at pH* 6.8. The inset shows the chemical shifts of the 4-H histidine signals as a function of pH*. The data are fitted to a pK_a of 8.8.

A further analogy with carbonic anydrase is worthy to be stressed here: in the latter enzyme the affinity of anions drops at high pH (12) consistently with competition of the negatively charged ligands with the hydroxyde ion. Figure 3 shows that this is indeed the case even in cobalt(II) carboxypeptidase A.

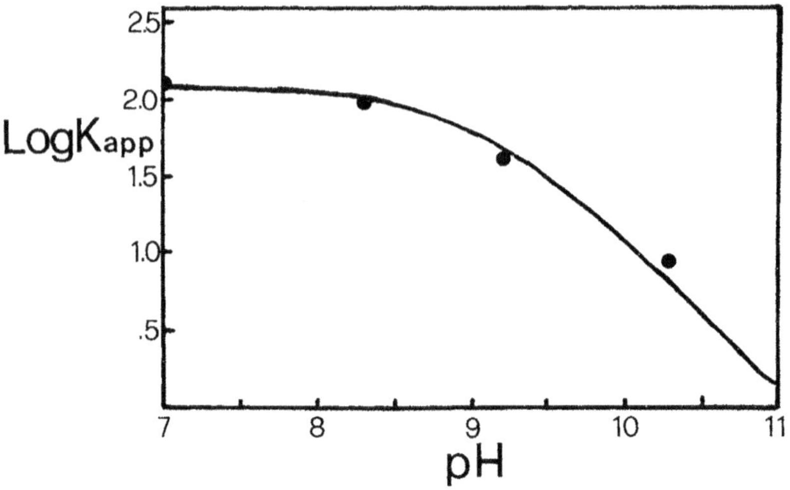

Figure 3. pH dependence of the apparent affinity constant (K_{app}) of N_3^- for cobalt(II) carboxypeptidase A. The data are best fitted to a pK_a of 9.1.

The body of experimental evidences discussed above leads us to suggest that a coordinated water molecule ionize to OH^- in cobalt(II) carboxypeptidase A with a pK_a of about 9. This does not imply that a coordinated hydroxide is required for the catalytic activity, which actually drops at high pH; nevertheless such a pK_a value may be lowered upon coordination of the substrate, allowing a proton transfer from water to another group in the active site. A possible mechanism takes advantage of the early proposal of the involvement of Tyr 248 in the catalytic pathway (4): the latter group would donate a proton to the nitrogen involved in the peptidic bond during the C-N bond cleavage: simultaneously, the carbonyl group, polarized by metal coordination, would undergo a nucleophylic attack by the newly generated hydroxide; the excess proton would finally neutralize the tyrosinate anion.

REFERENCES

(1) see chapter I in this book.
(2) see chapter VI in this book.
(3) Kuo, L.C., and Makinen, M.W.: 1982, J. Biol. Chem., 257, p. 24.
(4) Lipscomb, W.N.: 1980, Proc. Natl. Acad. Sci. USA, 77, p. 3875

(5) Rees, D.C., Lewis, M., Honzatko, R.B., Lipscomb, W.N., and Hardman, K.D.: 1981, Proc. Natl. Acad. Sci. USA, 78, p. 3408.
(6) Navon, G., Shulman, R.G., Wyluda, B.J., and Yamane, T.: 1970, J. Mol. Biol., 51, 15.
(7) Latt, S.A., and Vallee, B.L.: 1971, Biochemistry, 10, p. 4263.
(8) Bertini, I., Canti, G., and Luchinat, C.: J. Am. Chem. Soc., in press.
(9) Bertini, I., Canti, G., Luchinat, C., and Mani, F.: 1981, J. Am. Chem. Soc., 103, p. 7784.
(10) see chapter II in this book.
(11) see chapter IV in this book.
(12) Bertini, I., Luchinat, C. and Scozzafava, A.: 1982, Struct. Bonding, 48, 45.

COORDINATION PROPERTIES AND MECHANISTIC ASPECTS OF LIVER ALCOHOL DEHYDROGENASE

Michael Zeppezauer

Fachbereich 15, Analytische und Biologische Chemie
Universität des Saarlandes
D-6600 SAARBRÜCKEN, W. Germany

ABSTRACT

After a brief survey of current ideas on structure, kinetics and mechanism of action of liver alcohol dehydrogenase, the coordination chemistry of the metal binding centers is described following physical and chemical studies on native and metal substituted enzyme derivatives. Kinetic and spectroscopic studies with different metallo alcohol dehydrogenases are discussed which help to clarify the mode in which the catalytic metal ion participates in different steps of the catalytic cycle.

1. INTRODUCTION

1.1 Occurence and Function of NAD Dependent Alcohol Dehydrogenases

Alcohol dehydrogenases catalyzing the reaction

$$ {}^{1}R^{2}RCH\text{-}OH \; + \; NAD^{+} \rightleftharpoons {}^{1}R^{2}RC{=}O \; + \; NADH \; + \; H^{+} \qquad (1) $$

occur in the cells of numerous, if not all living species [1]. It is the aim of this review to illustrate the importance of metal ions for the catalytic performance of liver alcohol dehydrogenase. Major differences in oligomeric structure are found between the enzyme of mammalian liver and microorganisms, with 2 and 4 identical subunits, respectively. Still, there are common features such as sequence homology, conservation of residues important for the maintenance of tertiary structure, and a relative abundance of cysteine as compared with the average cysteine contents of the proteins of the species in question. Some of the alco-

I. Bertini, R. S. Drago, and C. Luchinat (eds.), The Coordination Chemistry of Metalloenzymes, 99–122.

hol dehydrogenases have been shown to contain Zn^{2+} , but a general relationship between zinc content and cysteine content or even more detailed information on active site components has not yet been available. Thus, from both chemical and biological points of view a more extensive comparison of alcohol dehydrogenases from different species seems worthwhile.

The physiological function of alcohol dehydrogenases is obvious in ethanol fermenting or heterofermentative microorganisms. A representative of the latter, Leuconostoc mesenteroides contains a tetrameric zinc alcohol dehydrogenase which in contrast to most other species, accepts $NADP^+$ more readily than NAD^+ [2]. However, the abundance of alcohol dehydrogenase in the liver of mammals, particularly of non-ruminants and its broad substrate specificity still have to be correlated with an adequate physiological role. This applies to both isoenzymes of horse liver, the ethanol active EE form and the steroid active SS form, since neither ethanol nor 3-β-hydroxysteroids are of major importance in mammalian metabolism. In addition, neither the esterase [3] nor the peroxidase [4] and isomerase activities of HLADH[a] have been placed in a proper metabolic context[5].

Most likely, the recent observation by Dutler and Ambar[6,7] that the EE isoenzyme catalyzes the oxidation of L-histidinol to L-histidinal and subsequently to L-histidine has opened the way towards an understanding of the biological function of this enzyme. It has also led to important ideas concerning the chemical principles underlying the catalytic action of metal ions in this type of metalloenzymes which are discussed in this volume [7,8].

1.2 Kinetic Mechanism

In the Theorell-Chance mechanism of alcohol oxidation [9] the formation of the binary complex $E \cdot NAD^+$ precedes the formation of the ternary complex $E \cdot NAD^+ \cdot$alcohol. Its interconversion into E•NADH•aldehyde is not rate limiting. Since also the dissociation of aldehyde from the ternary complex is rapid, no significant concentrations of the latter build up under steady state conditions. The rate of dissociation of NADH from the binary complex E•NADH determines the turnover rate, k_{cat}:

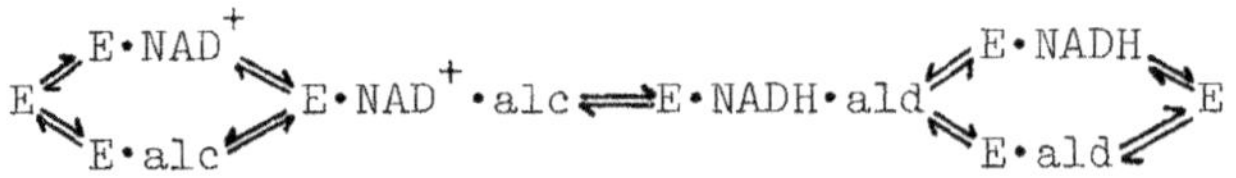

(2)

Although the ordered pathway sometimes does not hold for the oxidation of secondary alcohols when ternary complex conversion (i.e. "hydride transfer") becomes rate limiting, the Theorell-Chance mechanism, apart from its general importance for the development of enzyme kinetics, has been fundamental for the prediction of coenzyme-induced conformation changes in this class of proteins [10].

1.3 Molecular Structure and Conformational Transitions

General composition: The horse liver enzyme, molecular weight 80 000 daltons, is a dimer with 374 amino acids in each subunit. Fourteen of them are free cysteines. Each subunit contains two zinc ions and has one active site. Crystal forms: The coenzyme-free EE-form and binary complexes with various ligands (except coenzyme) crystallize in an orthorhombic space group, whereas most complexes containing either NAD^+ or NADH crystallize in a triclinic or monoclinic space group [1]. Examples are given in Table I.

Table I. Crystalline Complexes of Horse Liver Alcohol Dehydrogenase

Binary complexes orthorhombic	Ternary complexes triclinic or monoclinic
LADH (EE)	LADH•NADH• solvent
LADH•imidazole	LADH•NADH•isobutyramide
LADH•1,10-phenanthroline	LADH•NADH•DMSO
LADH•AMP	LADH•1,4,5,6 -tetrahydro-NAD•DACA
LADH•ADP-ribose• 1,10-phenanthroline	LADH•NAD^+fatty acid
LADH•salicylate	LADH•NAD^+•pyrazole
LADH•1,4,5,6-tetrahydro-NAD•solvent	
LADH•imidazole•NADH	
LADH•ADP-ribose	

High resolution studies of the orthorhombic [1] and the triclinic enzyme (viz. LADH•NADH•DMSO [60]) revealed significant differences in the protein conformation (see below). Molecular shape: The general shape of the molecule is that of a prolate ellipsoid of dimensions 45x60x110 Å. Each subunit is divided into two domains joined by a narrow neck region and separated by the deep active site cleft. One of these domains serves to bind the coenzyme and closely resembles the coenzyme binding domain of lactate, malate, and glycerinaldehyde-3-phosphate dehydrogenase [1]. The two zinc ions are bound within the second domain, called the catalytic domain. The latter is large and comprises 143 residues. The two subunits of the enzyme are joined together through their coenzyme binding domains which are separated from the catalytic domains on each side by the active sites [1].

Conformational States: Binding of coenzyme leads to a domain rotation. The catalytic domain moves relative to the coenzyme binding domain and the cleft between the two domains becomes narrower. Thus, after binding of coenzyme and substrate the active site becomes completely shielded from the exterior (Fig. 1).

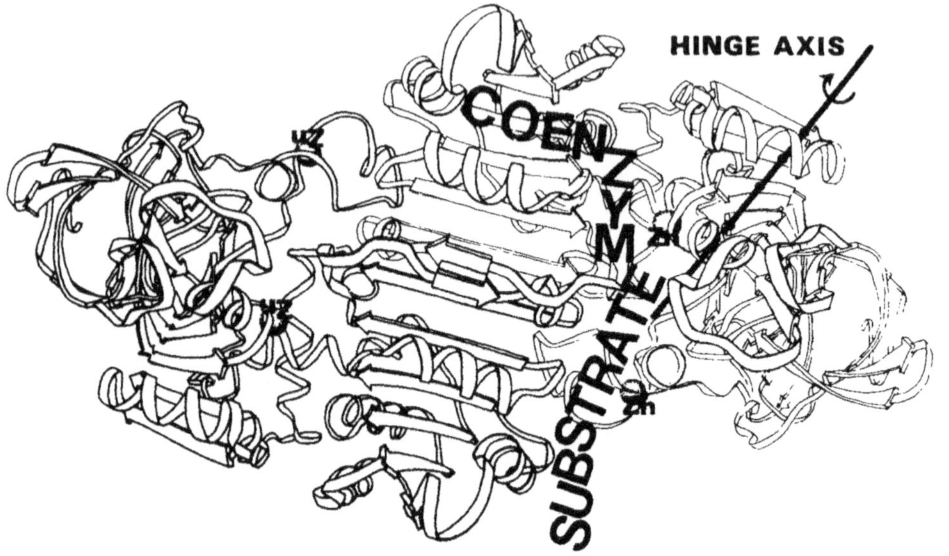

Figure 1. Polypeptide chain folding of horse liver alcohol dehydrogenase (HLADH). By courtesy of Elia Cedergren [59]

Metal ion environments: The catalytic zinc ion is situated at the bottom of the cleft between the coenzyme binding domain and the catalytic domain. It is accessible only through two channels, one which accomodates the coenzyme and another through which the substrate approaches the catalytic site. Three protein ligands are involved in binding the metal; i.e. cysteines 46 and 176, and histidine 67; the fourth ligand is a water molecule (Fig. 2). The geometry is distorted tetrahedral. 1,10-phenanthroline binds to the catalytic zinc ion, thereby replacing the water molecule and extending the coordination number to five.

The noncatalytic zinc ion is located in the periphery of the catalytic domain; the distance between both metal ions is 20 Å. The noncatalytic zinc ion is liganded by cysteines 97, 100, 103, and 111 in a tetrahedral geometry similar to that of iron in rubredoxin. The lobe which contains the noncatalytic metal binding sites exhibits striking similarities with regard to chain folding and spacing of the four cysteine residues as compared to the bacterial ferredoxins (Fig. 2).

1.4 Catalytic Mechanism

Enormous efforts have been undertaken to elucidate the catalytic mechanism of alcohol dehydrogenases by combining the tools of structural chemistry, enzymology, and physical chemistry (for recent reviews see references [6,11]). Major objectives have been the kinetic separation

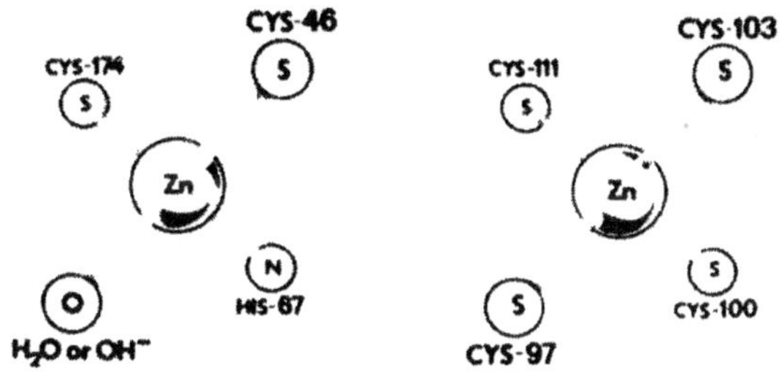

Figure 2. Metal ion ligands in HLADH. Left: Catalytic site, right: noncatalytic site. With permission from ref. [1].

of proton and hydride transfer steps, the identification of ionization equilibria along the kinetic pathway, their possible perturbations, and the nature of groups involved in proton transfer. A great deal of experimentation and thought has been devoted to the role the metal ion plays in catalysis, especially the mode of substrate binding and activation as well as the nature of metal-linked ionization equilibria, including the role of metal-bound water. Almost no attention has been paid so far to the question, how the metal may influence the thermodynamics and kinetics of coenzyme binding and the structural transition linked to this process. Concerning substrate binding and activation, two basically different ideas have evolved: (i) The substrate binds to the metal-bound water molecule, which acts as general acid-base catalyst (outer-sphere binding of substrate). This conclusion was drawn by A. Mildvan and associates from NMR experiments using cobalt substituted horse liver enzyme (see ref. [11]). The idea is principally attractive, since it places the catalytic mechanism of alcohol dehydrogenase in close neighbourhood to those of other NAD dependent dehydrogenases which perform general acid-base catalysis via protein side chains (notably histidine) [6]. (ii) The catalytic zinc ion functions as Lewis acid, it binds and activates the substrate by means of inner-sphere coordination. This idea has been more frequently advanced then (i) and has been pronounced in varying shapes. The simpler versions invoke only four-coordinate states of the metal ion. The water molecule is thought to be replaced by the substrate [1], alcohols are thought to coordinate and to react as alcoholate ions. Clearly, the proton abstraction from the substrate must be separable from hydride transfer in this phase of the catalytic cycle, which has been proven and inspired G. Pettersson to a "unified mechanism" of LADH catalysis [34]. Important features in these groups of mechanisms have been the assignment of the pK_a values of 9.2 to the ionization of unperturbed metal-bound water and of 7.4 to the ionization of this water after binding of and Coulomb-perturbation by NAD^+. In addition, attempts have been made to define the pK_a values of bound substrate (inhibitor) molecules [12].

Extended concepts have been developped invoking five-coordinated states, where the substrate molecule is bound as inner-sphere ligand without replacing the water molecule. In this way mechanisms have been formulated which do not necessarily depend on the formation of intermediate alcoholate ions [11]. Several of the proposed mechanisms provide logical sequences of chemical events which may very well be in accord with experimental, mostly kinetic data. However, all of them suffer from a severe drawback: Any reliable mechanism would necessarily have to include information about the concerted geometric and electronic rearrangements within the coordination sphere of the metal ion during the catalytic cycle. Only fragmentary information of this kind is available in the existing literature on alcohol dehydrogenases [36,38] as well as on other metalloenzymes. To our knowledge, so far only Dutler and Ambar have thoroughly analyzed the stereoelectronic aspects of the catalytic cycle of LADH in its different phases. Their concept of "Ligand Sphere Transitions" [7,8] will certainly apply great stimuli to both experimental and theoretical work.

2. COORDINATION CHEMISTRY OF METALLO ALCOHOL DEHYDROGENASES

2.1 Chemical Reactivity of and Metal Exchange at the Different Metal Binding Sites

2.1.1 Exchange by dialysis against divalent metal ions. It was early observed by Vallee and coworkers that anaerobic dialysis of LADH against divalent metal ions leads to the formation of metal - zinc-hybrid enzymes according to equation 3

$$2\ Me^{2+} + Zn_4LADH \rightarrow Me_2Zn_2LADH + 2\ Zn^{2+}$$
$$Me^{2+} = {}^{65}Zn^{2+},\ Co^{2+},\ Cd^{2+} \qquad (3)$$

within a few hours. Different conditions could be found, which either restrict the exchange to half of the Zn^{2+}, or permit the exchange of all four zinc ions. However, the second pair of metal ions is exchanged much slower. E.g., the complete exchange for Co^{2+} takes 72 hours [13]. The catalytic properties of the Me_2Zn_2LADHs prepared by dialysis techniques are virtually unchanged which demonstrates, that the noncatalytic zinc ions are replaced first. Raising the temperature should stimulate this process, which was proven using a column technique [14]. However, great losses of protein occurred, although the properties of the fully substituted cobalt enzyme prepared at 55°C were similar to those obtained by prolonged dialysis in the cold. It seems plausible, that the exchange at the noncatalytic site proceeds via binuclear intermediates:

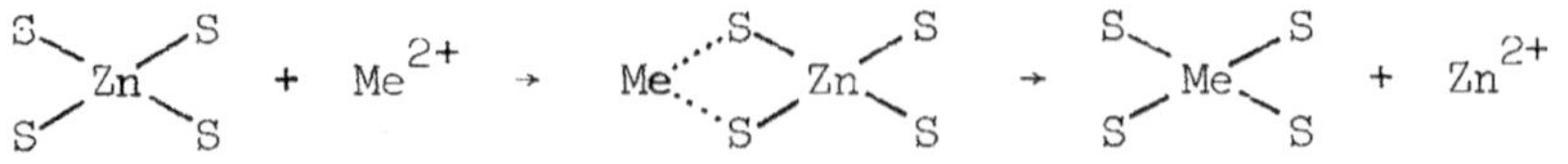

(4)

Similar pathways are known in the chemistry of low molecular weight metal complexes of similar structure [15]. The noncatalytic sites are close to the surface of the enzyme molecule and should therefore be easily accessible to this kind of reaction. This mechanism seems less probable for the exchange at the catalytic metal site, since the catalytic zinc ion is buried in a deep hydrophobic cleft. We have proposed, that for the exchange at the catalytic site to occur the dissociation of the metal ion from its site or a related process may be rate limiting.

2.1.2 Removal of zinc ions in solution. Since metal-chelating ligands can attack the catalytic zinc ion, polydentate ligands should be able to effectively compete with the protein for the catalytic metal. Indeed, the specific removal of one pair of Zn^{2+} with total loss of activity was achieved by treating soluble enzyme with dithiocarbamate [39]. However, it was not possible to reactivate this Zn_2LADH with Zn^{2+}. When soluble LADH is treated with the tridentate dipicolinic acid, the protein even precipitates [16]. No reactivation was reported of LADH deprived of all metal ions by treatment with EDTA [40] whereas LADH denatured and depleted of Zn^{2+} in 8 M urea could be reactivated [18]. Additional literature on this subject is rewiewed in [1].

2.1.3 Extraction and insertion of catalytic metal ions in crystalline LADH. This technique was originally pioneered by the late Björn Tilander, with the aim of preparing isomorphous heavy atom derivatives of carbonic anhydrase [7]. In the case of alcohol dehydrogenase it proved vital for the preparation of the demetallized species, $H_4Zn(n)_2$LADH, which can be reconstituted by either zinc or other divalent metal ions both in crystals and in solution [14]. It is interesting to note that reconstitution can be achieved at much higher metal concentration when working in crystals as compared to solutions of the enzyme. E.g., Zn^{2+} reconstitutes the crystalline enzyme at 1 mM concentration, whereas in solution concentrations > 20 µM lead to precipitation. For Co^{2+} the corresponding figures are 5 mM and 1 mM, respectively. This has been taken as evidence for a partial unfolding of the demetallized $H_4Zn(n)_2$ enzyme [14]. This is reasonable since the ligands of the catalytic metal ion stem from rather distant parts of the polypeptide chain. A certain flexibility of the $H_4Zn(n)_2$ enzyme cannot be excluded in solution, and this may lead to creation of "unnatural" metal binding sites. These sites may saturate at higher metal concentrations and then prevent the correct refolding. It is well known, that refolding of completely metal free LADH depends on the participation of zinc ions, although the mechanism and the pathways of this process are not yet understood [18]. It is noteworthy that dissolved $H_4Zn(n)_2$ enzyme can be recrystallized under the same conditions as native enzyme; it is reconstituted under conditions and with results identical to those when working solely in the crystalline state. This shows that the presumed unfolding of $H_4Zn(n)_2$-LADH is essentially reversible.

So far we have achieved reconstitution of $H_4Zn(n)_2$ enzyme in crystal suspension with the following metals: Zn^{2+}, Co^{2+} [14], Cu^{2+} [19], Cd^{2+} [20], Ni^{2+} [16], Pb^{2+}, Fe^{2+}, Fe^{3+} [16,21]. Under the usual conditions of assay [22] most metals seem to be catalytically active with the exception of Cu^{2+}, Fe^{3+} and Pb^{2+}. The Cu^{2+} and Fe^{3+} enzymes are bleached after exposure to oxygen, the resulting Cu^{2+} enzyme is active whereas the activity of the Fe^{2+} enzyme is low (5% that of the turnover of the native enzyme). The problem of correctly assessing "activity" (k_{cat}) is treated below. In our hands, the Pb^{2+} enzyme prepared in crystal suspension has shown no activity. In contrast, other authors have observed the recovery of enzymic activity upon incubation of dissolved $H_4Zn(n)_2$ enzyme (prepared according to ref. [14]) with Pb^{2+}, Hg^{2+} and Cd^{2+} ions [23]. Neither the metal contents nor proof for the location of the inserted metal ions were reported. During the experiments some precipitation of insoluble material seems to have occurred. No rate law describing the recombination of the $H_4Zn(n)_2$ enzyme with any of the heavy metal ions was presented. It is well known that noncatalytic zinc ions are readily exchanged by Cd^{2+} [15] and Hg^{2+} ions [24], and even the catalytic zinc ions seem to be replaced fairly easily both in solution and in crystals [24]. Also LADH precipitated after denaturation liberates zinc ions [25]. Thus several competing processes may occur, especially in solution, which can lead to the insertion of zinc ions as well as heavy metal ions in the catalytic site of $H_4Zn(n)_2$LADH parallel to insertion of heavy metals in the noncatalytic site. It will deserve careful analytical work to establish the chemical identity of the various metal hybrids prepared in solution, especially of species, where straightforward spectroscopic characterization is not feasible.

2.2 Kinetics and Mechanism of the Recombination of $H_4Zn(n)_2$LADH with Co^{2+} and Ni^{2+}

From the foregoing discussion it has become apparent, that the two metal binding sites in LADH exhibit considerable differences with respect to structure and chemical reactivity. A thorough understanding of their characteristic properties deserves, inter alia, knowledge of the kinetics and mechanism of metal exchange at noncatalytic sites, the metal extraction process and the recombination process at the catalytic sites. As indicated this kind of information is also a prerequisite for designing successful experiments for the preparation of true site specific hybrid metallo ADHs, especially where insertion rates at the catalytic sites may be comparable to exchange rates at the noncatalytic sites as it seems to be the case with the "soft" cations such as Pb^{2+}, Cd^{2+}, Hg^{2+}. Apart from its immediate practical usefulness, information on the kinetics and thermodynamics of metal binding in both sites would be of considerable biological interest. As mentioned above, the noncatalytic site is a biological model of the iron binding sites in rubredoxins, and the tetrahedral arrangement of two sulfurs, one nitrogen and one oxygen around the catalytic zinc puts this site in close neighbourhood to the blue copper proteins [14]. Therefore a comparison of the properties of the metal binding sites in alcohol dehydrogenases with those of related copper and iron proteins may help to elucidate the principles which govern the selectivity of binding sites and hopefully provide some clues as to their biosynthesis.

We have found that the recombination of Co^{2+}, Ni^{2+}, and Zn^{2+} ions could be studied on a time scale of minutes and by means of kinetic studies information could be obtained concerning stoichiometry, rate laws, activation parameters and the pathway of the metal into its binding site [26].

The formation rates were determined optically by activity measurements or by using the pH change during metal incorporation. Two protons per binding site are released during the binding of Zn^{2+} to $H_4Zn(n)_2LADH$ in the pH range 6.8 - 8.1, indicating that at least two of the ligands have a $pK_a > 8$. These two ligands are probably the -SH groups of cys 46 and cys 174. The recombination rates are also increased with pH, analogous to the observations for apocarboxypeptidase,-carbonic anhydrase, -superoxidedismutase, and -azurin. In contrast to available data on apocarboxypeptidase where metal binding occurs in a second-order reaction, the recombination of Zn^{2+}, Co^{2+} and Ni^{2+} ions with $H_4Zn(n)_2LADH$ is a consecutive process similar to that of the copper proteins azurin, stellacyanin and superoxidedismutase. The first step is the fast, reversible formation of an intermediate, characterized by the formation constant K_1. In the second step, the native (or the substituted) enzyme is reconstituted with the rate constant k_2 in a monomolecular process:

$$\frac{1}{2}\ H_4Zn(n)_2LADH + Me^{2+} \rightleftharpoons I \longrightarrow \frac{1}{2}\ Me(c)_2Zn(n)_2LADH + 2H^+ \qquad (5)$$

$$k_{obs} = k_2 \cdot k_1 \cdot [Me^{2+}]/(1 + K_1 \cdot [Me^{2+}]) \qquad (6)$$

The rate expression is valid for one metal binding site. No indication of cooperativity in the process of recombination was obtained. The observed rate showed the expected hyperbolic dependence on metal concentration in the case of Zn^{2+} and Co^{2+}, whereas it proved to be independent for Ni^{2+}. This would require a formation constant $K_1 > 10^5$ for the latter metal, which seems reasonable. The kinetic and activation parameters are given in Table II which gives a comparison with relevant data for the other proteins studied so far. This comparison shows that the reconstitution of $H_4Zn(n)_2LADH$ with both Co^{2+} and Ni^{2+} is characterized by exceptionally large negative activation entropies. Recombination of the other metalloproteins shows more or less large, positive activation entropies which partially compensate high activation enthalpies. This may also indicate some flexibility in the catalytic site of $H_4Zn(n)_2LADH$ which leads to a sterically restrictive transition state for the metal insertion pathway. Concerning the mechanism, two alternative models have to be discussed. One model considers the binding of the metal in its functional site as the first step and a slow conformational change, building up the final coordination sphere as the second step. The other possibility is the reversible binding of the metal ion at a peripheral site from which the ion moves to its functional site at a rate defined by k_2. These two models are not easily distinguished. However, using nuclear magnetic relaxation dispersion (NMRD), Andersson et al. could demonstrate the binding of Mn^{2+} ions to $H_4Zn(n)_2LADH$ at least at two different kinds of peripheral binding sites [28]. One type is characterized by high affinity (i.e. a dissociation constant of ~ 0.01 mM at 5 °C and pH 7.7 for Mn^{2+} and significantly smaller for Zn^{2+} and Cd^{2+}) and low relaxivity; the stoichiometry is about one site per molecule $H_4Zn(n)_2LADH$. The other type of site has low affinity (K_D ~ 1mM) and high relaxivity, the stoichiometry is uncertain. Zn^{2+} ions added to solutions of $H_4Zn(n)_2LADH$ so long as they are not in excess, bind initially to the new tight-binding sites before going to the catalytic sites. Mn^{2+} ions are set free and are detected by the NMRD signals of the aquo ions. When the zinc ions move to the catalytic sites the liberated Mn^{2+} can return to the low-relaxivity sites, and the decay of the NMRD signal can be used to determine the migration rates. The on-rate of Cd^{2+} ions to the catalytic sites is too fast to be followed with this technique; it is at least one order of magnitude greater than that of Zn^{2+}.

The pathway of the metal was revealed by studying the recombination in the presence of NAD^+ or NADH. Both forms of the coenzyme completely prevent the insertion of the metal ions. This demonstrates that hydrated Zn^{2+}, Co^{2+} and Ni^{2+} ions migrate exclusively through the coenzyme domain and not through the hydrophobic substrate channel.

Table II. Kinetic and Activation Parameters for the Recombination of Metal-Depleted Metalloproteins with Metal Ions

Protein	Rate Constant	Formation Constant of Intermediate (K_1 M^{-1})	$\Delta H^{\ddagger}$ ($kJ \cdot Mole^{-1}$)	$\Delta S^{\ddagger}$ ($kJ \cdot Mole^{-1} \cdot deg^{-1}$)	pH	Ref. 26
Carboxypeptidase	Zn^{2+} 7 $\cdot 10^5$ ($M^{-1}s^{-1}$)					
	Cd^{2+} 9 $\cdot 10^5$ ($M^{-1}s^{-1}$)					
	Mn^{2+} 1.7$\cdot 10^5$ ($M^{-1}s^{-1}$)					
	Co^{2+} 3 $\cdot 10^5$ ($M^{-1}s^{-1}$)					
	Ni^{2+} 5 $\cdot 10^3$ ($M^{-1}s^{-1}$)					
Carbonic Anhydrase	Zn^{2+} $\cdot 10^4$ ($M^{-1}s^{-1}$)		84.5	125.5		
	VO^{2+} $\cdot 10^8$ ($M^{-1}s^{-1}$)					
	Cu^{2+} 6 $\cdot 10^4$ ($M^{-1}s^{-1}$)		87.9			
	Co^{2+} 4.0 ($M^{-1}s^{-1}$)		100.4	142.3		
	Ni^{2+} 5.0 ($M^{-1}s^{-1}$)					
	Mn^{2+} 9.0 ($M^{-1}s^{-1}$)		83.7	117.2		
Superoxide Dismutase	Cu^{2+} 8.3$\cdot 10^{-4}$(s^{-1})		92		5.9	
Stellacyanin	Cu^{2+} 7.5$\cdot 10^{-3}$(s^{-1})		92	12.6	8.0	
Azurin	Cu^{2+} 4.0 (s^{-1})		83.7	37.7	6.0	
Horse Liver Alcohol Dehydrogenase	Zn^{2+} 5.2$\cdot 10^{-2}$(s^{-1})	1.2$\cdot 10^4$			6.9	
	Co^{2+} 1.1$\cdot 10^{-3}$(s^{-1})	2.5$\cdot 10^4$	51	-146	6.9	
	Ni^{2+} 2.2$\cdot 10^{-4}$(s^{-1})	10^5	48.5	-163	7.2	

2.3 X-ray Structural Investigation of Metallo Alcohol Dehydrogenases

We have undertaken X-ray studies of $H_4Zn(n)_2$LADH and $Co(c)_2Zn(n)_2$ LADH [27] and their complexes with coenzymes (unpublished results) to obtain answers to the following questions: (i) How flexible is the structure of the protein depleted of its catalytic metal ions? (ii) How is the binding of the coenzyme influenced by removal of the catalytic metal ions? (iii) Is the coordination geometry changed upon insertion of metals other than zinc into the catalytic site? The data on coenzyme binding to $H_4Zn(n)_2$LADH obtained so far are too preliminary, those on the unliganded species can be summarized as follows.

2.3.1 Structural Properties of $H_4Zn(n)_2$LADH:

(i) The electron density map ($F_{obs}-F_c$) shows a deep minimum at the position of Zn(c).

(ii) There is no change of electron density in the position of Zn(n).

(iii) The observations described under (i)-(ii) cover time periods of many weeks.

(iv) The side chain of cysteine 46 moves away from the metal site, the distance between cys 46 and cys 174 increases from 4 to 4.5 Å.

(v) Histidine 67 moves slightly away from the site.

(vi) No other changes in electron density are observed throughout the entire protein structure.

The conclusions are obvious: The protein depleted of the catalytic metal ions remains structurally rigid, at least when immobilized in the crystal lattice. It is noteworthy that the only observable changes are confined to the catalytic metal binding site. Evidently, no disulfide bridge is formed between cys 46 and cys 174 upon removal of the metal ion. Hence, no additional heterogeneity had been introduced with respect to the nature of the metal ligands which is important for the interpretation of spectroscopic and kinetic data. It is further evident from our X-ray study that no migration of Zn^{2+} occurs from the noncatalytic to the empty catalytic site upon prolonged storage of crystalline $H_4Zn(n)_2$ enzyme.

2.3.2 Structural Properties of $Co(c)_2Zn(n)_2$LADH: Here the only differences observed in comparison with the native zinc protein are very slight changes of the positions of the catalytic metal ion and of histidine 67. The coordination sphere is highly similar to that of the zinc enzyme which renders the cobalt enzyme a reliable model for the native protein (see Fig. 2a).

Summarizing the observations and extrapolating the conclusions obtained from structural studies on $H_4Zn(n)_2$LADH and $Co(c)_2Zn(n)_2$LADH, it seems as though the protein structure would be rigid enough to impose a tetrahedral geometry on other metal ions besides Zn^{2+} and Co^{2+}, which is corroborated by spectral studies of the Cu^{2+}, Ni^{2+}, and Fe^{3+} derivatives (see below).

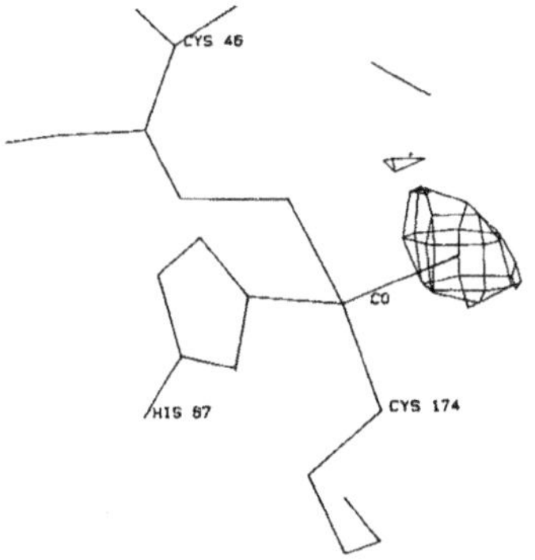

Figure 2a. Metal environment in $Co(c)_2Zn(n)_2$LADH from a difference Fourier electron density map $F_{obs}-F_c$, where F_{obs} denotes the observed F-values for the Co^{2+} enzyme and F_c the calculated structure factors from the refined model of the native protein. Since the native phases do not include the water molecule, it appears as a positive peak in the Co^{2+} enzyme [27].

2.4 Spectral Properties of Metallo Alcohol Dehydrogenases

Electronic and EPR spectra of alcohol dehydrogenases substituted with Co^{2+}, Cu^{2+}, Ni^{2+}, and Fe^{2+}, have provided useful information concerning the nature of the binding sites and their possible distortions when coenzyme and/or substrates (inhibitors) change the protein structure or interact directly with the metal binding sites. Additional information has been gained from Mössbauer spectra of the ^{57}Fe enzyme [21], Resonance Raman spectra of the Cu^{2+} enzyme [29] and Perturbed Angular Correlation spectra of the Cd^{2+} enzyme and its complexes [30].

2.4.1. Co^{2+} alcohol dehydrogenases: The first Co^{2+}enzyme which became available was the "green hybrid" enzyme, viz. $Zn(c)_2Co(n)_2$LADH in our nomenclature. It exhibits strong absorption bands at 340, 656, and 740 nm. The positions and intensities of these bands are comparable to known tetrahedral Co^{2+}-complexes with three or four thiolate ligands, e.g. Co^{2+}-rubredoxin and tetrathiophenolato cobaltate II [14]. In contrast, the "blue hybrid" enzyme, i.e. $Co(c)_2Zn(n)_2$LADH lacks the strong band at 740 nm and shows a weaker band at 520 nm, which is absent in the CoS_4 chromophores (Fig.3). Very similar spectra are found in the tetrahedral CoS_2N_2 chromophores of Co^{2+}-azurin, Co^{2+}-plastocyanin, and bis (2-mercapto-2,2-dimethylamino) cobaltate II [14]. These observations made it possible to identify the Co^{2+} binding sites, in accordance with assignments derived from the preparation routes for and the enzymic properties of both hybrid enzymes. In addition, this information permitted us to reexamine the assignment of a substituted yeast ADH containing four Co^{2+} ions per tetramer. This Co_4YADH was claimed to be substituted specifically at the catalytic sites [31]. Our conclusion has been that in Co_4YADH the Co^{2+} ions are statistically distributed over all eight binding sites, both the catalytic and noncatalytic ones [14].

Addition of certain ligands, i.e. N_3^-, SH^-, and CN^- shifts the 520 nm band of the blue hybrid to 570 nm and causes slight shifts of the intensive 650 nm band (Fig.4). In contrast, F^-, Cl^-, Br^-, I^- do not affect the electronic spectra of $Co(c)_2Zn(n)_2$LADH at all [14]. Increasing the pH has the same effect as the bound anions mentioned above; the apparent pK_a of this process is approximately 9.4 [32]. It would be tempting to assign this pK_a to the ionization of metal-bound water, since this process is believed to occur in the native enzyme with a pK_a of 9.2. So far, we do not know the reason for the spectral changes due to binding of certain anions near the Co^{2+}ions, either inner-sphere or outer-sphere. Therefore we are reluctant to ascribe the pH dependent spectral changes to the ionization of metal-bound water. Other reasons are given below, which render this process still less likely. The origin of the spectra of $Co(c)_2Zn(n)_2$LADH has not yet been elucidated completely. As discussed by Maret [19] it is reasonable to assume that the strong transition around 350 nm arises from ligand-to-metal charge transfer, whereas the visible bands occur in the region where d-d transitions are expected. The high intensities of the visible bands indicate considerable ligand-to-metal charge transfer character even for these transitions. Generally, the entire spectrum of the d-d transitions is redshifted in the sulfur containing metal binding sites of Co^{2+} ADHs as compared to other cobalt proteins which lack sulfur ligands.

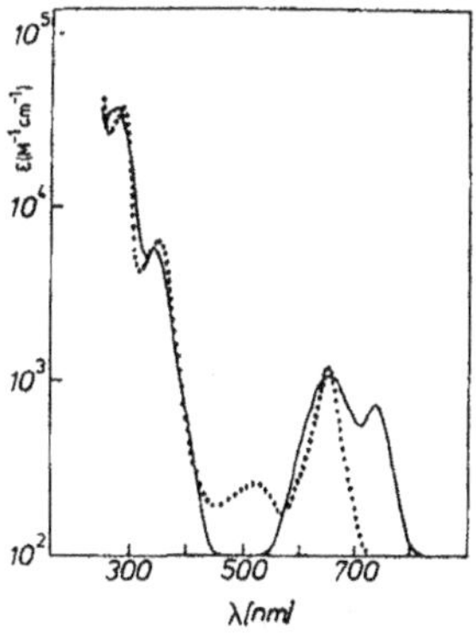

Fig.3 Absorption spectra of (+++) $Co(c)_2Zn(n)_2$LADH and (——) $Zn(c)_2Co(n)_2$LADH [14].

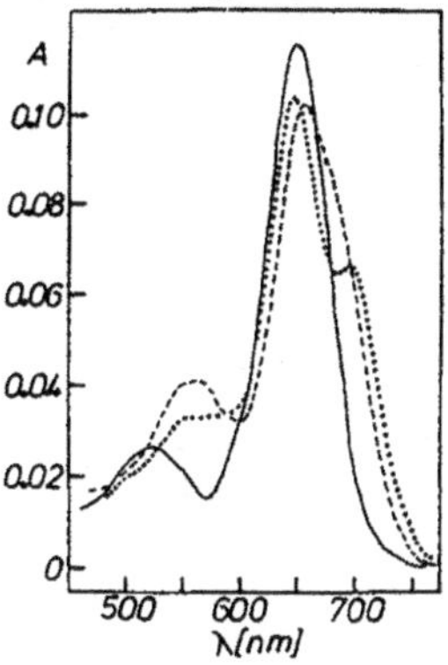

Fig.4 Absorption spectra of binary complexes of $Co(c)_2Zn(n)_2$LADH with HS^- (...) and N_3^- (---). From [14]

Maret has given a comparison of the spectra of binary and ternary complexes of $Co(c)_2Zn(n)_2$LADH with different inhibitors and/or coenzyme [19]. Ligands with coordinating nitrogen atoms influence the spectra both in the absence and presence of coenzyme, as demonstrated in Fig.5.

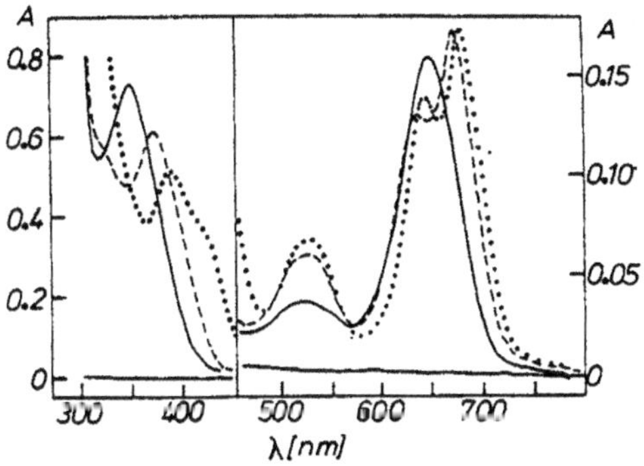

Fig.5 Absorption spectra of $Co(c)_2Zn(n)_2$LADH (——), its binary complex with pyrazole (---), and the ternary complex with pyrazole and NAD^+ (...) [19].

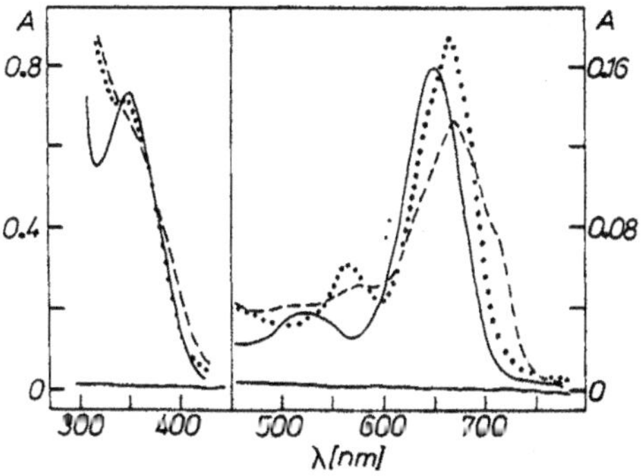

Fig.6 Absorption spectra of $Co(c)_2Zn(n)_2$LADH (——), the binary complex with NAD^+ (---), and the ternary complex with NAD^+ and acetate (...) [19].

Ligands with coordinating oxygen atoms affect the spectra of the catalytic Co^{2+}ion only in the presence of coenzyme (Fig. 6-8). A striking similarity is observed between the spectra of the ternary complexes of $Co(c)_2Zn(n)_2$LADH with NAD^+/acetate (Fig 6), NAD^+/trifluoroethanol (Fig.7) and NAD^+/ethanol (Fig.8). The latter is available as relatively short-

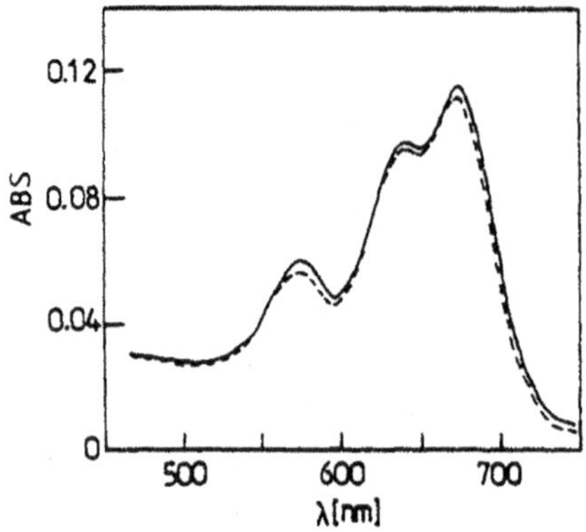

Fig.7 Absorption spectra of the ternary complex $Co(c)_2Zn(n)_2LADH\cdot NAD\cdot$2,2,2-trifluoroethanol at pH 4.95 (——) and pH 8.7 (---). From [32].

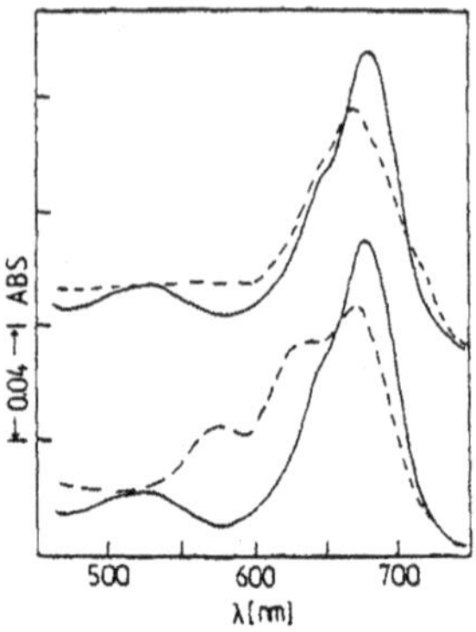

Fig.8 Absorption spectra of the abortive ternary complex $Co(c)_2Zn(n)_2LADH\cdot NADH\cdot$ethanol (——) and the productive ternary complex $Co(c)_2Zn(n)_2LADH\cdot NAD^+\cdot$ethanol (---) at pH 5.2 (upper) and pH 7.35 (lower). From [32].

lived transient species upon mixing of $Co(c)_2Zn(n)_2LADH\cdot NADH$ with acetaldehyde. The conclusion was drawn, that in all three species a negatively charged ligand is bound to the complex enzyme$\cdot NAD^+$. This would support the suggestion of Pettersson [33] that alcohol binds as alcoholate in these kind of ternary complexes. For the transient species $Co(c)_2Zn(n)_2LADH\cdot NAD^+\cdot$ethanol a pK_a of 6.2 was obtained from the pH dependence of the Co^{2+} spectra [32] which seems reasonable in view of the pK_a value of 6.4 postulated for the corresponding ternary complex formed with the Zn^{2+} enzyme [33]. On the basis of these isotropic optical spectra nothing definite can be said about possible changes in coordination number in addition to changes in coordination geometry which are very likely to occur. Polarized single crystal spectra [34] and EPR spectra of $Co(c)_2Zn(n)_2LADH$ and its complexes [35] have led Makinen to suggest a pentacoordinated Co^{2+} ion in ternary complexes containing NAD^+. This topic is discussed further in section 3.

2.4.2 Cu^{2+} alcohol dehydrogenase: The insertion of Cu^{2+} into the active site of LADH produces a "blue" (Type I) copper center [41]. Again, the EPR and optical spectra of Cu^{2+} LADH are best interpreted in terms of a tetrahedral geometry of the active site copper ion. An extension of the coordination number of the "blue" cupric ion to five has become plausible from NMRD data on the complex Cu^{2+} LADH$\cdot$pyrazole [42]. The cupric enzyme is inactive, but is able to bind coenzyme. It is evident from the spectral changes that the Cu^{2+} ion is sensing the coenzyme induced conformational changes [41,43].

2.4.3 Iron alcohol dehydrogenase: The Fe^{2+} enzyme prepared in crystal suspension is easily oxidized to Fe^{3+} enzyme. The latter shows a strong charge-transfer band at 560 nm and an EPR signal at g = 4.3 (77 K) typical of high-spin Fe^{3+}. The Mössbauer spectra of the Fe^{2+} enzyme show admixture of Fe^{3+} (Fig. 9 and 10). From the Mössbauer parameters penta-coordination was inferred for the Fe^{2+} enzyme [21].

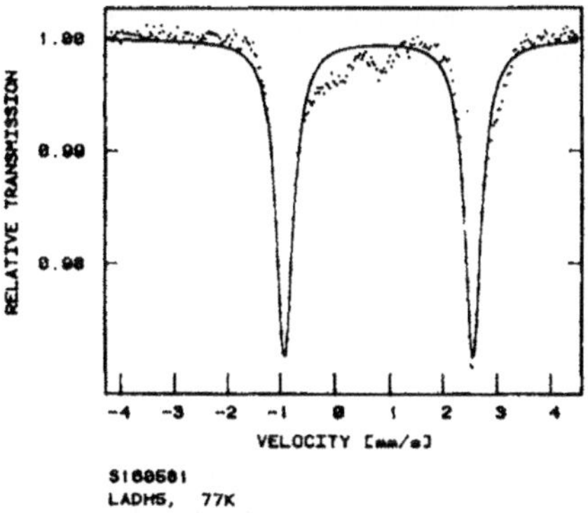

Fig.9 Mössbauer spectrum of Fe^{2+} LADH at 77 K

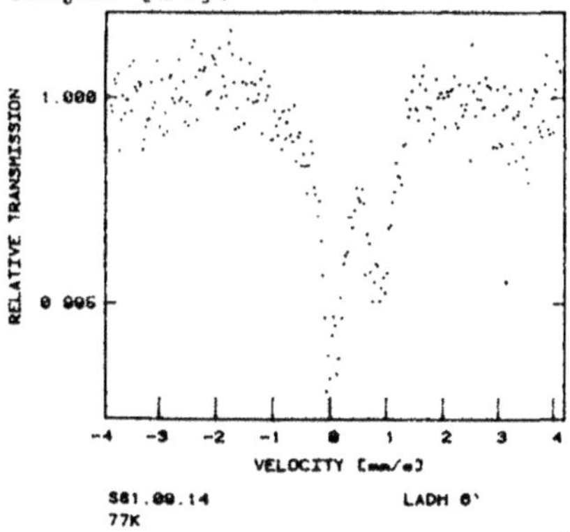

Fig.10 Mössbauer spectrum of Fe^{3+} LADH at 77 K

2.4.4 Cd^{2+} alcohol dehydrogenase: By means of the dialysis procedure and use of $^{109}Cd^{2+}$ it has been possible to define conditions for the preparation of Cd_4LADH and $Zn(c)_2Cd(n)_2$LADH [54]. Subsequently Andersson [20] has shown that the substitution technique in crystal suspension offers a route to the species $Cd(c)_2Zn(n)_2$LADH. Spectroscopically the formation of the cadmium derivatives is detectable due to LMCT transitions in the ultraviolet. The influence of Cd^{2+} on kinetic and spectral parameters of substrate and coenzyme binding is described below. ^{113}Cd NMR signals have been obtained from Cd_4LADH substituted by the dialysis exchange technique. The NMR signal of the catalytic Cd^{2+} could be distinguished from the other Cd^{2+} due to chemical shifts induced by complex formation [53].Perturbed Angular Correlation spectra were obtained from $^{111}Cd(c)_2$ $Zn(n)_2$LADH prepared in solution from $H_4Zn(n)_2$LADH. No ionization of a metal-bound ligand was evident below pH 9.5. The metal appeared to be tetracoordinated with and without NAD^+. In ternary complexes with NAD^+ both pyrazole and 2,2,2-trifluoroethanol seem to coordinate as fifth ligand without displacing water [30].

2.4.5 Ni^{2+} alcohol dehydrogenase: Until now, only LADH containing Ni^{2+} in the catalytic site has been available, which eliminates the problem of assignment [16,37]. As expected, the charge-transfer transitions occur at lower energies than those of Co^{2+} LADHs. Proposed assignments and comparison with other tetrahedral Ni^{2+} proteins e.g. Ni^{2+} azurin [57] are given in [37]. The enzyme forms binary and ternary complexes similar to those of $Co(c)_2Zn(n)_2$LADH. From their spectra it is evident, that all changes in the coordination sphere due to either ligand exchange at the metal or conformation changes in the protein are accompanied by geometric alterations in the metal binding site, which are reflected in the electronic absorption spectra.

3. MECHANISTIC ASPECTS

In this section experiments will be discussed which illustrate the usefulness of metal substitution in studying the properties of the metal -bound water, the interaction of substrate with the catalytic metal ion, the hydride transfer step, and the question in which way the metal ion is involved in the kinetics and thermodynamics of coenzyme binding.

3.1. Properties of the Metal-Bound Water Molecule

The interaction of water molecules with the active site and in particular, with the catalytic metal ion is still a largely unresolved matter. So far no satisfactory answers have been given to the following questions: Which is the pK_a of the metal-bound water molecule? Is it perturbed by bound coenzyme or other ligands? Does the substrate replace the water molecule as inner-sphere ligand or does it bind as outer-sphere ligand to the water? Can substrate and water bind simultaneously during the catalytic cycle?

3.1.1. Paramagnetic ions as relaxation probes. Cobalt alcohol dehydrogenases substituted to varying degrees by the dialysis exchange technique have been used to characterize the binding of substrate and inhibitor molecules by paramagnetic relaxation enhancement. However, the observed effects were extraordinarily small; both outer-sphere binding [55] and inner-sphere binding [58] was inferred. A recent NMRD study of $Co(c)_2Zn(n)_2$LADH has revealed that no significant differences exist between the water relaxation rates of the Co^{2+} and Zn^{2+} enzymes [42]. The data for the Co^{2+} substituted enzyme could be fitted to a theoretical curve calculated for a diamagnetic protein (Fig. 11). It is therefore not surprising, that addition of coenzyme and/or inhibitors had little or no effect on the observed relaxation rates (Fig. 12).

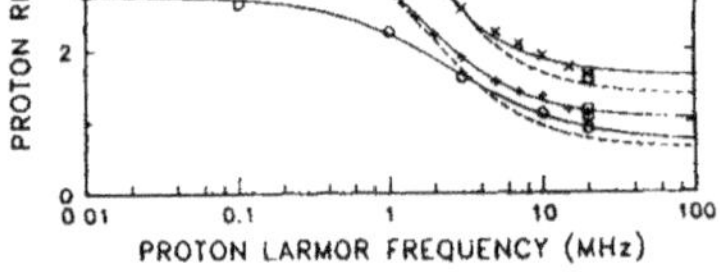

Fig. 11. NMRD spectra of $Co(c)_2Zn(n)_2$ LADH at 4 °C (x), 25 °C (+), and 39 °C (o) as compared to Zn_4LADH (---). The solid curves derive from a theoretical expression for relaxation by diamagnetic proteins [42].

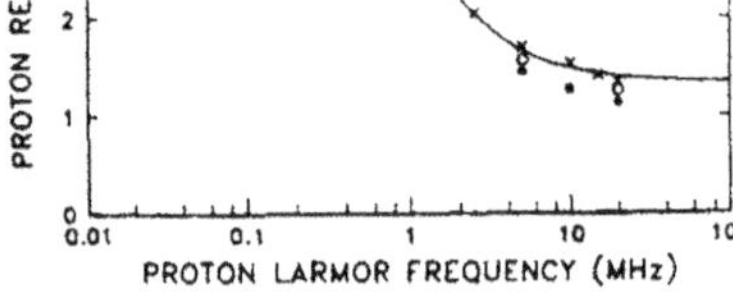

Fig. 12. NMRD spectra of $Co(c)_2Zn(n)_2$ LADH (x), after addition of NADH (o), and finally isobutyramide (*). The solid curve is calculated for relaxation by a diamagnetic protein [42].

In view of this unusual behaviour of $Co(c)_2Zn(n)_2$LADH it was concluded that the electron relaxation time of Co^{2+} is considerably shortened due to a strong spin-orbit interaction of the electronic spins of the Co^{2+} ions with their ligands, notably sulfur. The same applies to the relaxation behaviour of methanol protons. The Cu^{2+} enzyme showed a true paramagnetic contribution to the relaxation behaviour of solvent water or CH_3OD protons. However, it binds the substrate too weakly, which is the probable reason for its lack of activity. The final conclusion must be that Co^{2+} substitution is unsuitable as paramagnetic probe in order to gain information on the state of metal-bound water and substrate molecules in LADH.

3.1.2 Coenzyme-dependent perturbation of ionization equilibria. The ionization of metal-bound water with a pK_a of 9.2 and its perturbation by NAD^+ to 7.4 is one of the oldest ingredients of the catalytic schemes advanced so far [1]. The clearest commitment to the idea of perturbed water ionization is found in the papers by G. Pettersson and collaborators. In addition to both pK_a values already mentioned they assign a third pK_a of 11.2 to the ionization of water shielded by bound NADH [56]. In the native enzyme, the pK_a of 9.2 governs the association rates of NAD^+ and the pK_a of 7.6 the dissociation rates. In his thesis, H. Dietrich has compared these rates with the corresponding rates obtained for the demetallized enzyme, $H_4Zn(n)_2$LADH [16]. He found the association rates generally to be lower by an order of magnitude which results in weaker binding. Most important, the pK_a governing association of NAD^+ to $H_4Zn(n)_2$LADH is 9.5, that governing its dissociation is 7.7. Although the relatively weak binding of NAD^+ leads to considerable scatter of the data at low pH values two important conclusions can be drawn: (i) The binding of NAD^+ to demetallized HLADH is pH dependent. (ii) The pK_a values resemble closely those observed in the native enzyme. These are strong arguments against a participation of metal-bound water in the ionization processes related to the binding of coenzyme [44].

3.2 Interaction of Substrate with the Catalytic Metal Ion

3.2.1 Metal-alcohol interaction. From the data discussed in the preceding sections it has become apparent that the evidence in favour of outer-sphere binding of alcohols is weak. However, the available spectroscopic data (Figs. 7,8) are insufficient as definite proof of inner-sphere binding of alcohol or alcoholate. If indeed the alcohol ionizes in the ternary complex with NAD^+, then its inner-sphere binding is more likely for thermodynamic reasons. The dependence on substrate structure of pK_a values determined both kinetically [33] and spectroscopically [32] adds strength to the idea of substrate ionization. Clearly, more observations of this kind should be collected on different metallo alcohol dehydrogenases.

3.2.2 Metal-aldehyde interaction. The use of chromophoric aldehydes, notably trans-4-N,N-dimethylamino cinnamaldehyde (DACA) has provided detailed information on the dynamics of substrate binding and turnover. At high pH DACA forms a rather stable complex with LADH and NADH which breaks down upon lowering the pH to yield NAD^+ and the corresponding alcohol. Ternary complexes containing the coenzyme analog 1,4,5,6-tetrahydro NAD are stable at neutral pH. The kinetics of formation and breakdown of the complex derived from native LADH were studied by Dunn and collaborators [45, 46] who concluded that (i) hydride and proton transfer are kinetically separable, (ii) the substrate replaces the metal-bound water, and (iii) the coenzyme acts as noncovalent effector, since the binding of NADH must precede that of the substrate. In the absence of coenzyme, DACA is not bound at all. The large redshift of the bound chromophore was taken as strong evidence for inner-sphere coordination of the substrate to the zinc ion and as proof for the Lewis-acid behaviour of the catalytic metal [45]. This concept has been substantially refined by comparing the interaction of DACA with LADHs containing Cd^{2+}, Ni^{2+}, Co^{2+} and with $H_4Zn(n)_2$LADH [47, 16, 38]: (i) The presence of a metal ion is an absolute requirement for the binding of DACA, since $H_4Zn(n)_2$LADH does not form the ternary complex in spite of enhanced binding of NADH [47]. (ii) The magnitude of the redshift of bound DACA correlates positively with the expected Lewis-acid strength of the metal, i.e. the smaller the ionic radius of the metal ion the larger is its charge density and hence the polarization as well as the redshift induced in the chromophoric ligand. (iii) The dissociation constant of DACA in the ternary complex correlates with the Lewis-acid strength of the metal. These observations are summarized in Table III. Even more instructive is the comparison of association and dissociation rates of DACA in its ternary complexes with different metallo ADHs. (iv) The association rates are near-to-diffusion limited and independent of the nature of the metal ion. This means that dissociation of water is not rate-limiting for the binding of the aldehyde. Either the dissociation rate of water is increased several orders of magnitude in the binary complexes $Me(c)_2Zn(n)_2$LADH•NADH irrespective of the nature of the catalytic metal or the formation of the ternary complex is preceded by a pentacoordinated intermediate. (v) The rate of dissociation of DACA from the ternary complexes correlates with the expected Lewis acid strength of the catalytic metal ion. Unfortunately, the literature contains no data describing the kinetics of ligand substitution in tetrahedral complexes with one exchangeable water ligand. Thus the data shown in Table III seem to be the first of this kind assessing the Lewis acid strength of tetrahedral cobalt II. (vi) The rate of breakdown of the ternary complexes upon lowering the pH is also related to the nature of the catalytic metal in the expected manner (Table III). The kinetic data and the significantly lower isotope effect observed for the Cd^{2+}enzyme indicate that the nature of the transition state may be different as compared to the other metallo ADHs [38].

Both the binary complex formation between LADH and NADH (or 1,4,5,6-tetrahydro NAD) and the ternary complex formation lead to significant changes in the coordination sphere of the metal ion as can be seen from the spectra of $Co(c)_2Zn(n)_2$LADH (Fig. 13). The binary complex shows a redshift of the absorption band from 650 nm to 678 nm which most probably is due to a geometric distortion of the coordination sphere. As seen in Fig. 15 the binary complex shows no changes in the inner coordination sphere. There is still a considerable number of water molecules fixed within the catalytic site. Binding of DACA leads to a strong incre-

Table III. Correlation between Spectral and Kinetic Parameters for the Formation of Ternary Complexes (a) Enzyme•H_2NADH•DACA and (b) Enzyme•NADH•DACA and their Breakdown [16].

			Co^{2+}	Ni^{2+}	Zn^{2+}	Cd^{2+}
λ_{max} for DACA (a)			482 nm	479 nm	468 nm	-
λ_{max} for DACA (b)			478 nm	475 nm	464 nm	457 nm
Redshift of bound DACA (b) relative to free DACA (λ_{max} = 398 nm)			80 nm	77 nm	66 nm	59 nm
Association and dissociation rate constants (b)	$k_1 (M^{-1}s^{-1})$		$2.0 \cdot 10^7$	$1.6 \cdot 10^7$	$2.8 \cdot 10^7$	$2.5 \cdot 10^7$
	$k_{-1} (s^{-1})$		80	110	180	380
Dissociation constant (b)	$K_D = \frac{k_{-1}}{k_1}$ (μM)		4	6.9	7.0	15
Rate constant of hydride transfer	k_{app}^{max} (s^{-1})	NADH	11	6.9	6.9	0.21
		NADD	3.7	2.3	2.3	0.12
Primary isotope effect	k_H^{max}/k_D^{max}		3.0	3.0	3.0	1.8
Apparent pK_s of hydride transfer		NADH	5.6	6.0	6.1	≅6.5
		NADD	6.1	6.6	6.7	≅6.0

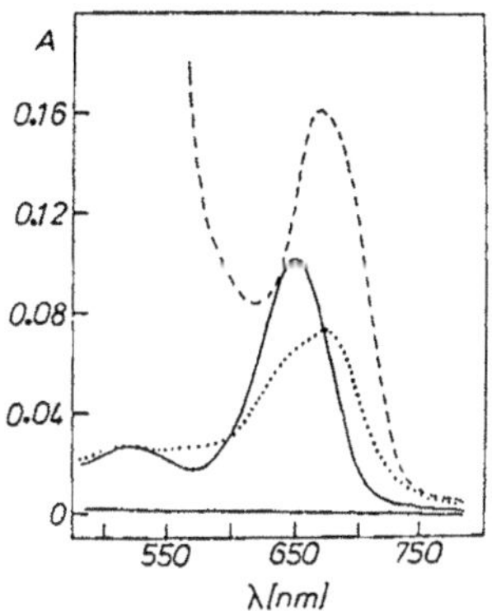

Fig. 13 Absorption spectra of $Co(c)_2Zn(n)_2$LADH(-), the binary complex with NADH(...), and the ternary complex with NADH and DACA (---). The high absorption below 600 nm is due to bound DACA [47].

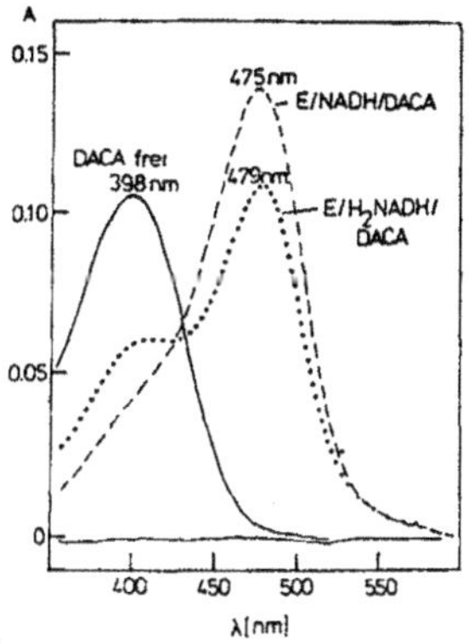

Fig. 14 Absorption spectra of DACA free(-) and bound to Ni^{2+} LADH·NADH(---) or Ni^{2+}LADH·H_2NADH (...) [16].

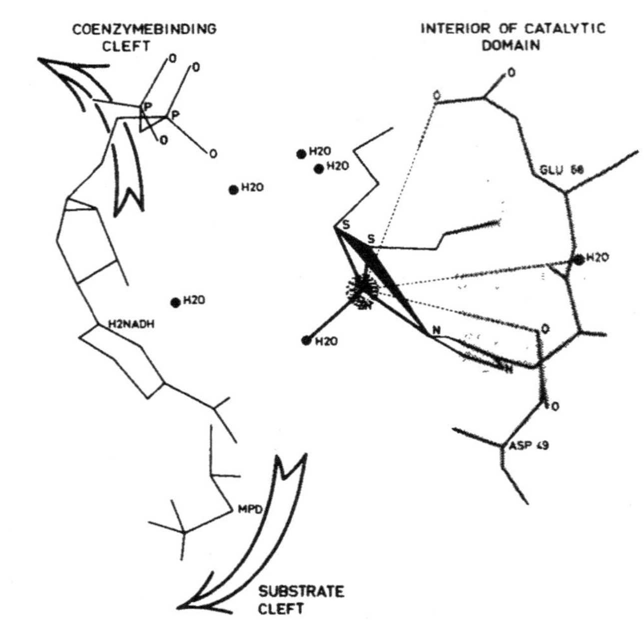

Fig. 15 Nearest neighbours to the active site zinc atom of the orthorhombic LADH·H_2NADH·MPD complex. The dotted lines indicate that two carboxylate groups and one water molecule are hidden inside the domain without direct contact to the surface of the protein. The distances between these groups and the center of the metal atom correspond to 5.5 - 6 Å. By courtesy of Elia Cedergren [59].

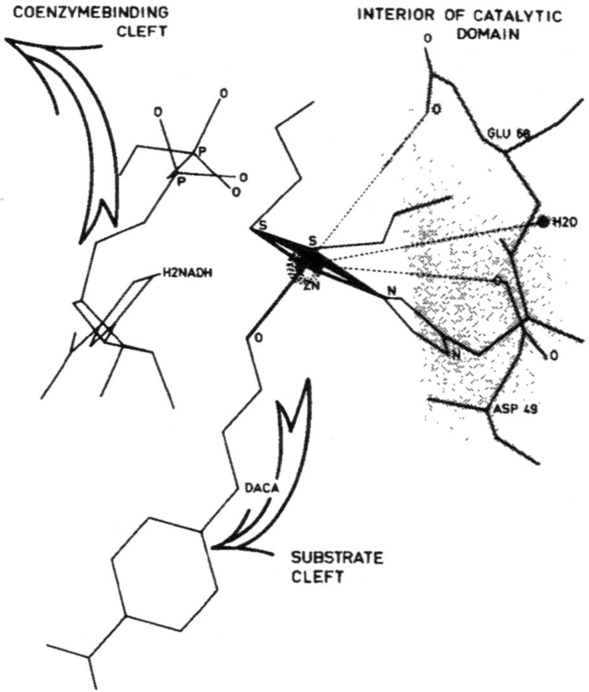

Fig. 16 Nearest neighbours to the active site zinc atom of the triclinic LADH·H_2NADH·DACA complex. The dotted lines are not bonds, they only demonstrate the presence of the same charged residues at the distal side of the coordination sphere as in Fig. 15. The internal water molecule was located as strong difference electron density in F_{obs}-F_c Fourier maps. The C4N, C5N and C6N atoms of the tetrahydronicotinamide ring are at van der Waals distance to the metal center and the cysteine ligands. By courtesy of Elia Cedergren [59].

ase in the intensity of the visible Co^{2+} band due to coordination with the highly polarizable, aromatic ligand (Fig. 13). Simultaneously, the water molecules are expelled from the catalytic site (Fig. 16). Comparison of the spectral data of complexes containing NADH with those containing 1,4,5,6-tetrahydro NAD shows but small systematic differences (Table III). We may conclude that the behaviour of the four metallo ADHs compared in this study is astonishingly similar. Most likely, this is a consequence of the rigidity of the protein structure which leads to highly similar metal environments in the catalytic site irrespective of the nature of the metal. Cd^{2+} indicates a tendency of developping a different transition state, which probably is a consequence of its significantly larger ionic radius as compared to Co^{2+}, Ni^{2+}, or Zn^{2+}.

3.3 Influence of the Catalytic Metal Ion on the Dynamics of Coenzyme Binding

Kinetic studies concerned solely with the native enzyme have so far provided few possibilities to grasp the importance of the catalytic metal ion for the dynamic interaction of the coenzyme with the protein. An investigation by Shore and Santiago [48] using fully substituted Co_4LADH presented the first clear evidence that both the hydride transfer rate and the turnover rate of ethanol oxidation are affected by the metal substitution. It was later found that the K_m values for ethanol vary considerably among the different metallo ADHs [49, 52]. Therefore, both the steady-state parameters and the rates of dissociation of NADH (i.e. the turnover rates or k_{cat} in the Theorell-Chance mechanism) will have to be compared as function of pH for the different metallo ADHs. So far we have obtained the following correlations: (i) The dissociation rate k_{-1} for NADH is strongly dependent on the presence and the kind of metal in the catalytic site. Thus, removal of the Zn^{2+} results in a thousandfold decrease in k_{-1} which is the reason for the much stronger binding of NADH to $H_4Zn(n)_2$LADH as compared to the native protein [16, 44]. The rate of dissociation of NADH from the Cd^{2+}enzyme is lowered tenfold in comparison with the Zn^{2+}enzyme [16, 51]. This shows clearly that the metal does influence the interior of the active site in a literally far-reaching way, since the coenzyme is no metal ligand. (ii) The turnover rate of ethanol oxidation is also strongly dependent on the nature of the catalytic metal. This is expected in all cases where the Theorell-Chance mechanism is valid. Relevant data are shown in Table IV.

Table IV. Relationship between the Kind of Catalytic Metal, the Rate of Coenzyme Release, k_{-1}, and Turnover, k_{cat}, of Ethanol Oxidation (pH = 10). From [16, 49, 52].

Species	k_{-1} (s^{-1})	k_{cat} (s^{-1})
$H_4Zn(n)_2$LADH	$5 \cdot 10^{-3}$	-
$Cd(c)_2Zn(n)_2$LADH	0.5	0.5
Zn_4LADH	5.5	5.5
$Co(c)_2Zn(n)_2$LADH	-	7.2

It is however necessary to establish (or to disprove) this validity in each case under consideration, usually by comparing the turnover rate, k_{-1}, and the hydride transfer rate at a given pH value. From the pH dependence of these parameters indications concerning the nature of the rate-limiting steps may be gained. This has been done so far with the Cd^{2+} enzyme [50, 51]. It was shown that hydride transfer is always considerably faster than the dissociation of NADH, but at low pH a third process becomes rate-limiting.

These findings emphasize the necessity of exploring the dynamics of enzyme-coenzyme interaction even in the absence of substrate. We may expect that comparison of different catalytic metal ions in this respect will reveal interesting aspects of structural dynamics in the protein.

Conclusion: Metal substitution in liver alcohol dehydrogenase has aided exploring structural and mechanistic principles underlying the catalytic action of this enzyme system. It is becoming clear that the catalytic cycle is performed by a subtle interplay of the four constituents of the system, viz. protein, metal, coenzyme, and substrate:

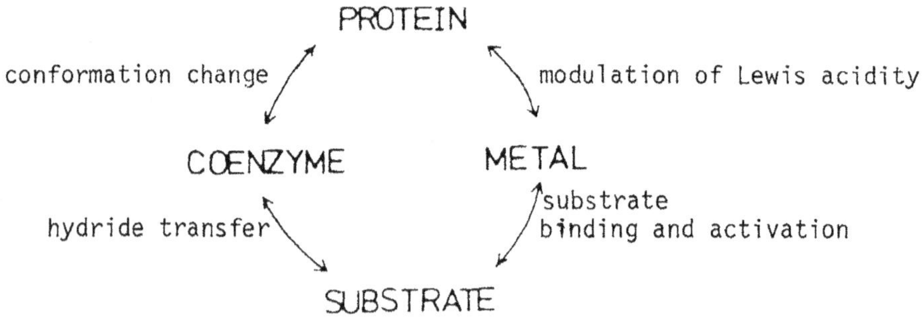

It is also becoming evident that the unique design of metal ion environments in proteins will disclose aspects of coordination chemistry which hitherto have been inaccessible by studying low molecular weight coordination compounds.

Acknowledgement: I am indebted to my coworkers both at home and abroad for stimulating collaboration; most of their names are cited below. Financial support was given by Deutsche Forschungsgemeinschaft, Fonds der Chemischen Industrie, EMBO, FEBS, Wissenschaftliche Gesellschaft des Saarlandes, and Vereinigung der Freunde der Universität.

Abbreviations: (H)(L)ADH, (Horse)(Liver)Alcohol Dehydrogenase; DMSO, dimethylsulfoxide; Me(c), catalytic metal ion; Me(n), noncatalytic metal ion; DACA, trans-4-N,N-dimethylaminocinnamaldehyde; H_2NADH, 1,4,5,6-tetrahydro NAD; NMRD, nuclear magnetic relaxation dispersion; LMCT, ligand-to-metal charge transfer; MPD, 2-methyl-2,4-pentanediol.

REFERENCES

[1] Brändén, C-I., Jörnvall, H., Eklund, H., and Furugren, B. (1975), in "The Enzymes" (P.D. Boyer, ed.) Vol 11, 103-190, Academic Press, New York.
[2] Schneider-Bernlöhr, H., Fiedler, H., Gerber, M., Weber, C., and Zeppezauer, M. (1981), Int. J. Biochem. 13, 1215-1224.
[3] Tsai, C.S., (1979), Biochem. Biophys. Res. Commun., 86, 808-814.
[4] Favilla, R., Mazzini, A., Fava, A., and Cavatorta, P., (1980), Eur. J. Biochem. 104, 217-222.
[5] Jeffery, J., (1980), in "Dehydrogenases", (J. Jeffery, ed.), 85-125, Birkhäuser Vlg. Basel
[6] ibid., p. 105
[7] Ambar, A., (1981), Dissertation, ETH Zürich, Nr. 6840
[8] see Chapter X in this book
[9] Theorell, H. and Chance, B., (1951), Acta Chem. Scand. 5, 1127-1144.
[10] Brändén, C.-I. (1965), Arch. Biochem. Biophys. 112, 215-217.
[11] Klinman, J.P. (1981), Crit. Revs. Biochem. 10, 39-78.
[12] Kvassman, J. and Pettersson, G. (1980) Eur. J. Biochem. 103, 557-564.
[13] Sytkowski, A. and Vallee, B.L. (1978) Biochemistry 17, 2850-2857.
[14] Maret, W., Andersson, I., Dietrich H., Schneider-Bernlöhr, H., Einarsson, R.,and Zeppezauer, M., (1979) Eur. J. Biochem. 98, 501-512.
[15] Sytkowski, A. and Vallee, B.L. (1976) Proc. Natl. Acad. Sci. USA 73, 344-348.
[16] Dietrich, H. (1980) Dissertation, Universität des Saarlandes, Saarbrücken.
[17] Tilander, B., Strandberg, B., and Fridborg, K. (1965) J. Mol. Biol. 12, 740-760.
[18] Rudolph, R., Gerschitz, J. and Jaenicke, R. (1978) Eur. J. Biochem. 87, 601-606.
[19] Maret, W. (1980) Dissertation, Universität des Saarlandes, Saarbrükken.
[20] Andersson, I. (1980) Dissertation, Universität des Saarlandes, Saarbrücken.
[21] see elsewhere in the book.
[22] Dalziel, K. (1957) Acta Chem. Scand. 11, 397-398.
[23] Skjeldal, L., Dahl, K.H., and McKinley-McKee, J.S. (1982) FEBS-Letters 137, 257-260.
[24] Eklund, H. personal communication
[25] Guckelmus, I. (1979) Staatsexamensarbeit, Universität des Saarlandes, Saarbrücken.
[26] Schneider, G. and Zeppezauer, M. (1982) J. Inorg. Biochem., in press
[27] Schneider, G., Eklund, H., Cedergren-Zeppezauer,E., and Zeppezauer, M., (1982) Proc. Natl. Acad. Sci. USA, submitted
[28] Andersson, I., Maret, W., Zeppezauer, M., Brown, R.D, III, and Koenig, S.H. (1980) Biochemistry 20, 3433-3438.

[29] Loehr, T. and Maret, W., unpublished data
[30] Andersson, I. and Bauer, R. (1982), Inorg. Chim. Acta, in press.
[31] Sytkowski, A.J. (1977) Arch. Biochem. Biophys. 184, 505-517
[32] Dietrich, H. and Zeppezauer, M. (1982) J. Inorg. Biochem., in press.
[33] Kvassman, J., Larsson, A., and Pettersson, G. (1981) Eur. J. Biochem. 114, 555-563.
[34] Kvassman, J. and Pettersson, G. (1980) Eur. J. Biochem. 103, 565-575
[35] Makinen, M.W.,Hill, S., and Zeppezauer, M. in preparation
[36] Makinen, M.W. and Yim, M.B. (1981) Proc. Natl. Acad. Sci USA, 78, 6221-6225
[37] Dietrich, H., Maret, W., Kozłowski, H., and Zeppezauer, M. (1981) J. Inorg. Biochem. 14, 297-311
[38] Dunn, M.F., Dietrich, H., MacGibbon, A.K.H., Koerber, S.C., and Zeppezauer, M. (1982) Biochemistry 21, 354-363.
[39] Drum, D.E., Li, T.-K., and Vallee, B.L. (1969) Biochemistry 8, 3783-3791
[40] Iweibo, K. and Weiner, H. (1972) Biochemistry 11, 1003-1010.
[41] Maret, W., Dietrich, H., Ruf, H.-H., and Zeppezauer, M. (1980) J. Inorg. Biochem. 12, 241-252.
[42] Andersson, I., Maret, W., Brown, R.D., III, and Koenig, S.H. (1981) Biochemistry 20, 3424-3432.
[43] Maret, W., Zeppezauer, M., Desideri, A., Morpurgo, L., and Rotilio, G. (1981) FEBS-Letters 136, 72-74.
[44] McGibbon, A., Dietrich, H., Dunn, M.F., and Zeppezauer, M. in preparation
[45] Dunn, M.F. and Hutchison, J.S. (1973) Biochemistry 12, 725-731.
[46] Dunn, M.F., Biellmann, J.F., and Branlant, G. (1975) Biochemistry 14, 3176-3182.
[47] Dietrich, H., Maret, W., Wallén, L., and Zeppezauer, M. (1979) Eur. J. Biochem. 100, 267-270.
[48] Shore, J.D. and Santiago, D. (1975) J. Biol. Chem. 250, 2008-2012.
[49] Pottmeyer, C. (1980), Diplomarbeit, Universität des Saarlandes, Saarbrücken
[50] Gerber, M., Weis, M., and Zeppezauer, M. in preparation.
[51] Weis, M. (1981), Staatsexamensarbeit, Universität des Saarlandes, Saarbrücken.
[52] Pfangert, U. (1982), Staatsexamensarbeit, Universität des Saarlandes, Saarbrücken.
[53] Bobsein, B.R. and Myers, R.J. (1981) J. Biol. Chem. 256, 5313-5316.
[54] Sytkowski, A.J. and Vallee, B. (1979) Biochemistry 18, 4095-4099.
[55] Sloan, D.L., Young, L.M., and Mildvan, A.S. (1975) Biochemistry 14, 1998-2008.
[56] Andersson, P., Kvassman, I., Lindström, A., Oldén, B., and Pettersson, G. (1981) Eur. J. Biochem. 113, 425-433.
[57] Tennent, D.L. and McMillin, D.R. (1979) J. Am. Chem. Soc. 101, 2307-2311
[58] Boccalon, G., Grillo, G., Baroncelli, V., Renzi, P., and Paretta, A. (1978) J. Mol. Cat. 4, 307-312.
[59] Cedergren-Zeppezauer, E., Samama, J.-P., and Eklund, H. (1982) Biochemistry, submitted.
[60] Eklund, H., Samama, J.-P., Wallén, L., Brändén, C.-I., Åkeson, Å., and Jones, T.A. (1981) J. Mol. Biol. 146, 561-587.

METAL-DIRECTED AFFINITY LABELLING AND PROMOTED ALKYLATION OF A THIOL LIGANDED TO THE CATALYTIC METAL ION IN LIVER ALCOHOL DEHYDROGENASE

Knut H.Dahl, Øystein Brandsnes, Lars Skjeldal and John S.McKinley-McKee

Department of Biochemistry, University of Oslo, Oslo, Norway

ABSTRACT. Cystine-46, which is a protein ligand to the catalytic metal atom in liver alcohol dehydrogenase, has been selectively alkylated with several alkyl-haloacids. When imidazole replaces H_2O as a free ligand to this metal atom, the alkylation rate increases. A thermodynamic transition-state analysis of enzyme alkylation showed that the effect of imidazole was due to a decreased enthalpy barrier. Imidazole also promoted or stimulated the alkylation with iodoacetate of the model-compound zinc-mercaptoethanol. Alkylation is promoted by imidazole donating electrons to the catalytic metal atom which distributes the increased electron density to the other ligands. Metal-directed affinity labelling with bromo-imidazolyl propionate has been carried out on liver alcohol dehydrogenase substituted with several metal atoms in the active site. Since the stability of the reversible complex and the nucleophilic reactivity of Cys-46 depends on metal type, metal-directed affinity labelling delineates different active site properties of different metalloenzyme species.

1. INTRODUCTION

Each of the two subunits of liver alcohol dehydrogenase (LADH) contains one catalytic and one structural zinc atom [1]. The catalytic metal atom has three protein ligands (one histidine and two cysteins), while in the fourth coordination position of the tetrahedron, H_2O/OH^- or the substrate can bind. The zinc atoms of the native enzyme have been replaced with several mono- and divalent metal ions to give new enzyme species with specific activities ranging from zero to almost that of the native zinc enzyme [2-8].

Cys-46 is one of the three protein ligands to the catalytic metal atom, and the sulphur of this particular

I. Bertini, R. S. Drago, and C. Luchinat (eds.), The Coordination Chemistry of Metalloenzymes, 123–134.

residue is seen to be selectively alkylated with iodoacetate and several other alkyl-haloacids [9,10]. As alkylation of Cys-46 inactivates the enzyme, alkylation can be followed by measuring the decrease in enzyme activity. Cys-46 is alkylated as a metal-bound thiol, which is generally less reactive than the corresponding thiolate anion ($R-S^-$). The alkylation of this metal-thiol with different reagents and the effect of other compounds on the nucleophilic reactivity of the metal thiol have been measured. The approach has given new insights into the interactions between a metal and the protein in a metalloprotein.

Recently attention was focused on the paradox that the Cys-46 metal-thiol is the more reactive thiol out of the total of 14 cysteine residues in each subunit [10]. In the alkylation of the enzyme with alkyl-haloacids, the shielding of cys-46 by the catalytic metal atom is counteracted by the formation of a reversible complex prior to alkylation.

Imidazole binds to LADH forming a binary complex, or in the presence of coenzyme a ternary complex [11,12]. Direct coordination of one of the nitrogen atoms to the active site metal was implicated in these complexes and in the ternary pyrazole complexes [13]. It has since been confirmed by crystallographic studies on the enzyme-imidazole complex [14] and by perturbation of the visible cobalt enzyme spectrum upon addition of imidazole [15]. Evans and Rabin [16] also observed that imidazole promoted or stimulated the alkylation of LADH with both iodoacetate and iodoacetamide. Since imidazole binds to the same metal atom as Cys-46, the binding of imidazole could be expected to effect the nucleophilicity of cys-46.

A detailed study of the effect of imidazole on the alkylation of cys-46 was undertaken in order to understand the role of the active site metal in the promoted alkylation [17]. Imidazole-promoted alkylation with iodoacetate was thus studied with LADH where the catalytic metal atom had been substituted with cobalt or cadmium.

In metal-directed affinity labelling of LADH with bromo-imidazolyl propionate (BIP), a reversible complex is formed prior to the irreversible alkylation of Cys-46 [15,18]. In the reversible complex, the imidazole part of BIP is liganded to the active site metal, while the carboxylic group interacts with the general anion-binding site,which also binds the pyrophosphate group of the coenzyme [1].

Inactivation of LADH with BIP has been used as a sensitive probe of the different properties of metalloenzymes where the catalytic zinc atom has been replaced with other metal atoms [8,15].

2. EXPERIMENTAL

Inactivation of LADH with alkyl-haloacids was studied as described in references [10,17], model alkylations as in reference [17], metal substitution as in references [7,8,15] and a thermodynamic transition-state analysis as in reference [19].

3. RESULTS AND DISCUSSION

3.1. Effect of Imidazole on Enzyme Inactivation by Alkylation

When iodoacetate, bromoacetate, 3-bromoacetate, 2-bromopropionate and 2-bromobutyrate were used to inactivate LADH, a reaction which was first-order with respect to enzyme was observed [17]. First-order kinetics were also observed in the presence of imidazole. Fig.1 shows a typical example of the semilogaritmic inactivation curves obtained. For all the alkyl-haloacids used, the inactivation rates gave a saturation effect both in the absence and in the presence of imidazole.

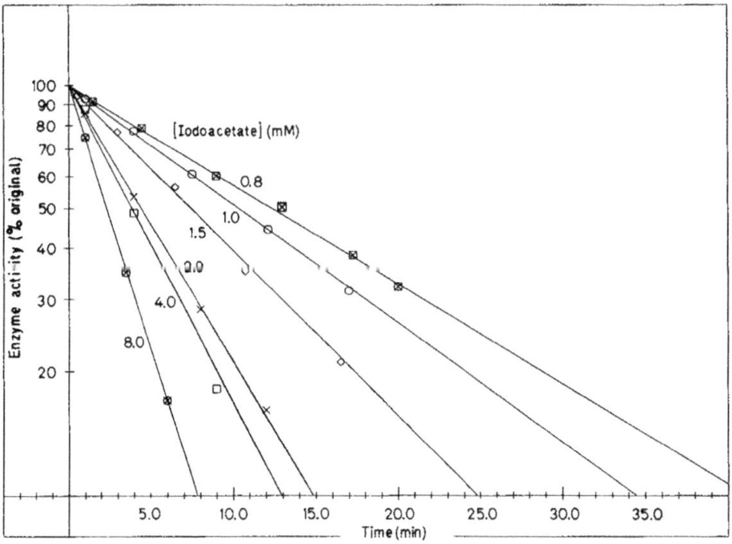

Fig.1. Semi-log plots for inactivation of cobalt alcohol dehydrogenase in 25 mM Pipes/Na^+ pH 7.0 in the presence of 20 mM imidazole.

The saturation effect is a result of a reaction mechanism where a reversible complex is formed prior to the irreversible alkylation. In the reversible enzyme•alkylhaloacid complex, the carboxylic group interacts with the general anion-binding site. Fig.2 shows double reciprocal plots of apparent rate constants for inactivation versus iodoacetate concentration in the presence of variable amounts of imidazole. It is evident that imidazole promotes inactivation with iodoacetate at all concentrations. Similar measurements have been performed with the other alkyl-haloacids with variable amounts of imidazole. Iodoacetate has also been used to inactivate cobalt and cadmium LADH in the presence of variable amounts of imidazole.

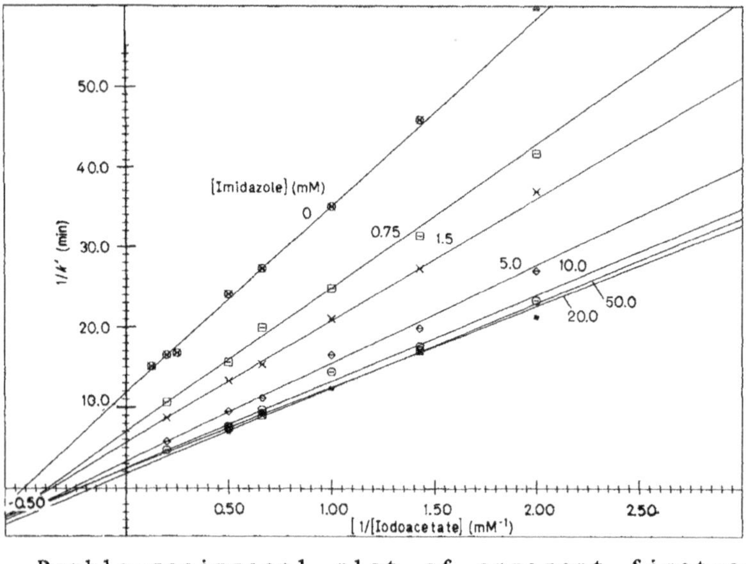

Fig.2. Double-reciprocal plot of apparent first-order rate constant versus iodoacetate concentration for the inactivation of native liver alcohol dehydrogenase in 25 mM Pipes/Na^+ pH 7.0 in the presence of increasing amounts of imidazole.

The observations indicated that the inactivation follows an affinity labelling mechanism, which in the presence of a promotor (like imidazole) can be represented as follows:

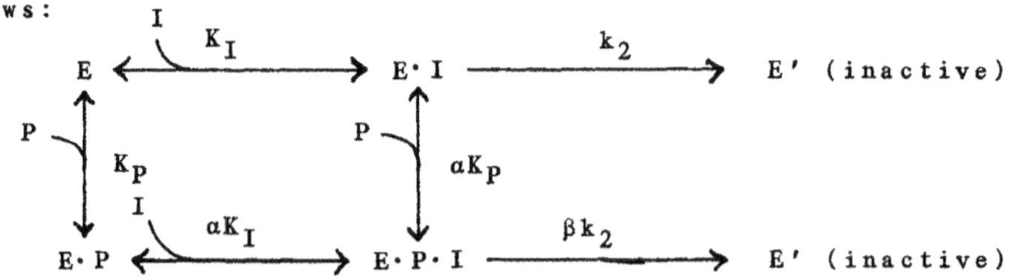

where E is free enzyme, I alkyl-haloacid, P promotor (imidazole), E·I and E·P reversible binary complexes, E·P·I reversible ternary complex and E′ inactivated enzyme. K_I and K_P are the dissociation constants for the reversible complexes and k_2 is the first-order rate constant for inactivation. The cooperativity factor α expresses the alteration in affinity of I for the enzyme due to bound P and vice versa. The promotion factor β denotes the effect of the promotor P on the irreversible alkylation rate and thus measures the promotion. The expression for the inactivation rate v, corresponding to the above mechanism is:

$$v = \frac{k_2 \cdot [I]\ (\alpha K_P + \beta [P])}{[I]\cdot[P] + \alpha(K_I \cdot [P] + K_P \cdot [I] + K_I \cdot K_P)}$$

Performing a nonlinear regression analysis on the kinetic data according to the above expression using v, [I] and [P] as variables, allowed the determination of the parameters k_2, α, K_P, β and K_I. The results of such regression analysis are summarized in table 1. In the case of 2-bromobutyrate where imidazole gave almost competitive protection of the enzyme, the regression analysis gave a high value for α, β could not be determined and the convergence criterion was not satisfied.

Table 1. Summary of the data for the imidazole-promoted inactivation of the different enzymes (native or fully substituted with cobalt or cadmium). The parameters are as defined in the text.

Enzyme/Reagent	$k_2 \cdot 100$ (min^{-1})	α	K_P (mM)	β	K_I (mM)
Zn_4/iodoacetate	7.9	3.5	1.8	7.6	1.7
Co_4/iodoacetate	8.5	3.1	2.5	9.3	3.3
Cd_4/iodoacetate	6.3	6.6	3.3	19.6	1.7
Zn_4/bromoacetate	17.3	1.7	1.1	4.15	4.9
Zn_4/2-bromopropionate	0.50	3.6	1.3	1.5	4.5
Zn_4/3-bromopropionate	0.37	6.0	1.0	3.0	3.4
Zn_4/2-bromobutyrate	1.0	>500	1.4	-	5.7

The observed K_I values indicate that the different alkyl-bromoacids have a similar affinity for the native enzyme, while iodoacetate has a slightly higher affinity. Iodoacetate has the same affinity for the zinc and cadmium enzyme, while its affinity is lower for the cobalt enzyme. The inactivation rate constant k_2 reflects the chemical reactivity of the reacting groups within the reversible complex. The value of k_2 varies for the different alkyl-haloacids being mostly due to their different chemical reactivities. Variations in k_2 are also observed when iodoacetate reacts with the zinc, cobalt and cadmium enzymes.

The dissociation constant for the binary enzyme-imidazole complex, K_P, is for the native enzyme on average 1.3 mM which agrees with previous results [11,20]. In the case of iodoacetate, K_P for the zinc, cobalt and cadmium enzymes follows the order Zn<Co<Cd as might be expected from the affinity of imidazole for the different metals [21].

The value of α varies considerably for the different alkyl-haloacids. As α measures the interaction between the alkyl-haloacid and imidazole, it is evident that their binding sites overlap to different extents. Generally the value of α increases with increasing size of the alkyl-chain, as might be expected for steric interactions.

The promotion factor β measures the rate enhancement of an alkylation reaction taking place within a reversible complex upon the additional binding of the promotor. The β factor will thus show the altered nucleophilicity of the metal-thiol, but the value of β will also depend on other factors, such as the leaving group and chemical reactivity of the alkyl-halide. To understand the promotion effect, it is necessary to realize that Cys-46 is alkylated not as a sulfhydryl or a thiolate anion, but as a metal-thiol. When imidazole binds to the active site metal, it is reasonable that the nucleophilic reactivity of Cys-46 is altered. In metal-coordination chemistry, imidazole is classified as a strong donator of both σ- and π-electrons. The σ-donating properties of imidazole are stronger than that of H_2O and electrons are thus pushed into the metal. The resulting increased electron density of the metal atom is to some extent distributed to the other metal ligands making the sulphur of Cys-46 more nucleophilic. This represents electrostatic interactions between different metal ligands in a metalloenzyme.

With imidazole promotion due to an inner sphere ligand field effect, metal type can be expected to influence the promotion factor β. The results with iodoacetate confirm this, with β for the cadmium enzyme much higher than for the cobalt enzyme which is slightly above that for the zinc enzyme.

To further characterize imidazole promotion, a transition-state analysis was performed on the bromoacetate inactivation of the native enzyme in the absence and presence of 10 mM imidazole [19]. A transition-state analysis involves measuring the reaction rates at various temperatures and calculating the activation parameters ΔH^*, ΔS^*

and ΔG^*. By comparing these for enzyme inactivation in the absence and in the presence of imidazole, the imidazole promotion may be attributed to a lowered enthalpy or entropy barrier.

Table 2. Thermodynamic parameters for enzyme alkylation. ΔH^* is the enthalpy of activation. $-T\Delta S^*$ is the entropy portion of ΔG^*, the free energy of activation.

	ΔH^*(kcal/mole)	$-T\Delta S^*$(kcal/mole)	ΔG^*(kcal/mole)
Bromoacetate	14.4	3.6	18.0
Bromoacetate + 10 mM imidazol	13.0	4.3	17.3

As is evident from table 2, the decrease in ΔG^* in the presence of imidazole is caused by a decrease in ΔH^* which is to some extent counteracted by a rise in $-T\Delta S^*$. Imidazole promoted alkylation of liver alcohol dehydrogenase with bromoacetate is thus seen to be an enthalpy effect, where imidazole influences the making or breaking of bonds. This agrees with imidazole promotion being an inner sphere ligand field effect.

3.2. Effect of Imidazole on the Alkylation of a Model Compound

Alkylation with iodoacetate of the model compound, zinc-mercaptoethanol in a 1:1 complex, was run in a 0.25M Mes buffer pH 5.7 with thiol and iodoacetate at the same concentrations [17]. Reaction was followed by determining unreacted thiol with DTNB. Fig.3 shows the effect of imidazole on the alkylation of the model-compound. In the presence of imidazole, the alkylation rate increases up to four fold. This promotion of the model-alkylation is considered due to imidazole forming a mixed complex with a stoichiometry depending on the zinc:thiol:imidazole ratio. In the mixed complex, the sulphur of mercaptoethanol is more nucleophilic due to the electron donating properties of imidazole. This confirms that the imidazole-promoted alkylation of the enzyme is caused by a displaced electron distribution in the active site coordination unit.

3.3. Metal-Directed Affinity Labelling.

Bromo-imidazolyl propionate (BIP) contains the reactive parts of bromoacetate and imidazole in one molecule. It inactivates the enzyme by alkylating Cys-46 in a reaction similar to those observed for the other alkyl-haloacids, with a reversible complex formed prior to irreversible alkylation [10,15,18,22].

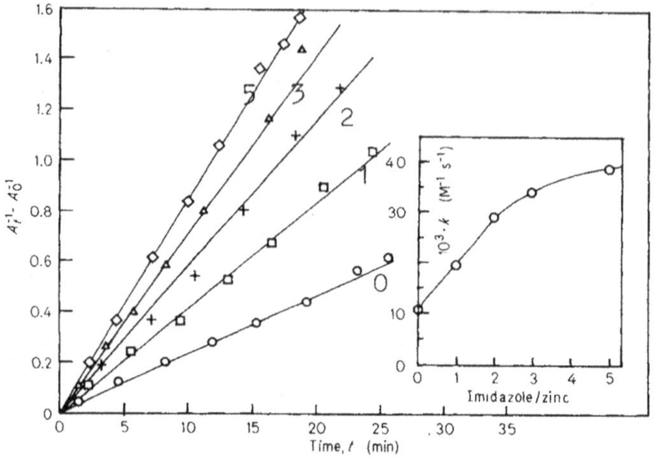

Fig.3. Second-order kinetic plots for the reaction of iodoacetate with the zinc-mercaptoethanol complex in 0.25 M Mes/Na^+, pH 5.7 in the presence of increasing amounts of imidazole. The reactions were followed by withdrawing aliquots for determination of unreacted thiol with DTNB. A_0 and A_t are absorbances at 412 nm in the assay with DTNB. The imidazole/zinc ratios were as indicated. The insert shows the resulting rate constant versus imidazole/zinc ratio.

$$E + BIP \overset{K_I}{\rightleftharpoons} E \cdot BIP \xrightarrow{k_2} E'(\text{inactive})$$

BIP interacts reversibly with the general anion-binding site which is implicated as arginines 47 and 369, but BIP is additionally also bound to the active site metal by the imidazole ring as shown in Fig.4. This mode of binding where the BIP molecule is anchored at both ends results in rigid positioning of BIP within the active site.

So far, BIP has only been available as a racemic mixture of its two enantiomers, but by using the pure enantiomers R-CIP and S-CIP of the corresponding chloro-compound it turned out that the reaction showed absolute enantioselective as only the S-enantiomer inactivated the enzyme [23]. Since both enantiomers of CIP could form a reversible complex, but only the S-enantiomer inactivated, the enantioselectivity was due to the mode of reversible binding of the alkylating reagent within the active site. This implies that the selectivity is at the irreversible step. This was confirmed by computer-assisted model-building where models of R- and S-CIP were built into the active site of an enzyme model [24]. In model-building it was easy to fit models of both enantiomers into the active site, but the following alkylation was possible only with the S-enantiomer.

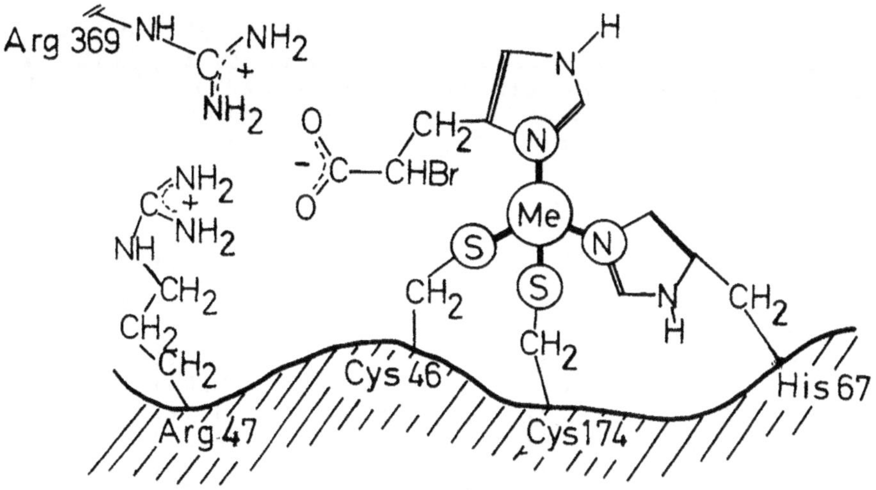

Fig.4. Schematic illustration of the reversible binding of BIP to the active site metal of LADH.

3.4. Affinity Labelling of LADH with different Metals in the Active Site

When the enzyme is inactivated with BIP $1/K_I$ measures the affinity of BIP for the enzyme, while k_2 measures the reactivity of the reacting groups within the active site. It seemed likely that the type of metal in the active site could influence both these parameters.

There are several approaches for substituting active site zinc in LADH with other metals. The results presented here are obtained with fully substituted cobalt and cadmium enzyme made by equilibrium dialysis of enzyme against an excess of the metal to be inserted [15]. More recently it has been shown to be more convenient to withdraw zinc from the catalytic center by dialysis of enzyme crystals against dipicolinic acid [4]. A variety of new enzyme species can now be formed by incubating in solution the metal depleted enzyme with a slight excess of the metal to be inserted. In this way reactivation of the inactive apo-enzyme has also been observed with the toxic heavy metals Pb(II) and Hg(II) [7] and also with the monovalent Ag(I) [8].

Using metal-directed affinity labelling with BIP, it has been shown that the active sites of the different enzyme species have different properties. From Table 3. it is evident that the inactivation kinetics differ between the zinc, the cobalt and the cadmium enzyme.

Table 3. Kinetic Data for BIP Inactivation at pH 7.0 and pH 8.2. K_I and k_2 are defined in the text, $t_{1/2}min$ is the minimum halftime of inactivation and k_2/K_I is the pseudo-bimolecular rate constant for inactivation. The Co and Cd enzymes are fully substituted, while Hg,Pb and Ag enzymes are apo-enzyme reactiviated with the respective metal ions.

Enzyme	K_I (mM)	k_2 (min^{-1})	$t_{1/2}min$ (min)	k_2/K_I ($min^{-1} mM^{-1}$)	pH
Zn(II)enzyme	0.47	0.077	9.0	0.163	7.0
Co(II)enzyme	1.25	0.150	4.6	0.120	7.0
Cd(II)enzyme	0.71	0.029	24.1	0.041	7.0
Zn(II)enzyme	0.68	0.127	5.4	0.189	8.2
Hg(II)enzyme	0.80	0.092	7.5	0.114	8.2
Pb(II)enzyme	0.97	0.090	7.7	0.092	8.2
Ag(I)enzyme	1.82	0.109	6.4	0.060	8.2

Table 3 also summarizes similar inactivation studies performed with the enzymes rectivated with Pb(II), Hg(II) and Ag(I). Also here significant differences are seen for the inactivation parameters for the different enzymes. Although the BIP inactivations have been done at somewhat different pH values, an overall comparison is possible for the different enzymes. The affinity of BIP for the enzyme, which is reflected in the parameter $1/K_I$ follows the order:

$$Zn > Hg \sim Pb \sim Cd > Co \sim Ag$$

On the other hand, the nucleophilic reactivity of the metal-bound thiol of Cys-46, which is reflected in the parameter k_2 follows the order:

$$Co > Zn > Ag > Hg \sim Pb > Cd$$

Metal-directed affinity labelling of LADH with different metals in the catalytic center is thus seen to be a very direct and quantitative way of delineating the active site properties of different metalloenzyme species. This is of particular importance for nonchromophoric metal atoms where UV or visible spectra do not enable characterization.

REFERENCES:

1. Brändén, C.-I., Jörnvall, H., Eklund, H. and Furugren, B. (1975)'The Enzymes' (Boyer, P.D., ed.) 3rd edn, vol.11, pp.104-190, Academic Press, New York and London.

2. Sytkowski, A.J. and Vallee, B.L. (1978) Biochemistry, 17, 2850-2857.

3. Sytkowski,A.J. and Vallee,B.L. (1979) Biochemistry, 18, 4095-4099.

4. Maret,W., Andersson,I., Dietrich,H., Schneider-Bernlöhr,H., Einarsson,R., and Zeppezauer,M. (1979) Eur. J. Biochem. 98, 501-512.

5. Dietrich,H., Maret,W., Kozłowski,H. and Zeppezauer,M. (1981) J.Inorg.Biochem. 14, 297-311.

6. Maret,W., Dietrich,H., Ruf,H.H. and Zeppezauer,M. (1980) J.Inorg.Biochem. 12, 241-252.

7. Skjeldal,L., Dahl,K.H. and McKinley-McKee,J.S. (1982) FEBS Lett. 137, 257-260.

8. Skjeldal,L., Dahl,K.H. and McKinley-McKee,J.S. (1982) submitted.

9. Reynolds,C.H. and McKinley-McKee,J.S. (1969) Eur.J. Biochem. 10, 474-478.

10. Dahl,K.H. and McKinley-McKee,J.S. (1981) Eur.J.Biochem. 118, 507-513.

11. Theorell,H. and McKinley-McKee,J.S. (1961) Acta Chem. Scand. 15, 1811-1833.

12. Sund,H. and Theorell,H. (1963) 'The Enzymes' (Boyer,P.D., Lardy,H. and Myrback,K. eds.) 2nd edn, vol. 7, pp. 25-33, Academic Press, New York and London.

13. Theorell,H. and Yonetani,T. (1963) Biochem.Z. 338, 537-553.

14. Boiwe,T. and Brändén,C.-I. (1977) Eur.J.Biochem. 77, 173-179.

15. Dahl,K.H. and McKinley-McKee,J.S.(1981)J.Inorg.Biochem. 15. 79-87.

16. Evans,N. and Rabin,B.R. (1968) Eur.J.Biochem. 4, 548-554.

17. Dahl,K.H. and McKinley-McKee,J.S. (1981) Eur.J.Biochem. 120, 451-459.

18. Dahl,K.H. and McKinley-McKee,J.S. (1977) Eur.J.Biochem. 81, 223-235.

19. Brandsnes,Ø., Dahl,K.H. and McKinley-McKee,J.S. (1982) Eur.J.Biochem. in press.

20. Reynolds,C.H., Morris,D.L. and McKinley-McKee,J.S. (1970) Eur.J.Biochem. 14, 14-26.

21. Sillén,L.G. and Martell,A.E.(1964) Stability Constants, Special Publication no. 17, The Chemical Society, London.

22. Dahl,K., McKinley-McKee,J. and Jörnvall,H. (1976) FEBS Lett. 71, 287-290.

23. Dahl,K.H., McKinley-McKee,J.S., Beyerman,H.C. and Noordam,A. (1979) FEBS Lett. 99, 313-316.

24. Dahl,K.H., Eklund,H. and McKinley-McKee,J.S. (1982) submitted.

LIGAND SPHERE TRANSITIONS: A NEW CONCEPT IN ZINC-ENZYME CATALYSIS

Hans Dutler and Abraham Ambar

Laboratorium für organische Chemie, ETH-Hönggerberg,
8093 Zürich

Functional studies with liver alcohol dehydrogenase led us to propose a new concept of zinc-ligand participation in reaction. The pentacoordinated zinc has square-pyramidal geometry and possesses a sixth ligand at longer distance from zinc below the pyramid base. This sixth distal ligand differs from the five proximal ligands in that it is more nucleophilic (or basic) and more flexibel. Any of the six ligands around zinc can occupy this distal position. Such ligand sphere transitions are induced either by the chemical changes at the ligands or by changes of protein conformation. Ligand sphere transition as well as increased flexibility of distal ligands provide ideal functional features from the point of view of molecular dynamics of reaction within the framework of the active site.

INTRODUCTION

Based on crystallographic evidence it had been proposed that the active-site zinc in liver alcohol dehydrogenase is tetracoordinated (1). In the free enzyme the four ligands are the sulfhydryls of Cys-46 and Cys-174, the imidazole of His-67 and water. In most mechanistic models hitherto emerging from these structural features the substrate oxygen is replacing the water oxygen and is directly hydrogen bonded to the hydroxyl of Ser-48 which is part of a proton relay including the 2'-hydroxyl of the nicotinamide ribose and the imidazole of His-51 (2).

From EPR measurements first indication was obtained for pentacoordinated zinc in enzyme-NAD^+ and in enzyme-NAD^+-pyrazole complexes (3). Pentacoordinated zinc in enzymatically active species was derived from the results of kinetic experiments (4). Mechanistically the proposed type of pentacoordination implied that both the substrate oxygen and the water oxygen are ligated to zinc and that the substrate oxygen is hydrogen bonded to the hydroxyl of Ser-48 via the water.

I. Bertini, R. S. Drago, and C. Luchinat (eds.), The Coordination Chemistry of Metalloenzymes, 135–145.

With respect to their value as an aid to explain the many conflicting results obtained with this enzyme mechanistic models based on tetracoordination and on the proposed type of pentacoordination are equally unsatisfactory. For example they do not provide a convincing explanation for the release of a proton upon binding of NAD^+ to the enzyme.

From crystallographic analysis of low-molecular weight zinc complexes (5) it appears that in pentacoordinated zinc the five ligands preferentially form a square pyramid with equal bond lengths (∿2.0 Å) and a sixth ligand binds below the pyramid base with a longer (∿2.6 Å) and variable bond length. This geometry of a ligand sphere with five proximal and one distal ligand provides a maximum of flexibility allowing distinctive ligand participation during reaction.

The new concept which we want to present in this chapter exploits the mechanistic potential of such ligand participation coupled with transitions occurring in the ligand sphere. The need for such a concept finds its source in our recent finding that liver alcohol dehydrogenase oxidizes histidinol to histidine (6), a reaction which most likely represents one of the physiological functions of this enzyme. This reaction requires, for example a cysteine sulfur to be involved in formation of a thiohemiacetal from histidinal intermediate. In this process the ligand cysteine sulfur must aquire the necessary nucleophilicity and must be able to adjust itself positionally to the given reaction path. Since the intermediate histidinal with great probability is fixed to the active-site zinc via its imidazole residue this reaction path is very strictly dictated by the given structural features.

The requirement for ligand flexibility in the sense of allowing proper adjustment to the given reaction path is also clearly discernible in reactions with rigid substrates studied kinetically in great detail in our laboratory (7). The results obtained from reduction of a large number of alkylcyclohexanones and oxidation of the corresponding alkylcyclohexanols demonstrate that the distinctive positional change that the substrate oxygen undergoes during reaction is in accord with the transitions occurring in the ligand sphere. These dynamical aspects are difficult to reconcile with an active-site zinc remaining tetracoordinated or pentacoordinated in the hitherto proposed mode throughout the entire reaction. Clearly the zinc must provide a more flexible ligand sphere in which it assumes a switch function which guarantees a smooth course of reaction.

In metalloenzyme-catalyzed reactions,where the substrate is rigidly fixed within the active site, the flexibility aspects are particularly important.Ligand sphere transitions represent ideal means of synchronizing the structural events on the side of the ligand sphere with those on the side of the substrate.In essence they have the effect of channelling molecular motion to within the paths required for reaction.

EXPERIMENTAL BACKGROUND

Histidinol-to-Histidine Oxidation

The reaction occurs in four steps (6): 1. Oxidation of histidinol to histidinal, 2. thiohemiacetal formation with histidinal, 3. oxidation of thiohemiacetal to histidine thioester, and 4. hydrolysis of the thioester.

Similar to the oxidation of ethanol, NAD^+ acts as the coenzyme. The occurence of histidinal as an intermediate is demonstrated by pH-jump experiments. The intermediate formed at pH 9 could be reduced back to histidinol with the available NADH by changing the pH to 7. In accordance with the known low stability of histidinal the intermediate showed a half-life of ca. 15 min. Evidence for covalent binding of the intermediate histidinal stems from stopped-flow experiments with NADH as the observable. At pH 7 a single turnover produced 2 mol of histidinal per mol of enzyme (2 active sites). At this pH no additional histidinal was produced and no oxidation to histidine took place. After rapid change of the pH to 9.3 the histidinal intermediate was oxidized to 2 mol of histidine in a second single turnover. Had the intermediate not been bound covalently, we would not have been able to observe two distinct single turnovers, each with an amplitude of 2 mol of NADH formed per mol of enzyme.

Histidinol as well as histidinal act as bidentate ligands. The involvement of the imidazole residue as a ligand to zinc was indicated by the effect free imidazole had on the rate of oxidation (6,8). Free imidazole was found at low concentrations (< 2.5mM) to promote alaninol and ethanol oxidation but not histidinol oxidation. The finding that free imidazole cannot act as a promoter when the substrate contains an imidazole residue can be interpreted to mean that free as well as substrate imidazole are ligands of zinc.

Further evidence for direct binding of the substrate-imidazole to zinc is provided by kinetic experiments (Table) with enzyme in which the active-site zinc was replaced by cadmium and cobalt (9). Relative to zinc enzyme histidinol oxidation is considerably slower with the cobalt-enzyme and considerably faster with the cadmium-enzyme than

Table. Oxidation rates for enzymes with different active-site metals. k_{cat} values (s^{-1}) were determined using the standard procedure (6). Buffer, 0.1 M glycine/KOH, pH 9.75; $[NAD^+]$, 3 mM; temperature, 25°C.

	Zn	Co	Cd
Ethanol	11.3	7.4	0.5
Histidinol	ca. 1.2	0.12	0.5

ethanol oxidation. This effect can only be explained when imidazole acts as a ligand for the active-site metal.

Finally, model building was of great help in deducing the nature of the covalent bond between enzyme and histidinal intermediate. With the imidazole residue ligated to zinc there was one single orientation of histidinal allowing formation of a covalent bond (Fig. 1). In this orientation the carbonyl carbon appears in an ideal position for nucleophilic attack by Cys-174 sulfur to form a thiohemiacetal or by ^{-}OH in the same position to initiate hydrolysis of the histidine thioester formed after oxidation of the thiohemiacetal.

In conclusion the results from histidinol oxidation allow us to develop the following scheme describing complex formation with the bidentate substrate histidinol. On starting with tetracoordinated zinc in free enzyme imidazole-$N^{\delta 1}$ and hydroxyl oxygen of histidinol are inserted between Cys-174 sulfur and water oxygen, the latter moving to the distal position. The ensuing involvement of Cys-174 sulfur is but an example of ligand participation during reaction. In the model which

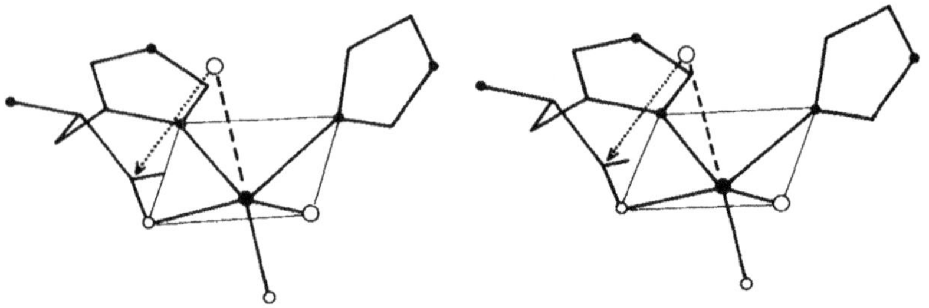

Fig. 1. Stereodiagram representing nucleophilic attack of Cys-174 sulfur on the carbonyl carbon of histidinal. With Cys-174 sulfur in distal position the direction of attack is at an angle of ca. 109° to the carbon-oxygen double bond. (●) Nitrogen, (○) oxygen, (◯) sulfur, and (⬤) zinc.

is to be developed from these results a number of additional, analogous modes of ligand participation can be anticipated and consequently follow the same rules. These include ester hydrolysis, NAD^+ binding, proton release, and oxidation-reduction.

Correlation Kinetics with Alkylcyclohexanones and Alkylcyclohexanols

Kinetic information on the reduction of alkylcyclohexanones was correlated with the three-dimensional structure of the enzyme (7). The substrates investigated were: 2-, 3-, and 4-alkylcyclohexanones with alkyl groups methyl, ethyl, i-propyl, and t-butyl. Kinetic studies establishing at which position of the cyclohexanone ring an alkyl group leads to fast, slow, or no reduction and at which position an increase in the size of the alkyl group leads to a decrease of the rate of reduction, allows one to deduce at which position an alkyl group leads to favorable or unfavorable interactions with groups of the enzyme. On the basis of the structure of the enzyme and on plausible assumptions regarding the arrangement of the reacting atoms, models were built of the enzyme-NAD^+-alkylcyclohexanol complexes formed during reduction. These models were analyzed with respect to favorable and unfavorable interactions. By changing the orientation of the cyclohexanol molecules in the complex it was possible to arrive at a structure in which the interactions observed in the model correlated well with those deduced from kinetic analysis. Similar studies were carried out for oxidation of the corresponding alkylcyclohexanols.

As a result, a probable structure of the enzyme-coenzyme substrate complex with productive substrate orientation was obtained. In this orientation the oxygen of the substrate appears to be directly bound to the active-site zinc. In particular these kinetic studies demonstrate that the path the substrate oxygen takes during oxidation coincides closely with the proximal-to-distal change of the accompanying ligand-sphere-transition process.

THE PROPOSED MODEL CONCEPT

The new concept proposes that ligand participation is made possible by typical positional changes of the ligands within the ligand sphere consisting of five proximal ligands forming a distorted square pyramid and one distal ligand. These ligand sphere transitions involve distal-to-proximal strictly coupled to proximal-to-distal changes of ligands. Which ligand takes which position primarily depends on bond strength and on topographical and dynamical features on the side of the protein. Chemical changes in which ligands are involved perturb the available order and cause the ligand sphere to readjust itself. Vice versa, changes in the ligand positions cause the ligands to become involved in reaction. In addition, these ligand sphere transitions are under control of protein conformation and, vice versa, protein conformation is under control of ligand sphere transitions. This interrelationship

Ligand involvement in reaction ⇄ Ligand sphere transitions ⇄ Protein conformational changes

demonstrates the central rôle of ligand sphere transitions. In the present discussion we want to restrict ourselves to the left part of the scheme concerning ligand involvement in reaction and ligand sphere transitions.

Ligand Sphere Transitions

Ligand positional changes can take place in two modes. When they involve apical ligands, intermediate states are formed in which the

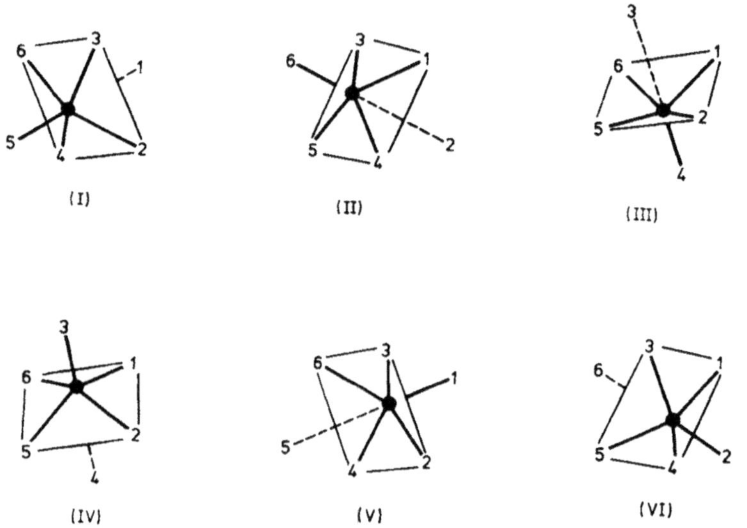

Fig. 2. Square pyramids with a sixth distal ligand. Depending on the involvement in reaction positions 1-6 are occupied as follows: 1, $N^{\varepsilon 2}$ (His-67); 2, S^- (Cys-46), O (H_2O); 3, S^- (Cys-174), O (H_2O), OH^-; 4, O (H_2O), OH^-; 5, O (substrate), O^- (substrate), O (H_2O); 6, O (H_2O), $N^{\delta 1}$ (substrate imidazole). Bond to the distal ligand in broken line, bonds to the proximal ligands in heavy lines, pyramid base in thin lines.

distal to proximal and the proximal to distal changing ligand form an approximately linear arrangement. This mode of transition leads to an inversion of the square pyramid. When they involve apical and basal ligands, the direction of approach and departure of the ligands is approximately at a right angle. In this mode the transition leads to a rotation of the pyramid.

Theoretically six types of square pyramids with a sixth distal ligand below the pyramid base are possible (Fig. 2). In the proposed model only five square pyramids (II-VI) are relevant; $N^{\varepsilon 2}$ (His-67) is not involved in reaction and remains proximal. The imidazole-N positions are designated as Ⓝ. In position 6 Ⓧ (equivalent to O (H_2O) or $N^{\delta 1}$ (substrate imidazole)) or Ⓝ undergo positional changes but are not involved in reaction.

Since the active-site metal and its ligands are buried in the interior of the protein and hence are completely shielded from solvent, the balance of charges has to be maintained at all times. When a negatively charged ligand changes from proximal to distal and is neutralized during reaction, the loss of negative charge within the ligand sphere needs to be compensated either by distal-to-proximal change of another negatively charged ligand or of a neutral ligand acquiring negative charge during reaction.

Ligand Involvement in Reaction

Ligands in distal position are more nucleophilic or more basic than in proximal position. Thus a proximal-to-distal change may allow a ligand to get involved, for example, in a nucleophilic addition reaction or in the uptake of a proton. Vice versa a distal-to-proximal change of a ligand may be coupled, for example, to the break-down of a tetrahedral intermediate or the release of a proton.

The aspect of flexibility of this type of metal coordination manifests itself in the ability of a distal ligand to undergo further positional changes. Exchange of ligands as for example in product-substrate exchange preferentially occurs in distal position. A negatively charged ligand can move further away from the distal position such as to act as counter ion for a positively charged reaction partner binding in the active site. Vice versa when a positive charge on such a reaction partner disappears during reaction, the counter ion moves back to its original position.

The positional changes of ligands, in particular proximal-to-distal and distal-to-proximal changes, allow reacting ligands to adjust themselves to the changing steric requirements as the reaction proceeds. This strict coupling of ligand sphere transition and reaction path is most clearly displayed in reactions where the steric course is known in detail as for example in nucleophilic additions to carbonyl (10,11).

Examples

1. NAD^+ Binding and Proton Release. Addition of NAD^+ to an enzyme species with the water oxygen in distal position leads to a transition of pyramid IV to II with Cys-46 S^- in distal position and water O^- in proximal position (Fig.3). Simultaneously Cys-46 S^- leaves this distal position and forms together with NAD^+ the ion pair $NAD^+ \cdot S^-$ (Cys-46).

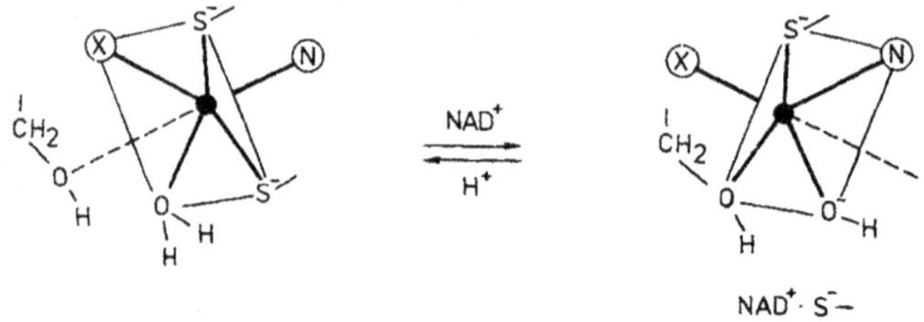

Fig. 3. NAD^+ binding and proton release.

The loss of negative charge in the zinc coordination sphere is compensated by formation of O^- from water oxygen. This release of a proton is the result of a pK change of water due to the distal-to-proximal change. From kinetic experiments it is known, that a pK change takes place from 9.6 to 7.6 (12) or 4.5 (13) depending on whether, respectively, NAD^+ binding occurs in absence or presence of substrate. The structure of the enzyme suggests that the proton is not directly released to solvent but is first transferred to a proton reley system consisting of Ser-48 OH, nicotinamide-ribose 2'-OH and His-51 imidazole. It is possible that the ion pair formation with the mobile Cys-46 S^- is more complex than anticipated in Fig. 3. This process may involve additional charged groups thus forming a loop between the groups to be neutralized.

2. Oxidation of Histidinol. The hydride transfer process in the ternary complex is accompanied by a transition of pyramid II with Cys-46 S^- acting as counter ion for NAD^+ to pyramid V (Fig. 4).After formation of neutral NADH, Cys-46 S^- moves back from its position as counter ion to NAD^+ first into the distal position of pyramid II and then, as a result of the ligand sphere transition, into the proximal position of pyramid V. Simultaneously the substrate hydroxyl releases its proton to neutralize the water O^- and turns over into substrate carbonyl which, being the weakest ligand, moves from proximal to distal.

During oxidation the tetrahedral hydroxyl carbon of the substrate is converted into a trigonal-planar carbonyl carbon. With substrates whose carbon skeleton is rigidly fixed in the active site, as with histidinol and with the alkylcyclohexanols, this geometrical change

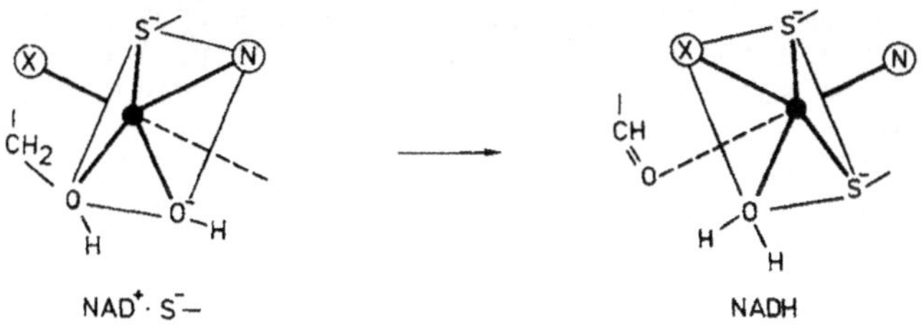

Fig. 4. Oxidation of histidinol.

implies a positional change of the substrate oxygen relative to zinc. With the aid of model building (Fig. 5) we made the interesting observation that this positional change closely coincides with that created by the ligand sphere transition converting pyramid II to V.

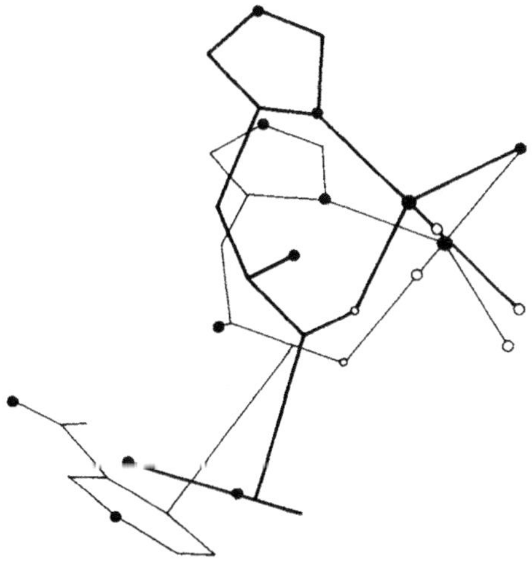

Fig. 5. Geometrical change during histidinol oxidation and ligand sphere transition from pyramid II to V. View perpendicular to the plane of positions 1,2,5, and 6. Histidinol, NAD+, and the ligand sphere of pyramid II in heavy lines; histidinal, NADH, and the ligand sphere of pyramid V in thin lines. Position 1 acts as pivot for the pyramid and the nicotinamide ribose as pivot for the coenzyme. (●) Nitrogen, (○) oxygen, (◯) sulfur, and (⬤) zinc.

3. Thiohemiacetal Formation. During nucleophilic attack of Cys-174 S^- on the substrate carbonyl carbon the coordination geometry of zinc changes from pyramid V to III. Cys-174 S^- moves from proximal to distal and thereby gains the necessary nucleophilicity. The developing negative charge on the substrate oxygen causes this ligand to undergo a distal-to-proximal change and compensates for the loss of the negative charge on Cys-174 sulfur. Analysis of the geometrical changes discloses again a close to perfect coincidence of ligand sphere transition and reaction path on the side of the substrate.

Fig. 6. Thiohemiacetal formation.

4. Hydrolysis of Thioester. The formation of the tetrahedral intermediate is a process analogous to the formation of the thiohemiacetal. The nucleophile is OH^- in place of Cys-174 S^-. For break-down of the tetrahedral intermediate the pyramid changes from III back to V. In this process Cys-174 S^- is liberated and moves back into its original proximal position (position 3).

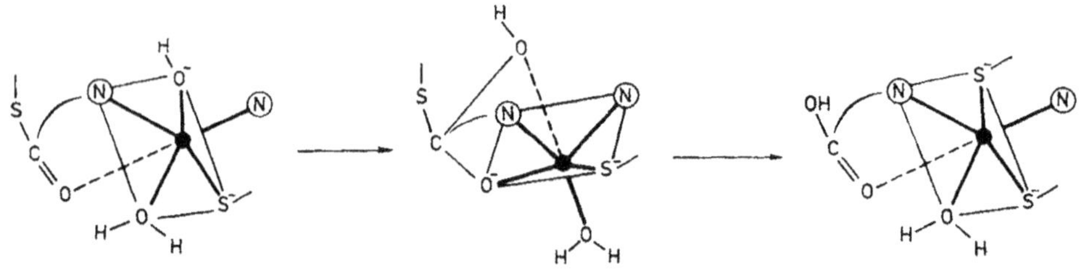

Fig. 7. Hydrolysis of thioester.

ACKNOWLEDGEMENTS

This work was carried out with financial support of the Schweizerischer Nationalfonds and Emil Barell-Stiftung of F. Hoffmann-La Roche & Co. A.G.

REFERENCES

(1) Bränden, C.J., Jörnvall, H., Eklund, H., and Furugren, B. (1975) in "The Enzymes" (Boyer, P.D.,ed.) 3rd ed., Vol. 11, Academic Press, New York.
(2) Bränden, C.I., and Eklund, H. (1980) in "Dehydrogenases Requiring Nicotinamide Coenzyme" (Jeffery, J., ed.), Birkhäuser Verlag, Basel.
(3) Makinen, M.W., and Yim, M.B. (1981) Proc. Natl. Acad. Sci. USA 78, 6221.
(4) Dworschack, R.T., and Plapp, B.V. (1977) Biochemistry 16, 2716.
(5) Freeman, H.C. (1967) in "Advances in Protein Chemistry" (Anfinsen, C.B., Anson, M.L., Edsall, J.T., and Richards, F.M., eds.) Academic Press, New York, London.
(6) Ambar, A. (1981) Diss. ETH, Nr. 6840.
(7) Dutler, H., and Bränden, C.J. (1981) Bioorganic Chemistry 10, 1.
(8) Theorell, H., and McKinley-McKee, J.S. (1961) Acta Chem. Scand. 15, 1148.
(9) Ambar, A., Gerber, M., Dutler, H., and Zeppezauer, M. (1982), unpublished results.
(10) Bürgi, H.B., Dunitz, J.D., and Shefter, E. (1973) J. Amer. Chem. Soc. 95, 5065.
(11) Bürgi, H.B., Dunitz, J.D., Lehn, J.M., and Wipff, G. (1974) Tetrahedron 30, 1563.
(12) Parker, D.M., Hardman, M.J., Plapp, B.V., Holbrook, J.J., and Shore, J.D. (1978) Biochem. J. 173, 269.
(13) Wolfe, J.K., Weidig, C.F., Halvorson, H.R., Shore, J.D., Parker, D.M., and Holbrook, J.J. (1977) J. Biol. Chem. 252, 433.

MOLECULAR MECHANISMS OF THE SUPEROXIDE DISMUTASE ACTIVITY OF THE CUPROZINC PROTEIN OF EUCARYOTIC CELLS (CuZn SUPEROXIDE DISMUTASE).

G. Rotilio

Department of Biology, 2nd University of Rome and CNR Center for Molecular Biology, Institute of Biological Chemistry, University of Rome, Rome, Italy.

1. MOLECULAR PROPERTIES OF THE PROTEIN

Cu Zn Superoxide dismutases are ubiquitary proteins of eucaryotic cells, made up of two identical subunits of 16 Kdal, each one carrying one Cu(II) and one Zn(II). Since they catalyze no other reaction than dismutation of O_2^- into H_2O_2 and O_2, they were named superoxide dismutase (1), although many Authors still prefer names related to their metal composition, as cytocuprein (2) or cuprozinc protein (3). The catalytic dismutation of O_2^- is actually a molecular function of this protein and in the past decade activity-structure relationship studies of superoxide dismutase have represented a major advance in the field of metalloproteins. Nevertheless, many Authors, including the author of the present review, hesitate to subscribe to the opinion that molecular function, even a very efficient and specific one, just equals the physiological role of any protein. However, this article has no commitment to discuss such a problem and will be dedicated to outline the basic mechanisms by which the cuprozinc protein of eucaryotic cells is a very active superoxide dismutase at the molecular level. In this context, a secondary message of this article will become apparent, that is a word of caution toward the use of any single methodological approach to structural-functional situations of proteins. In fact, comparative analysis of data obtained in the same experimental conditions, and rigorous control of all parameters of purified preparations are of primary importance in the study of molecular properties of proteins as much as the precision and the resolution power of advanced physical techniques of study.

The protein of bovine erythrocyte has been crystallized and its structure has been analyzed at 2 Å resolution (4). Each subunit is composed of 8 antiparallel β strands which form a flattened cylinder ("barrel") plus 3 external loops of non repetitive structure, which form almost half of the subunit. Two loops are larger and almost adjacent. They form a long channel connecting the solvent and the external surface of the barrel. The Cu(II) and Zn(II) are located at the bottom of this channel: the Zn is buried while the Cu is solvent accessible. However, when the Zn sites are made artificially vacant (5), Cu from one subunit

I. Bertini, R. S. Drago, and C. Luchinat (eds.), The Coordination Chemistry of Metalloenzymes, 147–154.

can migrate to an empty Zn site in another subunit.

The Zn(II) is bound to the ND1 (the proximal N) of His 61,69 and 78 and to the carboxylate oxygen of Asp 81. The geometry of coordination is tetrahedral with a strong distortion toward a trigonal pyramid with Asp 81 at the apex. The Cu(II) is coordinated by ND1 of His 44 and by NE2 (the distal N) of His 46,61 and 118, in a tetrahedrally distorted square planar geometry. His 61 thus bridges the Cu and Zn with the imidazole ring approximately planar to both metals, which lie 6.3 Å apart. The axial position of the Cu is much more open on the solvent side than on the protein side. The solvent peak is located about 3 Å from the Cu, suggesting that a water molecule is coordinated to copper in this position. These features match very well with the spectroscopic properties of the protein. The Cu(II) has a low energy band near 15,000 cm^{-1}, of moderately high intensity (ε = 150 $M^{-1}cm^{-1}$), and a rhombically distorted EPR spectrum (6) with parameters typical of five-coordinated Cu(II) (7). The Cu(II) relaxes efficiently the water protons (8). The U.V. spectrum is characterized by strong imidazolate-Cu(II) charge-transfer transitions (9). The Zn(II) can be substituted by tetrahedral Co(II) (10), and the Cu(II) by tetrahedral or five-coordinated Cu(II)(11). Co(II) and Cu(II) are strongly antiferromagnetically coupled (12,13).

In conflict with crystallographic and spectroscopic data, results of perturbed angular correlation of gamma rays in Cd-derivatives of yeast SOD have been interpreted as to exclude an imidazolate bridge between Cd (which substituted for Zn) and Cu(14). However it has been pointed out (15) that Cu does not bind to its native binding site in the procedure used by those Authors to reconstitute the protein with Cd and Cu from the apoprotein. This is an example of the use of sophisticated physical methods on samples not properly characterized.

2. GENERAL MECHANISM OF THE SUPEROXIDE DISMUTATION BY THE COPPER OF THE PROTEIN.

The Cu of superoxide dismutase has an unusually high redox potential (+ 400 mV; ref. 16). Therefore it is easily reduced by many biologically important molecules, including the product of superoxide dismutation H_2O_2 (16). The reduced protein is not reoxidized by O_2 at neutral pH. Kinetically, the reactions with O_2^- of either the Cu(I) or Cu(II) state of the protein are much faster than any other redox reactions, and the rate of the reduction and reoxidation processes of the Cu by O_2^- are the same and identical to the overall rate of turnover catalysis (17). This property makes the protein a superoxide dismutase; were the Cu(I) rapidly reoxidized by O_2, the protein would be a superoxide oxidase, by H_2O_2 a superoxide peroxidase. The bovine protein dismutates O_2^- at the same very high rate ($\simeq 2\times10^9 M^{-1}s^{-1}$) independently of pH between pH 5 and 10 (18). No other catalyst of this reaction displays such a property. Low molecular weight complexes of Cu(II) can even dismute O_2^- at a comparable rate, but a proton transfer rate-controlling step is evident in the kinetics of any dismutation reaction other than that of bovine Cu Zn superoxide dismutase.

3. INACTIVATION OF THE ENZYME

Inactivation is the irreversible, time-dependent, loss of activity by an enzyme. Three well-defined cases of inactivators of Cu Zn superoxide dismutases are H_2O_2, heat and diethyldithiocarbamate (DDC).

a) H_2O_2. This product of superoxide dismutation reduces the Cu(II) of the enzyme (16) in a slow reaction (19). The Cu(I)-enzyme is reoxidized by H_2O_2 in a Fenton-type of reaction, which destroys histidines of the copper environment (19). This inactivation does not occur when the enzyme is under turnover reaction with high O_2^-, as the active metal is always occupied by the substrate (17). However H_2O_2 inactivation can be of relevance in the biological environment, where H_2O_2 is produced in considerable amounts, while O_2^- is not. Many purified preparations of the enzyme may contain inactive enzyme molecules because of this reaction. The Cu(II) of the H_2O_2-inactivated enzyme has a more axial EPR spectrum (16) and is expected to behave differently from the native protein in its reactions with ligands.

b) Heat. Cu Zn Superoxide dismutases are very resistent to heat treatment at 70° C (20). This stability depends on the presence of copper and zinc bound at their native sites (20) as the metal-free protein is much less resistent. It has been recently discovered (21) that even the purest preparations contain at least three electrophoretic forms, with distinct pI's, and significantly different sensitivity to heat despite absolute identity of metal content and state. Moreover the form having intermediate electrophoretic mobility is dissociated by heat into non equivalent subunits, which reaggregate to homodimers with mobilities corresponding to those of the other two forms. This result has been interpreted as to indicate that "pure" Cu, Zn superoxide dismutase is a mixture of forms, borne out by asymmetrical distribution of charges in a parent heterodimer. This heterogeneity is likely to arise from post-translational reactions of amino acid side chains, as, for instance, deamidation. The finding that these forms differ as to heat sensitivity is a further warning against easy extrapolation of results obtained on uncontrolled samples.

c) DDC. This metal-chelating agent is currently used as an in vivo poison of Cu Zn superoxide dismutase (22). Its action has been explained as to be due to the formation of a ternary complex DDC-Cu-protein,which does not lead to removal of copper from the protein (23). More recently it has been shown that reaction with DDC actually removes the copper from the protein (24). The reaction can be conveniently manipulated, as it does not affect the zinc, and, under special conditions, leads to the isolation of a protein dimer carrying the copper on one subunit only (25). The importance of this derivative is that, when it is tested for activity, the copper appears to be more active than in the 2 Cu 2 Zn dimer. The dimer, where one subunit is coupled to another subunit carrying the copper inactivated by H_2O_2,is just half active as the native dimer (26). These results indicate a functional constraint on the active site of one subunit from another subunit, only when the latter one is metal-bound. This means that conformation and conformational flexibilities are different whether the subunits are bound to metal or are metal free. These results document how inactivating reactions can be managed in order to provide useful information on the structure-activity relationships of the enzyme.

4. INHIBITION OF THE ENZYME

Inhibitors are molecules that depress the enzyme activity by interacting with the active site via a mechanism that can be described by reversible chemical equilibria. Three situations apply to Cu, Zn dismutases, i.e. alkaline pH, anion binding and ionic strength. They are strictly interrelated to each other and the distinction made in this discussion refers to empiric experimental situations, not at all to theory or interpretation of the phenomena.

a) Alkaline pH. Many metallo proteins undergo alkaline transitions of spectroscopic properties and/or activity parameters. Such transitions are usually associated to the protonic equilibrium of a water molecule, which is bound to the metal coordination sphere. This seems to be the case for Cu Zn superoxide dismutase as well. The EPR spectrum changes from rhombic to axial (27), the nuclear magnetic relaxation rate of water protons increases (28) and the enzyme activity decreases (29), according to the same titration curve described by an apparent $pK'_a \simeq 11$. All changes are perfectly reversible and do not involve breaking of the imidazolate bridge (30). The alkaline transition is phenomenologically correlated with anion binding processes as the high-pH form binds anions less strongly. Also its interpretation is, by general agreement, that of special case of anion (OH^-) binding. More detailed discussion will therefore follow in the next paragraph b).

It should be recalled that Cu Zn superoxide dismutases also show an acid transition centered approximately at pH 4, which is again characterized by axialization of EPR spectra. This transition is irrelevant to the enzyme mechanism as below the pK'_a of O_2^- (pH = 4.8) enzymatic catalysis is no sense, since superoxide dismutation is very rapid in the presence of substantial amounts of $HO_2^{\cdot}$. The phenomenon is associated to breaking of the imidazolate bridge, due to protonation of the ND1 of His 61 (31). In fact, the Zn binding to the protein is labilized at low pH and dialysis at pH 3.0 is used to obtain the Zn-free protein (32). Therefore the acid transition, though reversible on its higher pH range, where the zinc is still bound to most of its protein ligands, is to be considered rather an inactivation, according to definition given above.

b) Anions. Singly charged anions interact with the enzyme in a fashion that may be considered as a rule for solvent accessible metal-binding centers of metalloproteins. However in the case of CuZn superoxide dismutase they are of special interest, as they represent strict substrate

-analogs. In fact, CN^-, N_3^-, OH^-, F^-, Br^- and Cl^- were shown (33,34) to inhibit the superoxide activity of the enzyme with a "competitive" type of pattern. Clearly, kinetics can no provide "positive" structural evidence. However the inhibition of activity is related to comparable degreesof inhibition of the water relaxation rate (34) suggesting that O_2^- and anions bind in the inner coordination sphere of the Cu at the same position, that is the water coordination position. This conclusion seems to be valid also for anions like NCO^- and NCS^-, if one is aware, in interpreting their reactions, of their behavior as non-"pure" metal ligands. In a recent paper (35) water proton relaxation rate values are reported to be 90% and 70% that of the native enzyme in the presence of 0.2 M NCS^- and NCO^- respectively. These values correspond to the decrease of activity in the same conditions, when the activity is normalized in terms of Cu(II), that is the only species "seen" by NMR. In particular, side effects on activity by ionic strength (see further on) and irreversible modifications of the protein part must be taken in account at such high concentrations of rather indiscriminate reagents. Further difficulties arise when NMR data are confronted with EPR spectra taken in frozen solutions (36). However when correlation is made between results obtained in the same conditions, the extent of EPR changes of the Cu parallels the magnitude of NMR effects (35) and of enzyme inhibition (34), that is $CN^- \simeq N_3^- > OH^- \simeq NCO^- > NCS^- \gg F^-$.

The degree of precision of such correlations is necessarily not very high, for two reasons: the different intrinsic sensitivity of variations techniques and the above-mentioned side effects on the protein by high salt concentration which may in turn affect the response of metal sites to ligands. Furthermore, affinity of inhibitors as calculated from the decrease of activity is higher than from spectroscopic measurements, because activity "sees" additional effects on the Cu(I) state of the enzyme (34).

The EPR spectra of the cyanide complex (6) and the ^{17}O NMR study of the high-pH form (37) demonstrate that anions bind equatorially. Crystal data (4) suggest axial coordination of water. This evidence does not conflict with the statement, made earlier in this section, that all anion bind to the copper by substituting the coordinated water molecule. Small anionic ligands, while entering the channel to the Cu and exchanging with the coordinated water, drastically alter the electrostatic interactions in the active site pocket, which is very sensitive to charge effects (see below). In the new arrangement the anion may lie in the plane formed by the copper and three equidistant histidines (which appear as equivalent N nuclei in the EPR spectrum of the cyanide adduct, see ref. 6), with the fourth histidine more apart at the apex of a regular square pyramid. In this picture all the protein ligands are still coordinated to the two metals, including His 61 (30), and only water is excluded from the first copper coordination sphere. This model is in line with three general principles: proteins are flexible objects that allow symmetry changes of coordinated metals (in this case from distorted tetrahedral to square pyramidal) with no change of coordination number; active site of enzyme can change conformation upon substrate binding ("induced fit" mechanism); in superoxide dismutase anions are substrate analogs.

Charge effects on the active site are suggested by the inhibition of the enzyme (38) at high concentrations of salts that, like $SO_4^{=}$ and ClO_4^{-}, neither coordinate the copper, as CN^{-} or N_3^{-} do, nor irreversibly alter the protein, as CNO^{-}. The plausible sites of these effects are to be identified with positively charged amino acid side chains, which would "guide" O_2^{-} into the inner coordination sphere of the Cu. Two types of evidence are available. Neutralization of the guanidine group of Arg. 141, a residue located at the gate of the active site channel, brings about a large decrease of activity, but alters the Cu geometry as well (39). This is a specific, but "mixed" effect, as it involves both a charged group and the Cu coordination. Succinylation of all the lysine residues of the protein leaves the Cu coordination unaffected but lowers the activity down to one tenth that of the native enzyme (40). This is a general, but "pure" charge effect. From these data the catalytic efficiency of Cu Zn superoxide dismutase appears to result from the right combination of the attitude of the Cu to bind anions and undergo rapid redox cycles with a small reducing anion, with à proper number and location of positively charged amino acids on the protein surface.

5. ROLE OF THE ZINC

In the absence of the Zn, the Cu changes its EPR spectrum into a more axial line shape and lowers its redox potential, as it is not reduced by H_2O_2 and ferrocyanide. Surprisingly enough, the Zn-free protein is as active as the CuZn protein (41). This result may indicate that the kinetic parameters of the enzyme reaction are not related to the unique geometry and redox potential of the Cu, thus emphasizing the influence of ionic strength and of the surface electrostatic potential of the protein in the catalysis. These considerations also lead to reevaluate the possible role of the Zn in the superoxide dismutase activity of the protein (42), and to restrict it to that of stabilizing the protein structure (43). This conclusion relates to possible physiological functions of the proteins other than superoxide dismutase. In fact, in the absence of the Zn, the Cu is mobilized from its site to vacant Zn sites of the other subunits (5); in the presence of Zn, on the other hand, the Cu-free protein is extremely avid of copper (44). More work on this line may revive the hypothesis that this protein is in some way involved in transport and control of intracellular copper.

REFERENCES

(1) Mc Cord, J.M. and Fridovich, I.: 1969, J. Biol. Chem. 244, pp.6049-6055.
(2) Carrico, R.J. and Deutsch, H.F.: 1969, J. Biol. Chem. 244, 6087.
(3) Valentine, J.S. and Pantoliano, M.W.: 1981 in: "Copper Proteins",

T. Spiro, ed., pp. 293-358.
(4) Tainer, J.A., Getzoff, E.D., Beem, K.M., Richardson, J.S. and Richardson, D.C., 1982, J. Mol. Biol., in press.
(5) Valentine, J.V., Pantoliano, M.W., McDonnell, P.J., Burges, A.R. and Lippard; S.T., 1979, Proc. Natl. Acad. Sci. U.S.A. 76: 4245-4249.
(6) Rotilio, G., Morpurgo, L., Giovagnoli, G., Calabrese, L. and Mondovì, B.: 1972, Biochemistry 11, pp. 2187-2192.
(7) Desideri, A., Morpurgo, L., Rotilio, G. and Mondovì, B.: 1979, FEBS Letters, 98, pp. 399-401.
(8) Fee, J.A. and Gaber, B.P.: 1972, J. Biol. Chem. 247, pp. 60-65.
(9) Fawcett, T.G., Bernarducci, E.E., Krogh-Jespresen, K. and Schugar, H.J.: 1980, J. Am. Chem. Soc. 102, pp. 2598-2604.
(10) Rotilio, G., Calabrese, L. and Coleman, J.E.: 1973, J. Biol. Chem. 248, pp. 3855-3859.
(11) Calabrese, L., Cocco, D. and Desideri, A., 1979, FEBS Letters. 106, pp. 142-144.
(12) Rotilio, G., Calabrese, L., Mondovì, B. and Blumberg, W.E.: 1974, J. Biol. Chem. 249, pp. 3157-3160.
(13) Desideri, A., Cerdonio, M., Magno, F., Vitale, S., Calabrese, L., Cocco, D., and Rotilio, G., 1975, FEBS Letters, 89, pp. 83-85.
(14) Bauer, R., Demeter, I., Hasemannn, V. and Johansen, J.T.: 1980, Biochem. Biophys. Res. Commun. 94, pp. 1296-1302.
(15) Rotilio, G. and Fielden, E.M. in "Copper Proteins", Vol. I, R. Lontie ed., in press.
(16) Rotilio, G., Morpurgo, L., Calabrese, L., and Mondovì, B.: 1973, Biochim. Biophys. Acta 302, pp. 299-235.
(17) Fielden, E.M., Roberts, P.B., Bray, R.C., Lowe, D.J., Mautner, G.N. Rotilio, G. and Calabrese, L.: 1974, Biochem. J. 139, pp. 49-60.
(18) Rotilio, G., Bray, R.C. and Fielden, E.M.: 1972, Biochim. Biophys. Acta 268, pp. 605-608.
(19) Bray, R.C., Cockle, S.A., Fielden, E.M., Roberts, P.B., Rotilio, G. and Calabrese, L.: 1974, Biochem. J. 139, pp. 43-48.
(20) Forman, H.J. and Fridovich, I.: 1973, J. Biol. Chem., 248, pp. 2645-2649.
(21) Civalleri, L., Pini, C., Rigo, A., Federico, R., Calabrese, L. and Rotilio, G.: 1982, Mol. Cell. Biochem., in press.
(22) Heikkila, R. and Cohen, G.: 1977, in "Superoxide and Superoxide Dismutase", Michelson, A.M. Mc.Cord J.M. and Fridovich, I. eds., Academic Press, London, pp. 367-373.
(23) Misra, H.P.: 1979, J. Biol. Chem. 254, pp. 11623-11628.
(24) Cocco, D., Calabrese, L., Rigo, A., Argese, E. and Rotilio, G.: 1981, J. Biol. Chem. 256, pp. 8983-8986.
(25) Cocco, D., Calabrese, L., Rigo, A., Marmocchi, F. and Rotilio, G.: 1981, Biochem. J. 199, pp. 675-680.
(26) Malinowski, D.P. and Fridovich, I.: 1979, Biochemistry 18, pp.237-244.
(27) Rotilio, G., Finazzi-Agrò, A., Calabrese, L., Bossa, F., Guerrieri, P. and Mondovì, B.: 1971, Biochemistry 10, pp. 616-621.
(28) Terenzi, M., Rigo, A., Franconi, C., Mondovì, B., Calabrese, L., and Rotilio, G.: 1974, Biochim. Biophys. Acta 351, pp. 230-236.

(29) Rigo, A., Viglino, P. and Rotilio, G.:1975, Anal. Biochem. 68, pp. 1-8.
(30) Calabrese, L., Cocco, D., Morpurgo, L., Mondovì, B. and Rotilio, G.: 1976, Eur. J. Biochem. 64, pp. 465-470.
(31) Calabrese, L., Cocco, D., Morpurgo, L., Mondovì, B. and Rotilio,G.: 1975, FEBS Letters, 59, pp. 29-31.
(32) Pantoliano, M.W., Mc Donnell, P.J. and Valentine, J.S.: 1979, J.Am. Chem. Soc. 101, pp. 6454-6456.
(33) Rigo, A., Viglino, P. and Rotilio, G.: 1975, Biochem. Biophys. Res. Commun. 63, pp. 1013-1018.
(34) Rigo, A., Stevanato, R., Viglino, P. and Rotilio, G.: 1977, Biochem. Biophys. Res. Commun. 79, pp. 776-783.
(35) Bertini, I., Borghi, E., Luchinat, C. and Scozzafava, A.: 1981, J. Am. Chem. Soc. 103, pp. 7779-7783.
(36) Bertini, I., Luchinat, C., and Scozzafava, A.: 1980, J. Am. Chem. Soc. 102, pp. 7349-7353.
(37) Bertini, I., Luchinat, C. and Messori, L.: 1981, Biochim. Biophys. Res. Commun. 101, pp. 577-583.
(38) Rigo, A., Viglino, P., Rotilio, G. and Tomat, R.: 1975, FEBS Letters 50, pp. 86-88.
(39) Malinowski, D.P. and Fridovich, I.: 1979, Biochemistry 18, pp. 5909-5917.
(40) Marmocchi, F., Mavelli, I., Rigo, A., Stevanato, R., Bossa, F., and Rotilio, G.: 1982, Biochemistry 21, pp. 2853-2856.
(41) O'Neill, P., Fielden, E.M., Cocco, D., Calabrese, L. and Rotilio, G., submitted to Biochem. J.
(42) McAdam, M.E., Fielden, E.M., Lavelle, F., Calabrese, L., Cocco, D. and Rotilio, G.: 1977, Biochem. J. 167, pp. 271-274.
(43) Rotilio, G., Calabrese, L., Bossa, F., Barra, D., Finazzi-Agrò, A., and Mondovì, B.: 1972, Biochemistry 11, pp. 2182-2187.
(44) Rigo, A., Viglino, P., Bonori, M., Cocco, D., Calabrese, L. and Rotilio, G.: 1978, Biochem. J. 169, pp. 277-280.

A COMMENT ON ANION BINDING TO SUPEROXIDE DISMUTASE

Ivano Bertini, Claudio Luchinat, Andrea Scozzafava
Istituti di Chimica Generale e Inorganica,
Università di Firenze,
Via Gino Capponi 7
50121 Firenze, Italy

As elegantly discussed in the previous chapter the interaction of anions with copper(II) in superoxide dismutase is rather complex. From X-ray data the copper(II) chromophore is essentially square pyramidal with the four histidine nitrogens in a puckered arrangement; the solvent exposed apical position is occupied by a semi-coordinated water molecule (1). Indeed, the ESR spectra give g and A values typical of chromophores which are referred to as distorted five coordinate (2).

Anions may be then thought to replace the apical water molecule. However, as outlined by Rotilio the physical data are not consistent with this simplified picture. For example, cyanide causes the removal of water, but also an ipsochromic shift of the absorption maximum of the d-d spectrum, axialization of the ESR spectrum, and an increase of $A_{\parallel}$ from 140 up to 190×10^{-4} cm^{-1}.

All of these data are consistent with binding in the tetragonal plane; a possibility considered by Rotilio is that cyanide displaces the water molecule causing a rearrangement of the histidine ligands and giving rise to a sort of elongated square pyramid, where the Cu-C bond axis essentially corresponds to the Cu-O axis in the native chromophore

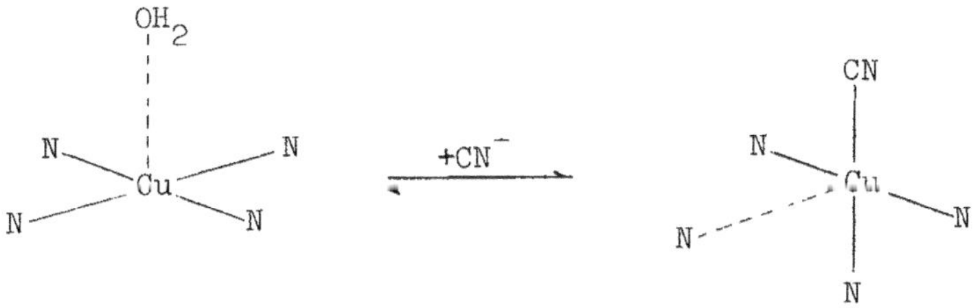

The investigation of the NCS^- derivative, however, allowed us to propose an alternative picture (3). ^{13}C and ^{14}N NMR measurements on the ligand nuclei (3,4) showed that the anion directly interacts with the copper center, owing to the large contact contribution to the ^{13}C and ^{14}N linewidths and to the short ^{13}C T_1 values which place the above

I. Bertini, R. S. Drago, and C. Luchinat (eds.), The Coordination Chemistry of Metalloenzymes, 155–158.

nucleus at a distance ($\lesssim$ 350 pm) typical of direct metal coordination. The ESR data at both liquid nitrogen (3) and room temperature (5) are consistent with the anion binding to the metal. Water proton NMR data on the NCS^- derivative have shown that water is still about 90% coordinated when copper(II) is 90% saturated with the above ligand (3). Therefore, owing to the non availability of a sixth coordination position, NCS^- can be thought to substitute a histidine basal nitrogen giving rise to a chromophore of the type

All the physical data on the cyanide derivative are also consistent with this picture: the ion removes one basal histidine nitrogen, water being no longer coordinated owing to the strong in plane ligand field.

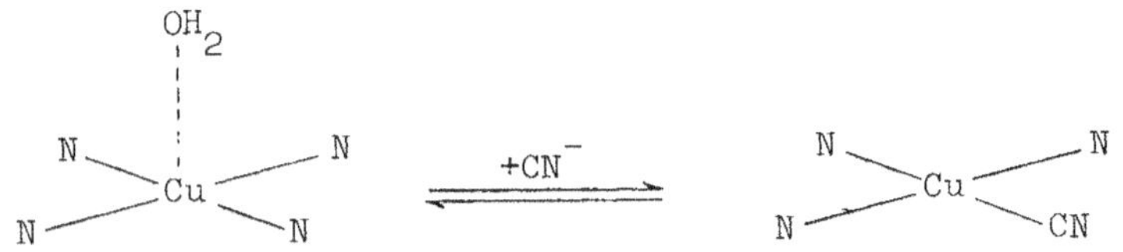

The azide ion, despite showing a more complex behavior, as well as cyanate and probably fluoride, can fit within the above scheme. The degree of displacement of the apical water in the various cases would be dictated by the in-plane ligand field strenght ($CN^- > N_3^- > NCO^- \approx NCS^- > F^-$, Table I).

The two pictures of i) anions replacing water and causing rearrangement of the protein in such a way to end up in a basal position, or ii) anions replacing one of the basal histidines, the presence of water being determined by the basal ligand field strenght, have both some appealing points and the discrimination between the two is rather a subtle matter. The former model is consistent with the general picture of flexibility of the active sites of enzymes, but does not satisfactorily account for the properties of the thiocyanate adduct; the latter model would require that fluoride is as strong a ligand as the histidine which is replaced. On the other hand, such replaced histidine can be weakly bound owing to stereochemical strain; coordination strain in CuN_4 moiety of the native chromophore is now widely accepted (6), and probably does not involve the bridging His 63 anion; it should be noted that F^-, like NCS^-, interacts

Table I. Stereochemistry around Copper(II) in some SOD Derivatives.

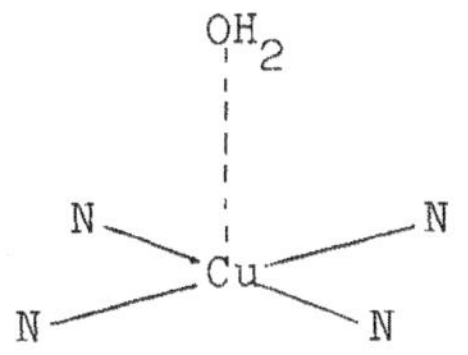
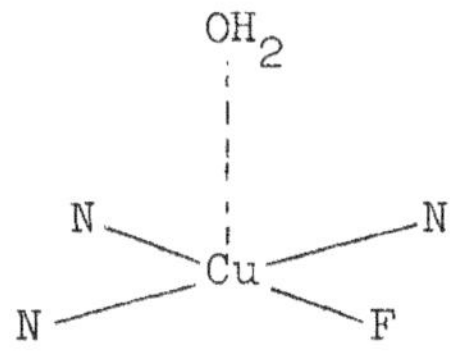
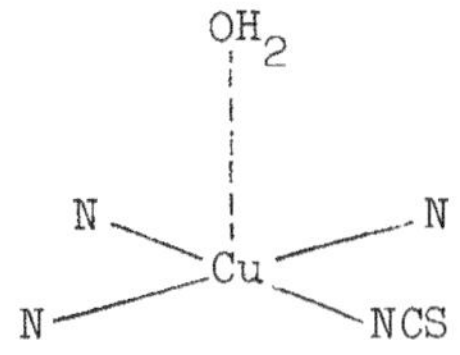

ν_{max} = 14.7 ($cm^{-1} x 10^{-3}$)	ν_{max} = 14.7	ν_{max} = 14.7
$g_{\parallel}$ = 2.26	$g_{\parallel}$ = 2.26	$g_{\parallel}$ = 2.25
$g_{\perp}$ = 2.07	$g_{\perp}$ = 2.07	$g_{\perp}$ = 2.06
$A_{\parallel}$ = 143 ($cm^{-1} x 10^{4}$)	$A_{\parallel}$ = 143	$A_{\parallel}$ = 148

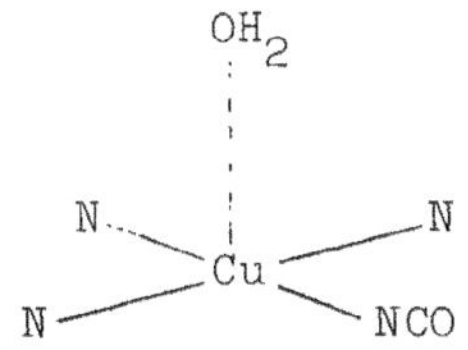
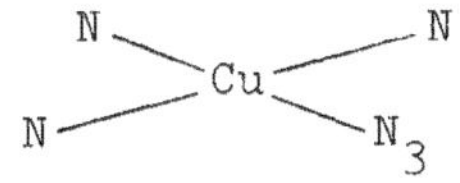
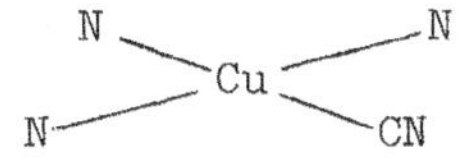

ν_{max} = 15.1	ν_{max} = 15.1	ν_{max} = 19.3
$g_{\parallel}$ = 2.26	$g_{\parallel}$ = 2.24	$g_{\parallel}$ = 2.21
$g_{\perp}$ = 2.05	$g_{\perp}$ = 2.04	$A_{\parallel}$ = 188
$A_{\parallel}$ = 158	$A_{\parallel}$ = 157	

with copper without displacing the coordinated water.

If it is accepted that fluoride has the same ligand field strenght of the strained histidine, then this model represents an appealing unifying approach to the anion binding mode. Furthermore, the pH dependence of the physical properties of superoxide dismutase could interestingly be framed within the model: ^{17}O NMR studies (7) show that a dramatic increase of the ^{17}O T_2^{-1} values occurs at high pH, with a pK_a of 11.3, paralleling ^1H NMR data (8,9). It is well established now that the increase in T_2^{-1} for the ^{17}O nucleus is largely due to contact interactions arising from direct Cu-O bond. The order of magnitude of such interaction is typical of equatorial binding of OH^-, just like the other anions. The semicoordinated water molecule could still be present with a higher pK_a.

Single crystal ESR measurements could in principle shed light on the orientation of the copper chromophore in the anion adducts. Indeed, recent studies have reasonably placed the z-axis close to the Cu-O bond direction in the native enzyme, while it is tilted somewhat in the cyanide derivative (10). Unfortunately, owing to the nature of the space group, the possible tilt angle can be either 30° or 60°. In both cases anion binding causes a tilting of the tetragonal plane, related to the value of the above angle; the smaller value would be quite consistent with model ii.

REFERENCES

(1) Tainer, J.A., Getzoff, E.D., Beem, K.M., Richardson, J.S., and Richardson, D.C.: 1982, J. Mol. Biol., in press
(2) Rotilio, G., Morpurgo, L., Giovagnoli, G., Calabrese, L., and Mondovì, B.: 1982, Biochemistry, 11, p. 2187
(3) Bertini, I., Luchinat, C., and Scozzafava, A.: 1980, J. Am. Chem. Soc., 102, p. 7349
(4) Bertini, I., Borghi, E., Luchinat, C., and Scozzafava A.: 1981, J. Am. Chem. Soc., 103, p. 7779
(5) Unpublished results from our Laboratory
(6) Valentine, J.S., and Pantoliano, M.W.: 1981, in "Copper Proteins", Spiro, T.G. ed., Wiley and Sons, N.Y., pp. 291-358
(7) Bertini, I., Luchinat, C., and Messori, L.: 1981, Biochem. Biophys. Res. Commun., 101, p. 577
(8) Terenzi, M., Rigo, A., Franconi, C., Mondovì, B., Calabrese, L., and Rotilio, G.: 1974, Biochim. Biophys. Acta, 351, p. 230
(9) Boden, N., Holmes, M.C., and Knowles, P.F.: 1979, Biochem. J., 177, p. 303
(10) Lieberman, R.A., Sands, R.H., and Fee, J.A.: 1982, J. Biol. Chem., 257, p. 336

KINETIC AND MAGNETIC RESONANCE STUDIES ON AMINE OXIDASES

Knowles, P.F.*, Lowe, D.J. †, Peters, J. †, Thorneley, R.N.F. †, and Yadav, K.D.S.*
Astbury Department of Biophysics, University of Leeds *
Agricultural Research Council Unit for Nitrogen Fixation, University of Sussex †

Abstract

The group of amine oxidases containing copper are widely distributed in nature. Most, if not all, contain two Cu^{2+} ions, and have two subunits, each of molecular weight 90,000 Da. Epr studies on the pig plasma enzyme have shown that the Cu^{2+} sites are non-identical. Reaction of the enzyme with azide or cyanide eliminates activity and makes the two Cu^{2+} sites indistinguishable. Water proton relaxation studies indicate that the coppers have one axial and one equatorial water coordinated. The equatorial, but not the axial water, can be displaced by azide or cyanide and this is suggested to be the molecular basis for the inhibition.

Quenched flow studies have indicated that ammonia is released after oxygen binding which supports an amino transferase type mechanism. Other kinetic studies, together with the magnetic resonance evidence, suggest that a role of $Cu^{2+} - H_2O$ in the amine oxidases might be as a source of hydroxyl ions at neutral pH, which would facilitate hydride ion transfer to oxygen. Analogies to other copper and zinc enzymes are discussed.

INTRODUCTION

Amine oxidases have been found in a variety of eukaryotic cells including those of fungal, plant and mammalian origin. They catalyse the oxidative deamination of mono-, di- and polyamines according to the equation:

$$RCH_2NH_2 + O_2 + H_2O \rightarrow RCHO + H_2O_2 + NH_2$$

The amine oxidases can be divided into "flavin-containing" and "non-flavin containing" groups. The former group is associated with mitochondria and will not be considered further in the present chapter. The latter group, designated amine-oxygen oxidoreductase (deaminating) E.C.1.4.3.6, has been subdivided into mono- and diamine oxidases, according to substrate specificity and the action of inhibitors. This subdivision is probably outdated, since "non-flavin containing" amine

I. Bertini, R. S. Drago, and C. Luchinat (eds.), The Coordination Chemistry of Metalloenzymes, 159–176.

oxidases have many structural and mechanistic features in common. Perhaps their most striking structural property is the presence of copper and this offers a convenient way to classify them. The main emphasis in this chapter will be to examine what is known about the structure and catalytic role of the copper centres in these enzymes.

Table 1 summarises structural features of selected copper-containing amine oxidases.

Source	Molecular Weight (in Daltons)	Subunits	Copper Content (mol/mol)	Carbohydrate Content (mol/mol)
Bovine plasma	170,000	2x87,000	2.2	7-8%
Pig plasma	186,000(±4,000)	2x95,000	2.1	∿10%
Pig kidney	179,000(±6,000)	2x90,000	2.0	not determined
Aspergillus niger	263,000(±11,000)	?x85,000	3.0	not determined

Table 1. Some properties of copper-containing amine oxidases (For references see (1))

From this table, it may be concluded that the amine oxidases probably have two subunits of molecular weight 90,000 Da. and contain two atoms of copper per molecule of protein. It has been reported that the enzymes from human placenta (2) and human pregnancy plasma (3) are active as monomers with correspondingly one atom of copper per molecule of protein. These claims need substantiating since they have important mechanistic implications. The omission of additional cofactors from Table I is deliberate. There has been considerable debate over the existence of a carbonyl cofactor in amine oxidases (1). Attempts to isolate such a cofactor have proved elusive. Further discussion of this topic will be given in the MECHANISTIC STUDIES section. For the present, it may be stated that the stoichiometry of inhibition by phenyl hydrazine, substrate addition and product release are all consistent with a single active site in the pig plasma enzyme (4).

The involvement of amine oxidases in the metabolism of regulatory amines has been reviewed (5,6). In addition, amine oxidases are implicated as catalysts of cross linking reactions in connective tissue (7). The enzyme responsible, termed lysyl oxidase, has been isolated from aorta (8) and embryonic cartilage (9); detailed studies on this enzyme have clear medical importance.

STUDIES ON THE COPPER COFACTOR

The copper atoms present in most (if not all) amine oxidases exist in the form of epr-detectable cupric. All epr studies have been made on polycrystalline solutions at cryogenic temperatures and reveal the presence of hyperfine interaction with ^{63}Cu and ^{65}Cu nuclei (I = 3/2). The epr parameters determined from measurements at 9 GHz frequency are shown in Table 2 and are typical of Type 2 copper (10).

Source	$g_{\perp}$	$g_{\parallel}$	$A_{\parallel}$ (mT)	Reference
Bovine plasma	2.06	2.29	16.1	11
Pig plasma	2.065	2.29	14.7	12,13
Pig kidney	2.063	2.294	16.0	14,15
Pea seedling	2.1	2.35	14.1	16
Aspergillus niger	2.07	2.31	16.2	17

Table 2. Epr spectral parameters of amine oxidases

These parameters suggest 5-coordinate Cu^{2+} in a site close to square-pyramidal symmetry (18, 19).

Study of the pig plasma enzyme by 35 GHz epr revealed that the two Cu^{2+} sites are chemically distinct (13). Reasonable simulation of the spectra could be achieved by mixing equal proportions of components with axial and rhombic symmetry (Table 3 and Figure 1). Comparison with the epr spectral parameters from copper complexes of known structure (20) allowed the tentative conclusion that the ligand atoms were nitrogen and oxygen.

Enzyme Form	Component	g_{xx}	g_{yy}	g_{zz}	A_{xx}	A_{yy}	A_{zz}
Native enzyme	Axial	2.078		2.286	1.0		15.5
	Rhombic	2.039	2.065	2.294	1.0	1.0	14.0
Enzyme with 23mM NaN_3	Axial	2.055		2.257	1.0		16.5
Enzyme with 52mM KCN	Axial	2.054		2.218	1.0		15.8

Table 3. Parameters used in the simulation of the 35 GHz epr spectra (Figures 1 and 2) of pig plasma monoamine oxidase. The hyperfine splittings refer to ^{63}Cu and ^{65}Cu nuclei. The values given between the x and y columns correspond to $\perp$ values in the case of axial symmetry.

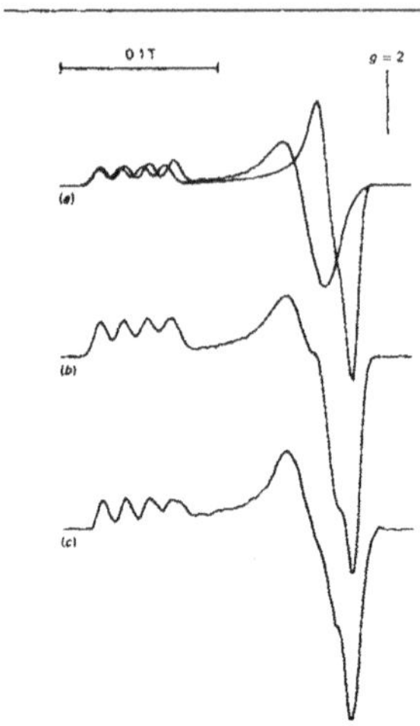

Figure 1. 35 GHz epr spectrum of monoamine oxidase from pig plasma at 150K. (a) Calculated axial and rhombic components; (b) sum of spectra in (a) equally weighted; (c) experimental spectrum.

Similar 35GHz epr studies on other amine oxidases have not been attempted though from water proton relaxation studies on the enzyme from pig kidney, Kluetz and Schmidt (21) also concluded that the two coppers were not identical.

Chemical intuition would suggest that copper plays a redox role in the amine oxidases; thus, the amine substrate might be expected to reduce Cu^{2+} to Cu^{+} whilst oxygen would restore the copper to its divalent state. However, the results from all epr investigations on amine oxidases show that the vast majority of copper remains divalent following anaerobic reduction with amine substrate. There have been reports (14, 22) that these conditions lead to changes in line shape; this point will be taken up in the section describing MECHANISTIC STUDIES.

Good evidence to show that the copper has a functional role has come from studies on the pig plasma enzyme where first Lindström et al. (23) and later Barker et al. (13) observed that azide both modified the epr spectrum and inhibited enzyme activity; similar results were found with cyanide. The effect of cyanide titration of the 35 GHz epr spectrum of the enzyme are shown in Figure 2. The epr spectra, which include simulation based on the parameters in Table 3, show that both coppers in the cyanide (and azide) complexes have identical symmetry. This suggests that the distinct chemical nature of the two coppers is a requirement for activity.

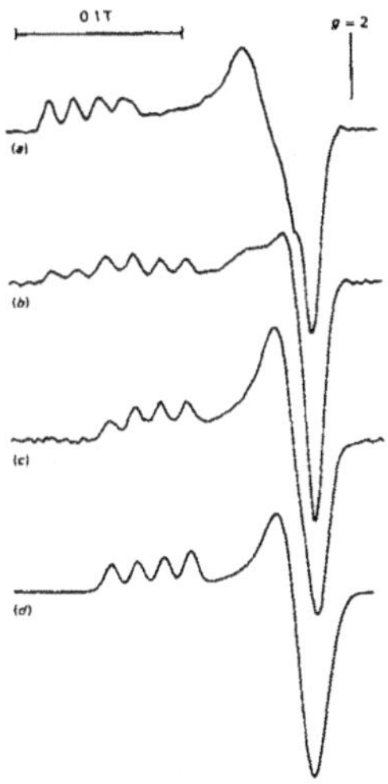

Figure 2. Effect of potassium cyanide on the 35 GHz epr spectrum of pig plasma monoamine oxidase at 150K. (a) no cyanide; (b) 4.5mM cyanide; (c) 52mM cyanide; (d) simulation of spectrum c.

It is not possible to draw further conclusions from these epr studies on the nature of ligands coordinated to copper in the amine oxidases. Such information has come, however, from nmr studies.

Water proton relaxation measurements (24) demonstrate clearly that the Cu^{2+} sites in monoamine oxidase (and superoxide dismutase) are accessible to solvent (Figure 3); contrast the behaviour of blue copper proteins which do not affect the water proton relaxation behaviour.

A general picture of water exchange at the Cu^{2+} sites of monoamine oxidase is shown in Figure 4.

It can be seen that several parameters of biochemical interest might be derived from water proton relaxation measurements; the exchange rate (τ_M^{-1}) of water between solvent and coordinated environments, the Cu-H_2O distance (r) in the site and the number of waters coordinated to each Cu^{2+} (q).

The effect of temperature and Larmor frequency on the normalised paramagnetic contributions, $(p_M T_{2p})^{-1}$ and $(p_M T_{1p})^{-1}$ to the water proton relaxation rates are shown in Figure 5.

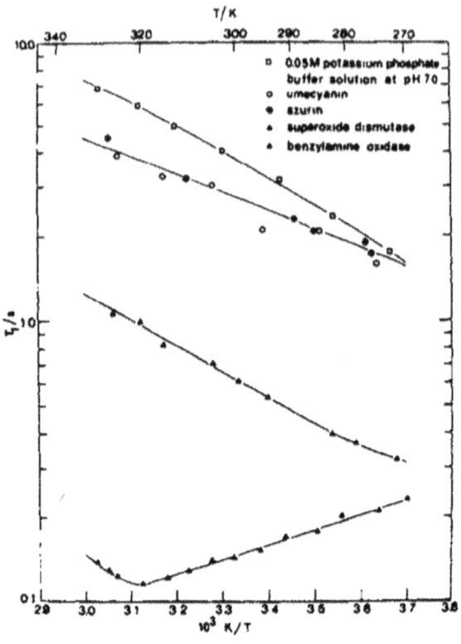

Figure 3. Water proton T_1 values measured at 30 MHz in solutions of various copper proteins.

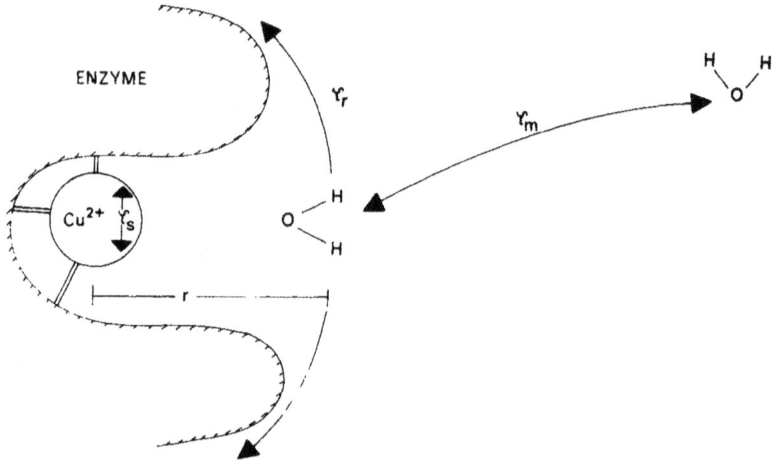

Figure 4. Water exchange at the Cu^{2+} sites of monoamine oxidase.

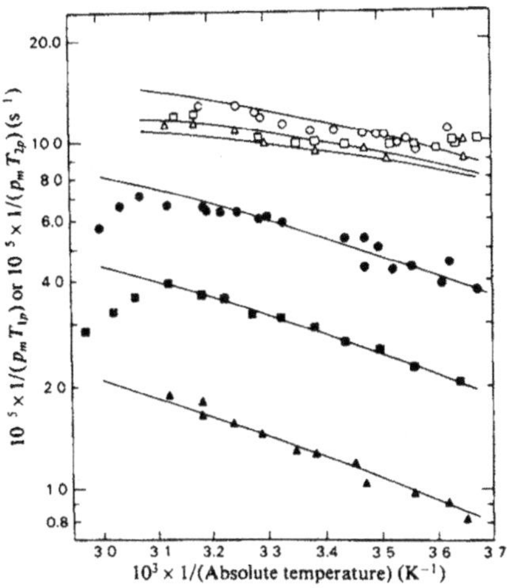

Figure 5. Temperature dependence of the normalised paramagnetic contributions to the water proton relaxation rates measured in solutions of pig plasma monoamine oxidase.

●, ■, ▲, T_{1p} data at 10, 30 and 60 MHz respectively;

○, □, △, T_{2p} data at 10, 30 and 60 MHz respectively.

Interpretation of the $(p_M T_{1p})^{-1}$ behaviour is possible using the Luz-Meiboom equation (25).

$$(p_M T_{1p})^{-1} = q/(\tau_M + T_{1M})$$

where T_{1M} is the spin lattice relaxation time of bound water protons. It can be seen from Figure 5 that $(p_M T_{1p})^{-1}$ shows a marked dependence on Larmor frequency which implies $\tau_M < T_{1M}$. Hence we can write $(p_M T_{1p})^{-1} = q/T_{1M}$.

T_{1M} is given by the Solomon-Bloembergen equations (26,27) which in the present case take the form

$$\frac{1}{T_{1M}} = \frac{2}{15} \frac{\mu_{eff.}\ \gamma_I^2}{r^6} \left[\frac{3\ \tau_c}{1+\omega_I \tau_c^2}\right]$$

where ω_I and γ_I are respectively the Larmor frequency and magnetogyric ratio of the proton, μ_{eff} is the effective electronic magnetic moment of Cu^{2+} and τ_c is the correlation time for the dipolar interaction. τ_c is given by

$$\tau_c^{-1} = \tau_r^{-1} + \tau_m^{-1} + \tau_s^{-1}$$

where τ_r is the rotational correlation time and τ_s is the electron spin lattice relaxation time.

Two explanations for the behaviour of $(p_M T_{1p})^{-1}$ are suggested

(i) $\omega_I \tau_c < 1$; $\tau_c = \tau_s$ where τ_s increases as the temperature increases

(ii) $\omega_I \tau_c > 1$; τ_c has contributions from τ_r, τ_s or τ_M

The observed behaviour of $(p_M T_{2p})^{-1}$ (see Figure 5) is consistent with explanation (ii) being valid (13).

Quantitative interpretation of the data in Figure 5 has been attempted on the basis of the above theory leading to the solid lines (computer fits) through the data points. It has been assumed to a first approximation that both Cu^{2+} sites are identical, since otherwise the number of variable parameters becomes intractable. The values for some of the parameters derived from the computer fits to the data are shown in Table 4.

	Frequency		
Quantity	10 MHz	30 MHz	60 MHz
q/r^6 (nm)$^{-6}$		$(3.8 \pm 0.4) \times 10^{-3}$	
τ_c(s)	$(8.0 \pm 0.8) \times 10^{-8}$	$(2.6 \pm 0.3) \times 10^{-8}$	$(1.8 \pm 0.2) \times 10^{-8}$
τ_M(s)		$(1.6 \pm 0.6) \times 10^{-6}$	

Table 4. Values of parameters shown in Figure 4 obtained from water proton relaxation studies on pig plasma monoamine oxidase.

Further information on the structure of the Cu^{2+} sites in the pig plasma enzyme and their involvement in the catalytic mechanism can be obtained by examining how the water proton relaxation rates are affected by changes in pH or by addition of inhibitors (azide and cyanide). Figure 6 shows that the relaxation rates decrease by a factor of 2 as the pH is raised from 7 to 9. Figure 7 describes how the enhancement ratio (28) changes as the enzyme is titrated with cyanide; similar results were obtained with azide. It can be seen that the enhancement ratio at pH 7.0 extrapolates to a value of 0.5 at infinite concentrations of inhibitor.

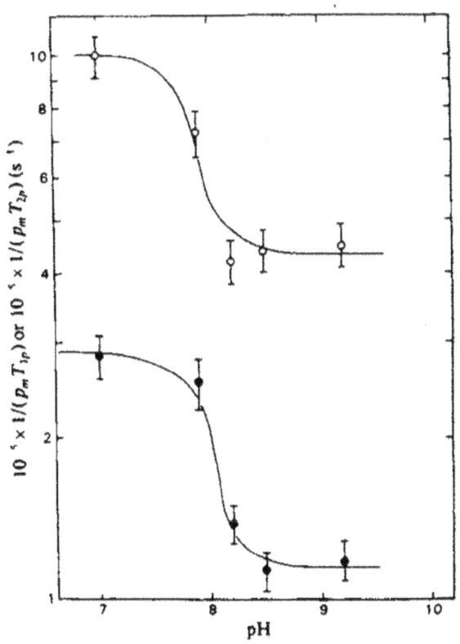

Figure 6. Variation of $(T_{1p})^{-1}$ and $(T_{2p})^{-1}$ with pH in solutions of pig plasma monoamine oxidase.
●, T_{1p}; O, T_{2p}

Values for the dissociation constants of the azide and cyanide complexes (K_D) can be derived from this data (28) and are given in Table 5.

	K_D (mM)	
Ligand	From kinetics	From nmr
Azide	84±16	48±15
Cyanide	0.76±0.21	0.7±0.3

Table 5. Comparison of values of the dissociation constant for enzyme-inhibitor complexes obtained from nmr and steady state kinetics at pH 7.

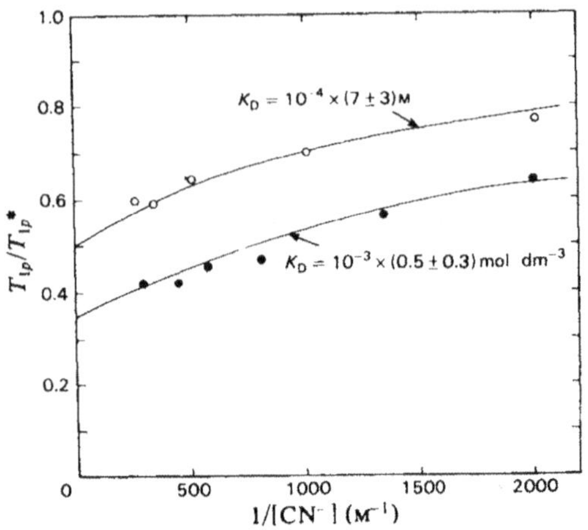

Figure 7. Attenuation of the water proton enhancement ratio from pig plasma monoamine oxidase as a function of [KCN]. O, pH 7.0; ●, pH 9.2.

The observed effects of pH and inhibitors may be rationalised in terms of changes in the number of protons from water molecules coordinated to Cu^{2+}. Two explanations for the pH effects are possible; firstly, displacement of one half of the bound water molecules by a neighbouring group on the protein and secondly, deprotonation of all bound water molecules to OH^-. These alternatives may be distinguished through the results of titration studies by cyanide at higher pH (Figure 7) and suggest that deprotonation of all coordinated waters is probably correct.

Considering the inhibitor effects in more detail, it is immediately clear that not all the water molecules coordinated at the two Cu^{2+} sites are displaced by the inhibitors. This suggests that either water molecules coordinated to one only of the two Cu^{2+} ions can be displaced or that each Cu^{2+} ion has two coordinated waters differing in their reactivity with the inhibitor. The first explanation implies that one of the two Cu^{2+} sites is accessible to solvent water but not to the inhibitors; there is no chemical basis for this assertion. The second explanation is consistent with the known properties of axial and equatorial water molecules coordinated to Cu^{2+} (29,30). If we postulate one equatorial and one axial water per Cu^{2+} with no fluxion between these positions, only the equatorially-coordinated waters would be substituted by the inhibitors. Support for this postulate comes from comparison of the water exchange rate (τ_M^{-1}) in monoamine oxidase (Table 4) with that in $(Cu(tren)H_2O)^{2+}$ where the coordinated water has equatorial character [2×10^5 $\ell mol^{-1} s^{-1}$] (29).

The information on the Cu^{2+} sites in pig plasma monoamine oxidase derived from epr and nmr measurements can be summarised as follows. The two Cu^{2+} sites are chemically distinct and have square pyramidal symmetry with three nitrogens (from the protein) and two waters (one axial, the other equatorial) as ligands. At this point, it is interesting to note that Marwedel et al. (31) have concluded from epr and ^{19}F nmr relaxation studies that the single type 2 copper in galactose oxidase has one axial and one equatorial water coordinated.

MECHANISTIC STUDIES

Two mechanisms for the action of amine oxidases have been proposed. The first, advocated most strongly by Pettersson and co-workers see (32) and references therein, can be termed an amino transferase mechanism and is consistent with a carbonyl cofactor. (Scheme 1)

$$E\text{-}Pyr\text{-}CHO + RCH_2NH_2 \rightarrow E\text{-}Pyr\ CH_2NH_2 + RCHO$$

$$E\text{-}Pyr\ CH_2NH_2 + O_2 \rightarrow E\text{-}Pyr\text{-}CHO + H_2O_2 + NH_3$$

Scheme 1. Postulated amino transferase mechanism for amine oxidases. Pyr-CHO and Pyr-CH_2NH_2 represent pyridoxal 5-phosphate and pyridoxamine 5-phosphate respectively.

It can be seen that aldehyde would be the only product liberated under anaerobic conditions and evidence in favour of this has been presented (4).

An alternative mechanism has been proposed by Abeles and his colleagues (33) (Scheme 2). This can be described as an imine mechanism where the imine, $\underset{\displaystyle NH}{\overset{\|}{CH}} - R$, is hydrolysed under anaerobic conditions to liberate ammonia and aldehyde products. Evidence has been presented that these products are, indeed, released anaerobically (34); other evidence has come from studies using suicide inhibitors and isotopically labelled substrate amine (33,35).

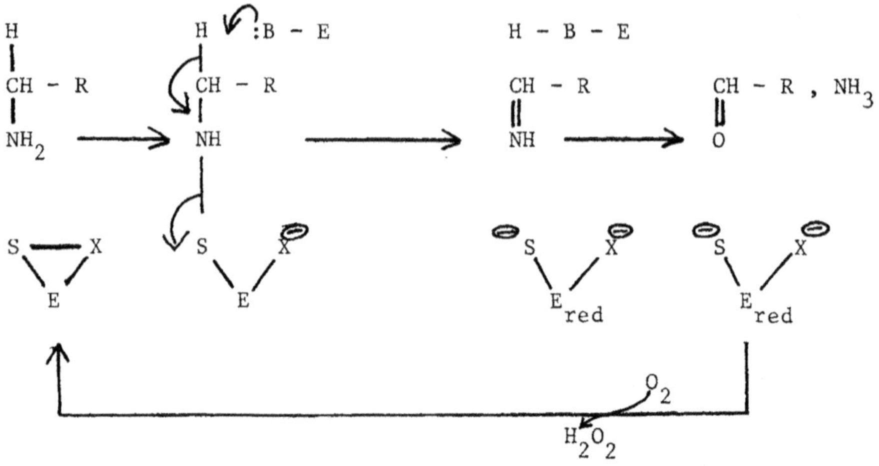

Scheme 2. Postulated mechanism for bovine plasma monoamine oxidase (32). B - E represents a basic group (b) bound to the enzyme and X is an unidentified grouping on the enzyme. E and E_{red} are oxidised and reduced forms of the enzyme.

The time course of ammonia release should provide unambiguous evidence to resolve these opposed views. It will be seen that a minimal scheme

$$E \xrightarrow{\text{Amine}} E_{reduced} \xrightarrow{O_2} E$$

is common to both mechanisms and the rate constants for these two steps in the pig plasma monoamine oxidase mechanism are known from the stopped flow studies of Pettersson et al. (35,37). In fact, the kinetics of re-oxidation of E reduced have been studied in more detail (37); oxygen was shown to react by a second order process to give an intermediate (termed E oxidised) which re-converts to E in a first order step. Under aerobic conditions, the oxygen independent reaction represents the rate limiting step at pH 9.0 as shown in Scheme 3. The pseudo first order rate constant for the step E reduced $\longrightarrow$ E oxidised is based on conditions of O_2 saturating.

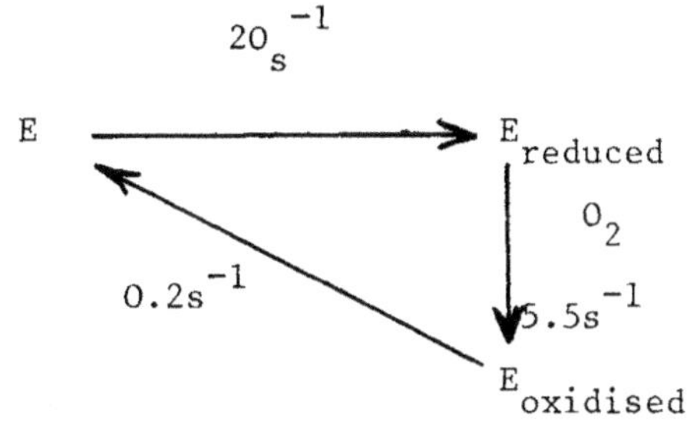

Scheme 3. Abbreviated mechanism for pig plasma monoamine oxidase at pH 9.0 based on stopped flow studies (36,37).

Using techniques developed at the Unit of Nitrogen Fixation to study the pre-steady state kinetics of ammonia production by nitrogenase, we have followed the time course of ammonia release from monoamine oxidase under conditions of O_2 saturating at pH 9.0.

A rapid quench apparatus designed by R. C. Bray (38) was utilised as described by Eadie et al. (39). Pig plasma monoamine oxidase (92μM) was dialysed extensively against 0.1M sodium phosphate buffer, pH 9.0, to remove ammonium sulphate, then saturated with oxygen prior to transfer to the syringe of the rapid quench apparatus. Benzylamine (10mM) in the same buffer was similarly oxygenated then transferred to the quench apparatus. The experiment was conducted at 25°C. After different reaction times (10ms - 20s), the reaction mixture was quenched by extrusion into Eppendorf vials containing 0.2mls of 1 N HCl;control experiments had shown that these conditions irreversible inactivate the enzyme. Ammonia in the quenched samples was assayed using a radio chemical method, based on that of Kalb et al. (40), which is sensitive to ammonia concentrations of less than 1 nmol. Six ammonia assays were performed for each reaction time. Simulation of the time course for ammonia release according to Scheme 3 were conducted on a PDP11/34 computer utilising a program based on NAG subroutines E04JBF (least squares fitting) and D02ABF (solution of differential equations).

The results are shown in Figs 8 and 9. The clear evidence for a burst in ammonia production eliminates the possibility of its release at the rate limiting rate ($0.2s^{-1}$). Statistical assessment of the data using an F-test established that release of ammonia in the step E reduced → E oxidised was 10x more likely than its release in the step E → E reduced. This provides firm evidence in favour of the aminotransferase mechanism (Scheme 1) over the imine mechanism (Scheme 2).

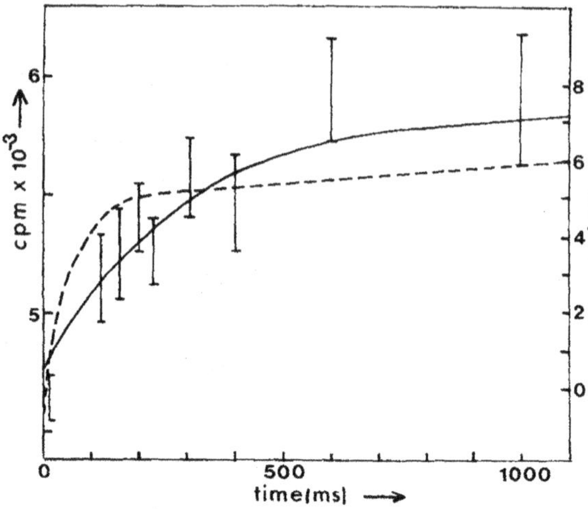

Figure 8. Quenched flow studies of the time course (0-1s) of ammonia release during aerobic reaction of benzylamine with pig plasma monoamine oxidase at pH 9.0. The dashed and solid lines are the calculated values for ammonia release in the steps E → E reduced and E reduced → E oxidised respectively of Scheme 3. Error bars represent standard deviations (6 determinations).

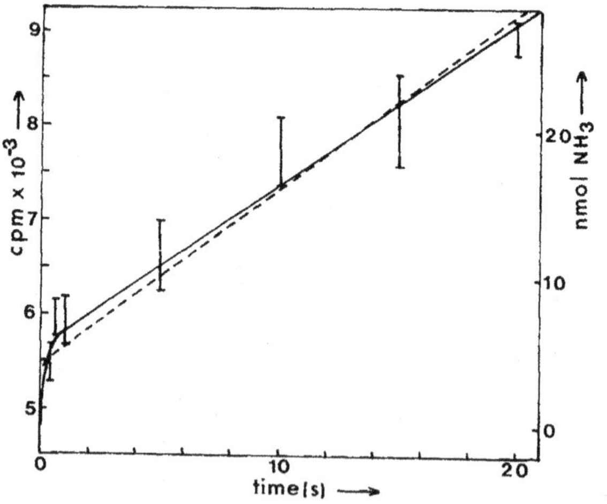

Figure 9. Quenched flow studies of the time course (0-20s), of ammonia release; other details as for Figure 8.

Further details on the catalytic mechanism of pig plasma monoamine oxidase have come from stopped flow and steady state kinetic studies. Azide has been shown (13) to be a competitive inhibitor against oxygen binding to the enzyme. The inhibitor pattern with cyanide is more complex but again has been interpreted in favour of competition against oxygen binding. Thus these inhibitors must affect the catalytic step involving re-oxidation (by oxygen) of the E reduced species. From the water proton relaxation studies discussed earlier, it was concluded that azide and cyanide displaced the equatorial (but not the axial) water coordinated to Cu^{2+} in the enzyme. There is reasonable quantitative agreement between dissociation constants of the azide and cyanide complexes determined from the kinetic and proton relaxation data (Table 5). Thus, taking the two sets of data together, it may be concluded that water coordinated equatorially to Cu^{2+} is involved in re-oxidation of the E reduced species. The same conclusion can be drawn from steady state and stopped flow studies into the mode of action of azide, ammonia and imidazole on the enzyme (23, 41, 42).

A catalytic mechanism for pig plasma monoamine oxidase has been proposed (42) which is consistent with all the stopped flow kinetic data from Pettersson et al, as well as the magnetic resonance and quenched flow kinetic data discussed in the present chapter (Scheme 4).

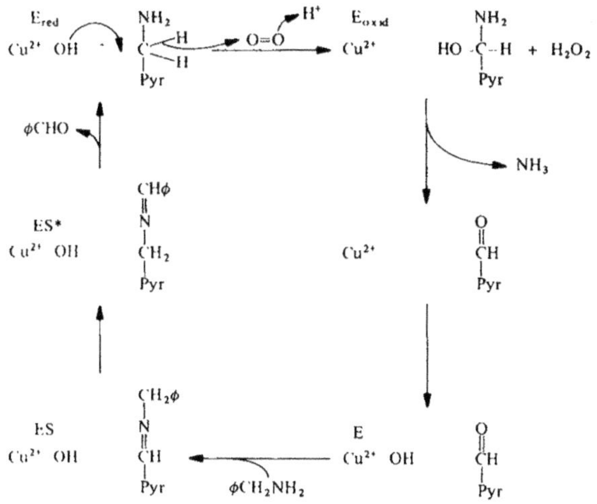

Scheme 4. Postulated catalytic mechanism for monoamine oxidase. The Cu^{2+} and pyridoxal phosphate (Pyr) co-factors are protein-bound. Ligands to the Cu^{2+} other than hydroxyl (or water) are not shown.

It can be seen that hydroxyl coordinated to Cu^{2+} is postulated to act as a nucleophile facilitating hydride ion transfer to oxygen in the step E reduced → E oxidised. No valence change in the copper would be required for this role of Cu^{2+}-OH as a nucleophile which is consistent with all epr studies on amine oxidases, including those on intermediates in the catalytic cycle (22). Clearly hydride ion transfer (or electron and hydrogen atom transfer as separate steps) to oxygen (a triplet species) would not occur unless the π^* orbitals on the oxygen were modified; coordination of the oxygen to the other copper present in the active site of the enzyme might have this effect (43). A stimulating article by Hamilton (44) gives alternative views on the role of copper in monoamine oxidases and also speculates on the nature of the carbonyl cofactor.

There is some analogy between this postulated role of Cu^{2+}-OH as a nucleophile in monoamine oxidases and that of Zn^{2+}-OH in carbonic anhydrase (45). It has also been demonstrated that Zn^{2+}- H_2O plays a role in the redox enzyme alcohol dehydrogenase (46,47). Metal depletion and substitution studies on the amine oxidases might therefore be most informative.

Finally, it may be noted that evidence for water (hydroxyl) coordinated to Cu^{2+} has been presented for several other copper enzymes involved in oxidative reactions. These include the Type 2 Cu^{2+} of galactose oxidase (3), laccase (48) and Cu_B of cytochrome oxidase (49). Thus a nucleophylic role for Cu^{2+} in addition to its well-established redox roles may be widespread.

Acknowledgements

We thank Dr R. C. Bray for use of his quenched flow apparatus and V. Blakeley, G. Clifford and M. Reeve-Fowkes for skilled technical assistance.

References

(1) Knowles, P.F. and Yadav, K.D.S. in "Copper Proteins"; Lontie, R., editor; CRC Press, publishers. In press.

(2) Crabbe, M.J.C., Waight, R.D., Bardsley, W.G., Barker, R.W., Kelly, I.D. and Knowles, P.F.: 1976, Biochem.J. 155, pp.679-687.

(3) Baylin, S.B. and Margolis, S.: 1975, Biochim.Biophys.Acta. 397, pp.294-306.

(4) Lindström, A. and Pettersson, G.: 1978, Eur.J.Biochem. 83, pp.131-135.

(5) Zeller, E.A.: 1963, "The Enzymes", 2nd edition; Boyer, P.D., Lardy, H.A. and Myrbäck, K. Editors; Academic Press (New York): 8, pp.313-335.

(6) Blaschko, H. ibid., pp.337-351.

(7) Carnes, W.H.: 1971, Fed.Proc., 30, pp.995-1000.

(8) Buffoni, F., Marino, P. and Pirisino, R.: 1976, It.J.Biochem. 25, pp.191-203.

(9) Stassen, F.L.H.: 1976, Biochim.Biophys.Acta. 438, pp.49-60.

(10) Vänngård, T.: 1972, "Biological Applications of Electron Spin Resonance", Swartz, H.M., Bolton, J.R. and Borg, D.C., editors; Wiley Interscience (New York), pp.411-447.

(11) Suzuki, S., Sakurai, T., Nakahara, A., Oda, O., Manabe, T. and Okuyama, T.: 1980, FEBS letters, 116, pp.17-20.

(12) Buffoni, F., Della Corte, L. and Knowles, P.F.: 1968, Biochem.J., 106, pp.575-576.

(13) Barker, R., Boden, N., Cayley, G., Charlton, S.C., Henson, R., Holmes, M.C., Kelly, I.D. and Knowles, P.F.: 1979, Biochem.J., 177, pp.289-302.

(14) Mondovì, B., Rotilio, G., Costa, M.T., Finazzi-Agrò, Z., Chiancone, E., Hansen, R. and Beinert, H.: 1967, J.Biol.Chem., 242, pp.1160-1167.

(15) Kluetz, M.D. and Schmidt, P.G.: 1977, Biochem.Biophys.Res.Comms., 76, pp.40-45.

(16) Nylén, U., and Szybek, P.: 1974, Acta.Chem.Scand. 28B, pp.1153-1160.

(17) Yamada, H., Adachi, O. and Yamano, T.: 1969, Biochim.Biophys.Acta, 191, pp.751-752.

(18) Hathaway, B.J. and Billing, D.E.: 1970, Coord.Chem.Rev. 5, pp.143-207.

(19) Bencini, A., Bertini, I., Gatteschi, D. and Scozzafava, A.: 1978, Inorg.Chem., 17, pp.3194-3197.

(20) Barbucci, R. and Campbell, M.J.M.: 1976, Inorg.Chim.Acta, 16, pp.113-120.

(21) Kluetz, M.D. and Schmidt, P.G.: 1980, Biophys.J., 29, pp.283-293.

(22) Grant, J., Kelly, I.D., Knowles, P.F., Olsson, J. and Pettersson, G.: 1978, Biochem.Biophys.Res.Comms., 57, pp.1216-1224.

(23) Lindström, A., Olsson, B. and Pettersson, G.: 1974, Eur.J.Biochem. 48, pp.237-243.

(24) Boden, N., Holmes, M.C. and Knowles, P.F.: 1974, Biochem.Biophys. Res.Commun., 57, pp.845-848.

(25) Luz, Z. and Meiboom, S.: 1964, J.Chem.Phys., 40, pp.2686-2692.

(26) Solomon, I.: 1955, Phys.Rev., 99, pp.559-565.

(27) Bloembergen, N.: 1957, J.Chem.Phys., 27, pp.572-573.

(28) Dwek, R.A.: 1973, Nuclear Magnetic Resonance in Biochemistry; Clarendon Press, (Oxford).

(29) Cayley, G.R., Cross, D. and Knowles, P.F.: 1976, J.Chem.Soc. (Chem.Commun.), pp.837-838.

(30) Cayley, G., Kelly, I.D., Knowles, P.F. and Yadav, K.D.S.: 1981, J.Chem.Soc. (Dalton), pp.2370-2372.

(31) Marwedel, B.J., Kosman, D.J., Bereman, R.D. and Kurland, R.J.: 1981, J.Amer.Chem.Soc., 103, pp.2842-2847.

(32) Olsson, B., Olsson, J. and Pettersson, G.: Eur.J.Biochem., 87, pp.1-8.

(33) Suva, R.A. and Abeles, R.H.: 1978, Biochemistry, 17, pp.3538-3545.

(34) Berg, K.A. and Abeles, R.H.: 1980, Biochemistry, 19, pp.3186-3189.

(35) Battersby, A.R., Staunton, J., Klinman, J.P. and Summers, M.C.: 1979, FEBS letters, 99, pp.297-298.

(36) Lindström, A., Olsson, B., Olsson, J. and Pettersson, G.: 1976, Eur.J.Biochem., 64, pp.321-326.

(37) Olsson, B., Olsson, J. and Pettersson, G.: 1977, Eur.J.Biochem., 74, pp.329-335.

(38) Gutteridge, S., Tanner, S.J. and Bray, R.C.: 1978, Biochem.J. 175, pp.887-897.

(39) Eady, R.R., Lowe, D.J. and Thorneley, R.N.F.: 1978, FEBS letters, 95, pp.211-213.

(40) Kalb, V.F., Donohue, T.J., Corrigan, M.G. and Beruldur, R.W.: 1978, Anal.Biochem., 90, pp.47-57.

(41) Kelly, I.D., Knowles, P.F., Yadav, K.D.S., Bardsley, W.G., Leff, P. and Waight, R.D.: 1981, Eur.J.Biochem., 114, pp.133-138.

(42) Yadav, K.D.S. and Knowles, P.F.: 1981, Eur.J.Biochem., 114, pp.139-144.

(43) Drago, R.S. and Corden, B.B.: 1980, Acc.Chem.Res., 13, pp.353-360.

(44) Hamilton, G.A., 1981, "Copper Proteins"; Spiro, T.G. editor; Wiley, pp.205-218.

(45) Werber, M.M.: 1976, J.Theor.Biol., 60, pp.51-58.

(46) Evans, S.A. and Shore, J.D.: 1980, J.Biol.Chem., 255, pp.1509-1514.

(47) Andersson, P., Kvassman, J., Lindström, A., Oldén, B. and Pettersson, G.: 1981, Eur.J.Biochem., 113, pp.425-433.

(48) Goldberg, M., Vuk-Pavlović, S. and Pecht, I.: 1980, Biochemistry, 19, pp.5181-5189.

(49) Karlsson, B. and Andréasson, L.-E.: 1981, Biochim.Biophys.Acta., 635, pp.73-80.

METAL COORDINATION AND MECHANISM OF BLUE COPPER-CONTAINING OXIDASES.

Bengt Reinhammar

Department of biochemistry and biophysics.
Chalmers Institute of technology and
university of Göteborg
Göteborg, Sweden.

1. INTRODUCTION

The blue copper-containing oxidases; laccase, ceruloplasmin and ascorbate oxidase comprise a rather small group of enzymes. They are the only oxidases, which together with cytochrome *c* oxidase, can couple the one-electron transfer from reducing substances to the full reduction of dioxygen to water in the general reaction: $O_2 + 4e^- + 4H^+ \rightarrow 2H_2O$.

The blue oxidases contain rather unique types of copper ions, one of which is responsible for the strong blue colour of these proteins. Extensive mechanistic studies, in particular with laccase, have revealed that all the metal sites are involved in the reduction or reoxidation processes. The laccases are less complicated than ceruloplasmin and

I. Bertini, R. S. Drago, and C. Luchinat (eds.), The Coordination Chemistry of Metalloenzymes, 177–200.

ascorbate oxidase, as concerns their catalytic mechanism, but all three enzymes are believed to function in similar ways.

Owing to their important role as efficient catalysts in the reduction of dioxygen and also to their unique spectroscopic properties these enzymes have attracted much interest by various researchers in the last two dacades. This is documented in a number of review articles, some of which are given in the reference list (1-6). It is therefore not the intention of this chapter to give a complete bibliography of all literature available, but to concentrate on the recent knowledge on the structure and mechanism of these enzymes. The laccases have been more thoroughly studied than ceruloplasmin and ascorbate oxidase. They are therefore better inderstood at least concerning their metal coordination and catalytic mechanism. The discussion will therefore mainly concern some recent studies of these oxidases.

2. CHEMICAL COMPOSITION AND REACTIONS CATALYZED

2.1. Enzyme Distribution and Function.

Both laccase and ascorbate oxidase are widely distributed in the plant kingdom. For example, ascorbate oxidase has been prepared from cucumber and squash and has also been found in some algae and bacteria (7). Laccase was first discovered as early as in 1883 (8) in the sap of the japanese laquer trees and this laccase has been extensively studied. Laccase is also produced by many fungi and the enzyme is particularly abundant in the white-rot fungi which are able to degrade lignin (9-11). Ceruloplasmin can be prepared from the blood plasma from several mammalian species (6).

All blue oxidases exhibit a rather low specificity and they can therefore rapidly oxidize a number of quite different substrates. Laccase and ceruloplasmin show similar substrate specificity. The best substrates are different phenols, such as p-diphenols and related substances, such as aminophenols and diamines. Both *o*-phenols and *m*-phenols as well as some monophenols are oxidized at substantial rates (6,12-14). Some inorganic compounds, for example hexacyanoferrate(II) is rapidly oxidized by fungal laccase (15) and one of the best substrates for ceruloplasmin is Fe^{2+} and this enzyme has therefore been classified as a ferroxidase (16).

The best substrate for ascorbate oxidase is ascorbate but the enzyme can also oxidize a number of substituted polyhydric- and aminophenols (7).

2.2. Molecular Properties and Structure.

All blue oxidases are glycoproteins with varying carbohydrate contents. Laccases contain between 10 and 45 % carbohydrates (3), ceruloplasmin about 7 % (6) and ascorbate oxidase about 2 % carbohydrates (17).

The laccases possibly consist of a single polypeptide chain of about 500 amino residues (18). The large variation in reported molecular weights, 64 000-140 000, is probably a reflection of different carbohydrate contents (3). Ceruloplasmin, with a molecular weight of about 132 000, is found to consist of a single polypeptide chain of about 1050 amino residues (19). Despite the great difficulties involved to examine such a long polypeptide chain most of the protein sequence has recently been solved (20).

The molecular weight of ascorbate oxidase is about 140 000 (21). It is reported to consist of two identical subunits, which in their turn are composed of two different peptide chains of about 30 000 and 40 000 which are linked by disulfide bridges. The presence of three type 1 and one type 2 copper ions (see below) is, however, not consistent with such a symmetric structure.

There are only very limited information about the structures of these oxidases. Despite great efforts by several groups there has been no success in producing protein crystals which are suitable for X-ray analysis. It is believed that one of the difficulties to get well ordered crystals is due to the high carbohydrate content.

The amino acid sequences have been determined of two large proteolytic fragments of human ceruloplasmin(20). These fragments show strong internal homology suggesting evolutionary replication of at least two smaller units. They also exhibit significant homology around a cysteine residue with the copper binding sites in azurin and plastocyanin as shown in Figure 1. These small proteins contain a single type 1 copper and the three dimensional structures have recently been determined for both proteins (22,23). It is therefore known that the metal is coordinated to two histidine, one cysteine and one methionine residue. Since these amino acid residues are conserved in the two fragments of ceruloplasmin it is suggested that the two type 1 coppers present in this protein are bound in similar ways as in the small proteins (20).

A small peptide sequence around the single free cysteine residue in *Polyporus* laccase has also been determined (24). It contains histidine residues on both sides of the cysteine residue in similarity to the 19 000 fragment of ceruloplasmin. This structure is not found in the small blue proteins and in the 50 000 fragment of ceruloplasmin. It was therefore suggested that these histidine residues are ligands to other copper ions. Since both laccase and ceruloplasmin contain only one type 2 copper it was thought that this metal is bound to these histidine residues. This hypothesis gains some support from kinetic studies which suggest a conformational interaction between types 1 and 2 copper ions (section 5).

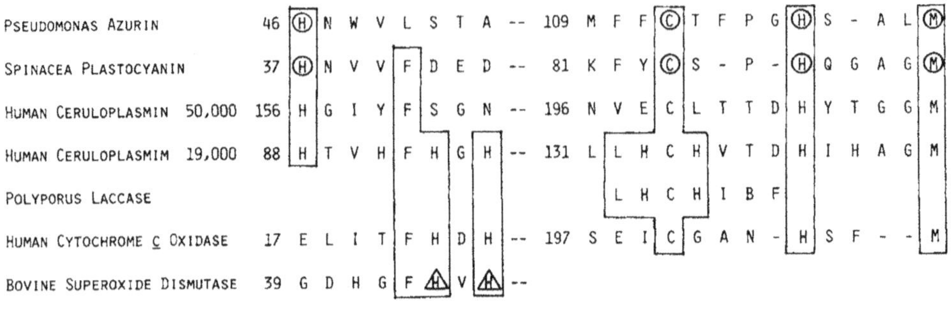

Fig. 1 Amino acid sequences around the copper-binding sites in plastocyanin and azurin (circles) and superoxide dismutase (triangles).

The sequences around the two proposed type 1 copper sites in ceruloplasmin show another interesting point. The 19 000 fragment exhibits a striking homology, which is close to the first copper ligand (histidine 88), to the known site for binding of the copper ion in bovine superoxide dismutase (20). A similar homologous region is found in subunit 2 in human cytochrome *c* oxidase (20) but is absent in the small blue proteins and in the 50 000 fragment of ceruloplasmin. As discussed in section 3.2. a new EPR signal has recently been detected in both *Polyporus* and *Rhus* laccases and also in cytochrome *c* oxidase (25,26). These signals were shown to come from the half-reduced type 3 copper sites in the laccases and the hitherto EPR-silent Cu_B in cytochrome *c* oxidase. Rather similar EPR signals are also reported from Cu(II) in the native copper site of bovine superoxide dismutase (27). It is therefore possible that at least one of the type 3 copper ions in the blue oxidases and the Cu_B in cytochrome *c* oxidase are bound in these peptide sequences in the different proteins. The lack of this sequence in the 50 000 fragment in ceruloplasmin might then support the view that this enzyme contains only one type 3 copper pair (19).

2.3. The Three Types of Copper Sites.

In the first studies of a blue copper oxidase, made as early as in 1894 by Bertrand, it was reported that the *Rhus* laccase is a metalloenzyme which contains manganese (28). However, in later studies by Keilin and Mann (29) with highly purified enzyme it was demonstrated that this enzyme contains copper but not manganese. Subsequent studies with the other blue oxidases have established that the only metal detected in stoichiometric amounts in these proteins is copper.

The copper content in blue oxidases varies from four to eight atoms per molecule and the metals are present in at least three diffferent coordination environments. They are classified according to the following definition:

1. Type 1 Cu(II) is characterized by an intense blue colour, with an extinction coefficient higher than 10^3 M^{-1} cm^{-1}, and an EPR spectrum with unusually narrow hyperfine structure in the g_z region with A_z smaller than 100 gauss.
2. Type 2 Cu(II) exhibits normal EPR parameters and appears to have only weak absorption in the visible and near ultraviolet regions. This metal binds certain inhibitory anions e.g. F^-, N_3^- and CN^- very strongly.
3. Type 3 Cu(II) ions are usually not detected by EPR since they form strongly antiferromagnetically coupled pairs. This complex has a strong absorption band at about 330 nm which disappears on reduction.

The first attempt to determine the state of the copper ions was made with the *Rhus* laccase. The results were interpreted to show that there were four Cu(II) ions, all of which contributed to the blue colour (30). A later investigation with the *Polyporus* laccase and ceruloplasmin showed, however, that only about 50 % of the copper ions were detected by EPR (31). Another EPR study of the *Rhus* enzyme also demonstrated that about 70 % of the copper ions are in the cupric state (32).

The early investigations also showed that the EPR detectable fraction of the metal ions gave rise to superimposed signals from Cu(II) in different coordination environments. One of the signals showed characteristics of type 1 Cu(II). The other EPR signal was believed to originate from denaturated molecules since the type 1 Cu(II) signal changed to a similar signal on denaturation of the protein. Later studies with the *Polyporus* laccase showed, however, that this enzyme contains equal amounts of Cu(II) in two different sites (33). Subsequent studies of other laccases, ceruloplasmin and ascorbate oxidase have now established that all blue oxidases contain at least three types of copper sites.

Several blue oxidases have been carefully examined concerning the state if their copper ions. Table 1 gives what we consider the correct values for the distribution of the types of copper ions in these proteins. It shows that all laccases contain one each of types 1 and 2 and two type 3 copper ions. An exception is the *Podospora* laccase I which is a tetrameric protein with 16 copper ions. It is, however, likely that each monomere contains the same numbers of the three types of copper ions as the other laccases.

According to copper analysis and molecular weight determinations ceruloplasmin would contain only six copper ions (19). Since there are two type 1 and one type 2 coppers there would be three EPR non-detectable copper ions to be consistent with the EPR results. Besides only one type 3 pair there would be a type 4 copper which must be in the Cu(I) state constantly since it is not detected by EPR. As discussed in section 2.2. there could be only one type 3 pair in this enzyme. Another interpretation would be that there are two type 3 copper pairs and that the molecular weight is either too low or that some molecules have lost their copper ions resulting in a copper content which is too low. There has also been considerable disagreements on the metal stoichiometries in ascorbate oxidase. Several recent studies agree, however, with the data shown in Table 1.

TABLE 1

Distribution of Copper Centers in Blue Oxidases.

Protein	Type 1	Type 2	Type 3	References
Laccase:				
Rhus	1	1	2	34
Polyporus	1	1	2	35
Podospora[a]	1	1	2	36
Neurospora	1	1	2	37
Ceruloplasmin	2	1	4	38
Ascorbate oxidase	3	1	4	17,39

[a]Copper ions per monomer

The EPR non-detectable copper ions were originally suggested to be in the Cu(I) state since they do not contribute to the magnetic susceptibility of laccase and ceruloplasmin made at room temperature (40) or low-temperatures (41,42). However, later redox studies showed that these oxidases can accept as many electrons as there are copper ions present (43-45) and that there is a cooperative two-electron acceptor in two laccases in addition to the type 1 and 2 Cu(II) ions (44,45). Since no new EPR signals appear on reduction it was suggested that the EPR silent metal ions consist of strongly antiferromagnetically coupled Cu(II). Recent careful studies between 40-300 K show that only types 1 and 2 Cu(II) ions contribute to the magnetic susceptibility (46,47). Therefore, if the type 3 copper ions are exchange coupled Cu(II) pairs the coupling constant - J must be at least in the order of 300 cm^{-1} (47). The idea that the type 3 copper ions are in the Cu(II) state is supported by the recent discovery that it is possible to reduce one of the metals which makes the other EPR detectable (see section 3.2. and ref. 48). An increase in the magnetic susceptibility is also observed in the peroxo-complex of tree laccase and this is interpreted as a decrease in the antiferromagnetic coupling of the type 3 copper pair on binding of peroxide (49).

3. SPECTROSCOPIC DATA

3.1. Optical Properties.

The blue copper oxidases exhibit an intense absorption at about 600 nm (with a molar absorptivity of about 5 000 per type 1 Cu(II)) and at least two weaker absorption bands at about 800 nm and 450 nm as illustrated in Figure 2. These bands are assigned to the type 1 Cu(II) since they are also present in small blue proteins, which contain only this type of copper, and can be made to disappear on selective reduction of this metal in laccases (50).

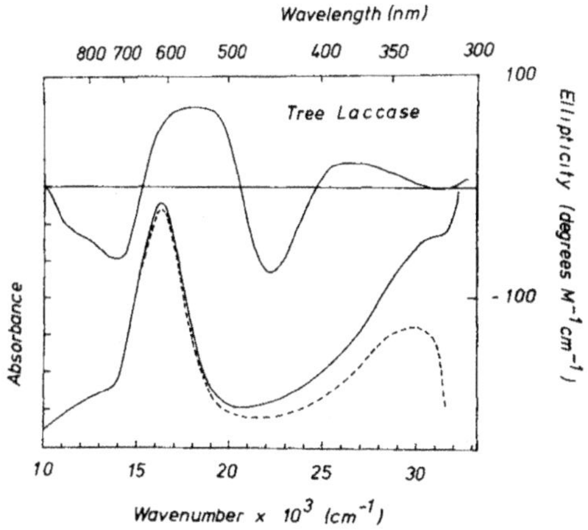

Fig. 2 Optical absorption of oxidized (full line) and oxidized-reduced (dashed line) as well as CD spectrum of *Rhus* laccase.

The absorption bands were earlier assigned of d-d transitions in a distorted tetrahedral Cu(II) center (51). Later CD studies of laccase and ceruloplasmin showed, however, that there are at least six bands associated with this chromophore as illustrated in Figure 2. Therefore, charge transfer and also ligand transitions are probably involved (50). The CD studies have recently been extended into the infrared and two more bands were detected below 11 000 cm^{-1} in two laccases, ceruloplasmin and ascorbate oxidase as well as in the small blue proteins stellacyanin, azurin and plastocyanin (52). As an illustration to the complicated electronic structure of the type 1 copper chromophore Table 2 shows the energies of the CD transitions which are assigned to type 1 Cu(II) in different blue proteins.

A number of other spectroscopic and chemical methods have been used in the studies of the blue chromophore. The results, which have recently been summarized (52), are consistent with a distorted tetrahedral coordination of type 1 copper. The intense absorption band at 600 nm is

TABLE 2

CD Band Position and Transition Assignments in Blue Proteins.

Protein	Band position in wave numbers (cm^{-1})								References
Polyporus laccase	7250		11000	13200	15800	19000	22500		50,52
Rhus laccase	6000	9500	11700	14000	16800	19000	22400		50,52
Ceruloplasmin	6100	10000	11500	13500	16500	18700	22300		50,52
Ascorbate oxidase	5800	9100		13250	16530	18200	21275	23810	52
Stellacyanin	5250	8750	11000	12800	16500	19000	22400	33000	50,52
Transition assignment	$^2B_2 \longrightarrow$ ^2E	$^2B_1$	$^2A_1$	¶S	σS	σS*	¶N	not assigned	52
					$\longrightarrow d(x^2 - y^2)$				

likely due to S(Cys)→Cu(II) transition. Table 2 shows the transition assignments of the different CD bands in the blue oxidases and also in stellacyanin for a comparison. Three near-infrared bands are assigned to d-d transitions and the bands with higher energies to charge transfer transitions from cysteine sulfur and histidine nitrogen ligands to the $d(x^2-y^2)$ orbital of Cu(II). The band at about 19 000 cm^{-1}, present in all blue proteins, have tentatively been assigned to σS*→$d(x^2-y^2)$ transition where S* is a thioether-type sulfur atom (52). Methionine sulfur isa metal ligand in plastocyanin (22) and azurin (23) and probably in ceruloplasmin (Figure 1) and in tree laccase (52). Since methionine is absent in stellacyanin (53) this CD band might be the result of a transition from a disulfide to the metal in this protein (52).

The strong absorption band at about 330 nm is only observed in the oxidases and it can therefore not originate from type 1 copper. Reductive titrations show that it is associated with a cooperative two-electron acceptor which is believed to be the type 3 copper pair (44,45). Selective reduction of type 1 Cu(II) in fungal laccase demonstrate that there are no strong CD activity associated with the types 2 or 3 sites (50). Weak absorption and CD bands have, however, recently been detected in fungal and tree laccases on selective reduction of type 1 Cu(II) with NO (54). The absence of near-infrared electronic transitions appears to rule out a tetrahedral coordination for the types 2 and 3 sites but supports a tetragonal coordination.

3.2. EPR Spectra.

The blue oxidases contain various amounts of copper ions in at least three different coordination environments. The EPR spectra of the resting enzymes show the presence of superimposed signals from two different types of copper as shown in the experimental and simulated EPR spectra of *Neurospora* laccase in Figure 3. One of the signals exhibit an unusually small hyperfine splitting constant and is associated with the type 1 Cu(II). Table 3 summarizes the EPR parameters of this copper ion in some well characterized oxidases. The spectra show deviation from axial symmetry which is consistent with a distorted tetrahedral coordination. The low g-values and small hyperfine coupling constants might depend on a delocalization of the unpaired spin onto the metal ligands or to a tetragonally flattened tetrahedral geometry, which allows for a mixing of 4p character into the primary 3d orbital levels (52).

The EPR parameters of type 2 Cu(II) in the different oxidases are summarized in Table 4. These are typical of copper coordinated to nitrogen and oxygen ligands in tetragonal complexes. This metal binds certain anionic inhibitors, such as F^-, N_3^- and CN^- very strongly. For example, F^- binds extremely strongly to the fungal laccase and displaces an OH^- which shows that H_2O or OH^- (depending on the solution pH) is a ligand to this metal (59). Another F^- binds somewhat weaker and both anions perturb the type 2 Cu(II) spectrum.

The blue oxidases also contain 2-4 type 3 copper ions. They are EPR

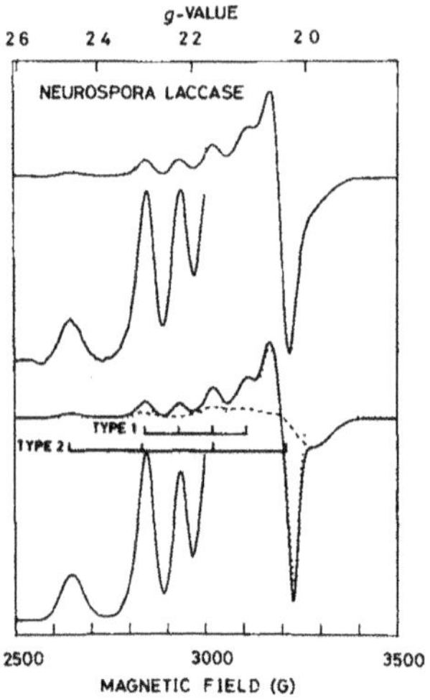

Fig. 3 Experimental (upper spectrum) and simulated EPR spectra at 9 GHz of *Neurospora* laccase.

TABLE 3

EPR Parameters of Type 1 Copper.

Protein	g_x	$g_\perp$	g_y	g_z	A_z^a	References
Laccase						
Polyporus	2.033		2.050	2.190	9.0	35,55
Rhus v.	2.030		2.055	2.298	4.3	25,34
Podospora	2.034		2.050	2.209	8.0	36
Neurospora		2.04		2.190	9.2	37
Rhus s.		2.045		2.204	7.6	56
Ceruloplasmin		2.05 2.06		2.206 2.215	7.4 9.5	57
Ascorbate oxidase	2.036		2.058	2.227	5.8	17,39

[a] A_z values in Tables 3-5 are in units of $10^{-3}cm^{-1}$

TABLE 4

EPR Parameters of Type 2 Copper.

Protein	$g_{\perp}$	$g_{\parallel}$	$A_{\parallel}$	References
Laccase				
Polyporus	2.036	2.243	19.4	35,55
Rhus v.	2.053	2.237	20.6	25,34
Podospora	2.046	2.246	17.6	36
Neurospora	2.04	2.23	19.4	37
Rhus s.		2.217	18.4	56
Ceruloplasmin	2.060	2.247	18.9	57
Ascorbate oxidase	2.053	2.242	19.9	17,39

non-detectable under most experimental conditions. However, an EPR signal with a new shape was recently observed during reoxidation of native and type 2 depleted tree laccase and it was demonstrated that it originates from the type 3 copper pair (25,26,60). A somewhat similar signal (Figure 4C) has also been produced on reduction of type 2 copper-depleted fungal

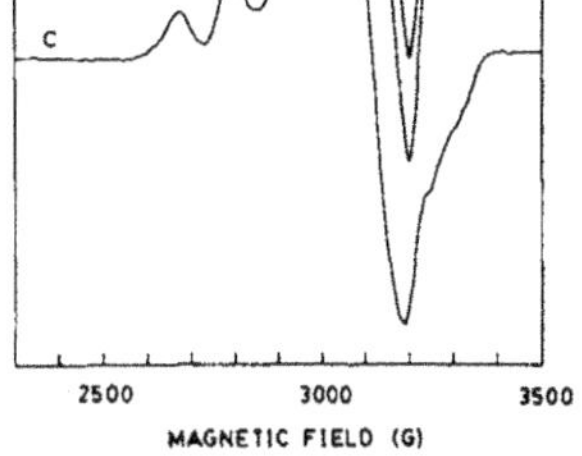

Fig. 4 EPR spectra at 9 GHz of native (A), type 2-depleted (B) and type 3 Cu(II) *Rhus* laccase.

and tree laccases (25,48). Since type 1 is reduced and type 2 is absent this new EPR signal must come from one of the type 3 Cu(II) ions which becomes detectable when the other is reduced. The EPR parameters for these signals are given in Table 5. These signals show strong rhombic character and appear to be identical. The Cu(II) ions are therefore possibly bound in similar sites in both laccases. Of particular interest is that very similar rhombic Cu(II) EPR signals have also been detected in other proteins containing a bimetallic center, such as superoxide dismutase, half-met hemocyanin and cytochrome *c* oxidase. In the latter protein the new signal is attributed to Cu_B which is EPR non-detectable in the resting oxidase (26). As discussed in section 2.2. there is a pronounced amino acid sequence homology between a blue oxidase (ceruloplasmin) and cytochrome *c* oxidase with the copper-binding site on bovine superoxide dismutase (Figure 1). This similarity between Cu(II) signals and protein sequences in the different proteins might then be a reflection of similarities also in copper coordination.

TABLE 5

EPR Parameters of Cu(II) from Bimetallic Centers in Various Proteins.

Protein	g_x	g_y	g_z	A_z	References
Polyporus laccase, type 2-depleted	2.025	2.148	2.268	13.2	26
Rhus laccase, type 2-depleted	2.03	2.15	2.277	13.2	48
Half-Met hemocyanin	2.04	2.09	2.30	13.2	61
Bovine superoxide dismutase	2.025	2.103	2.257	13.9	27
Cytochrome *c* oxidase	2.052	2.109	2.278	10.8	26
Native *Rhus* laccase	2.05	2.15	2.305	8.4	25,26

Since it is possible to obtain this signal on reduction of type 2 copper-depleted tree and fungal laccases the type 3 copper ions must be oxidized in the resting proteins (26,48). This result is therefore in accordance with redox titration and kinetic studies (section 4. and 5.) which suggest that the type 3 copper ions consist of a pair of antiferromagnetically coupled Cu(II) ions.

3.3. Endor Spectroscopy.

The technique of pulsed EPR has been applied to study the coordination of types 1 and 2 Cu(II) in laccase and type 2 in ceruloplasmin. In all cases the nuclear modulation patterns indicate that imidazole is a metal ligand (62). The type 3 Cu(II) site shown in Figure 4C has also recently been studied with ^{14}N endor and the region for nitrogen resonances is shown in Figure 5 (63). The data at different spectrometer frequencies are consistent with at least three distinct ^{14}N ligand atoms. The Larmour-split doublets, centered at $A_x/2$, are indicated by bars in the figure. A fifth peak at about 6 MHz (not shown) is tentatively

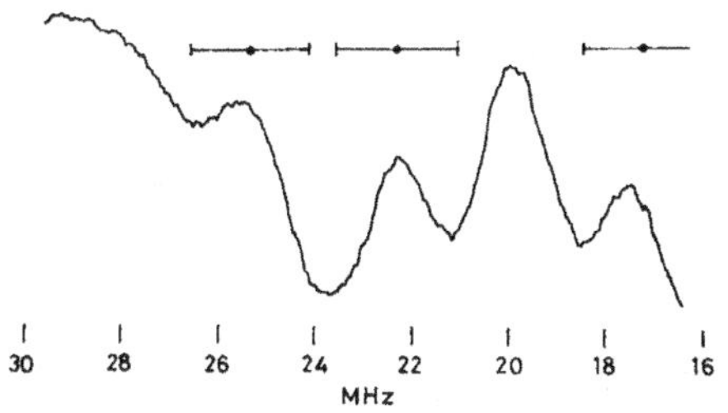

Fig. 5 ^{14}N endor spectrum of the type 3 Cu(II) site in *Rhus* laccase.

assigned to the remote nitrogen of one or more imidazole rings. The three nitrogens observed have hyperfine couplings that are significantly larger or smaller than the coupling constants for $(Cu(Imidazole)_4)^{2+}$ which indicates a low symmetry for the coordination site of type 3 Cu(II) in this protein.

The proton endor (not shown) with H_2O or D_2O demonstrate that most proton peaks are not affected by solvent deuteration indicating that they are associated with structural protons. However, at least one pair of resonances is eliminated by D_2O. It is therefore at least one exchangeable group in close proximity to the copper site.

This EPR-detectable type 3 copper ion and its set of coordinating ligands may then be characterized as (CuN_3X) where at least one of the nitrogenous ligands appears to be a histidyl imidazole. The X would represent a non-nitrogenous fourth ligand, which could be a water or hydroxyl molecule occupying the coordination site X which in its turn is normally occupied by a bridge between the two coppers of the type 3 center.

4. OXIDATION-REDUCTION PROPERTIES

The redox properties of the metal centers have been examined in two laccases and in ceruloplasmin. The methods used are based on optical absorption and EPR measurements in combination with potentiometric studies of the metal sites as they are successively reduced or reoxidized. The advantages of the different methods used and the difficulties involved in the interpretation of the data have recently been reviewed (35).

The estimated best potential values for these proteins are summarized in Table 6. The experiments to determine the potential of type 2 copper are most difficult to perform since EPR measuremnts and rather high concentrations of enzymes are required. The potential of this site

TABLE 6

Reduction Potentials of Copper Sites in Blue Oxidases.

		Potentials in mV			
Protein	pH	Type 1	Type 2	Type 3	References
Rhus laccase	7.5	394	365	434	44,45
+ 10 mM NaF	7.5	390	390	390	
+ $Fe(CN)_6^{4-}$	7.5	434		483	
Polyporus laccase	5.5	785		782	44
+ 1 mM NaF	5.5	780		570	
Ceruloplasmin	5.5	490,580			38

is therefore known only in the tree laccase. The 330 nm chromophore, which is probably associated with the type 3 copper pair, titrates as a cooperative two-electron acceptor in both fungal and tree laccase but titrates with a Nernst coefficient of only 0.5-1 in ceruloplasmin, a result which is not easily understood.

The inhibitor F^-, which binds strongly to type 2 Cu(II), affects the potential of both types 2 and 3 copper sites but not the type 1 site. Thus the potential of the two-electron acceptor in the *Polyporus* laccase decreases by as much as 210 mV in the presence of this inhibitor. Also $Fe(CN)_6^{4-}$ seems to interact with the type 2 copper in the *Rhus* enzyme and affect the potentials of both type 1 and 3 copper sites.

5. THE CATALYTIC MECHANISM

There have been several attempts to examine the role of the different metal sites in the electron transfer from one-electron reducing substances to dioxygen performed by the blue oxidases. These enzymes are efficient catalysts. For example, the type 1 Cu(II) in the fungal laccase is reduced with rate constants of about 10^6-10^8 $M^{-1}sec^{-1}$ by such different reductants as $Fe(CN)_6^{4-}$, hydroquinone and hydrated electrons (15,64-67). The reaction with dioxygen is also rapid with rate constants in the order of 10^6 $M^{-1}sec^{-1}$(section 5.2.).

The reduction and reoxidation reactions are very complex even in the presence of only reductants or oxidants. In order to simplify the interpretations it has been necessary to use only one reactant at a time. Although the presence of both substrates would reflect the normal catalytic events it has been shown that steady state results are in accordance with the transient state experiments which are performed with only one reactant at a time. The reduction and reoxidation mechanisms, which will be described in this section, are therefore considered valid for the overall catalytic mechanism of laccases and probably also for the more complicated enzymes ceruloplasmin and ascorbate oxidase.

5.1. The Reduction Reactions.

Extensive mechanistic studies have been performed with the laccases, in particular with the *Rhus* laccase (59,68-70). In order to examine the redox behaviour of all three electron acceptors in these oxidases a combination of stopped-flow and rapid-freeze EPR techniques were used. The anaerobic reduction patterns of the *Rhus* enzyme at pH 6.5 and 7.4 are summarized in Figure 6 (59,69). At pH 6.5 both types 1 and 2 Cu(II) ions are reduced in similar reactions. They are characterized by an

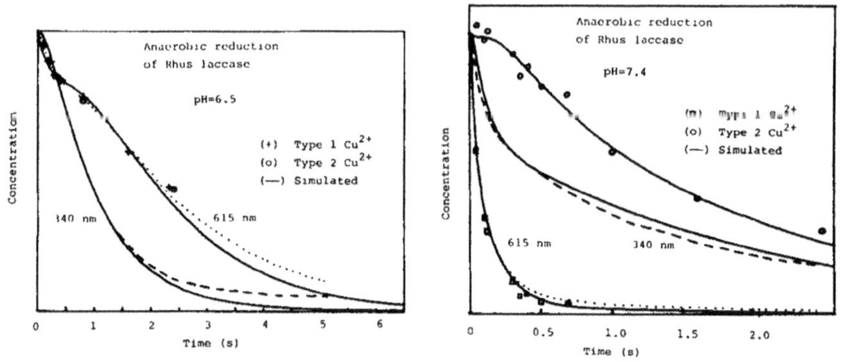

Fig. 6 Experimental (symbols, dashed and dotted lines) and simulated (full lines) reduction curves of types 1-3 copper sites in *Rhus* laccase.

initial rapid reduction phase followed by a slowing down and then a final reduction phase. The 340 nm absorption, which is associated with the oxidized type 3 copper pair, shows an initial lag phase and is then apparently reduced faster than both types 1 and 2 Cu(II). The presence of this lag phase indicates that electrons are transferred from types 1 and 2 Cu(I) to the two-electron accepting type 3 site.

The simultaneous reduction of types 1 and 2 copper sites would suggest that they are reduced independently of each other with the same rate constants or that type 2 is reduced by electron transfer from type 1 Cu(I). However, an analysis with 0.1 to 10 mM hydroquinone, as shown in Figure 7, demonstrates that neither of these possibilities are plausible. For example, an independent reduction of these copper ions would result in a much longer lag phase in the type 3 copper reduction.

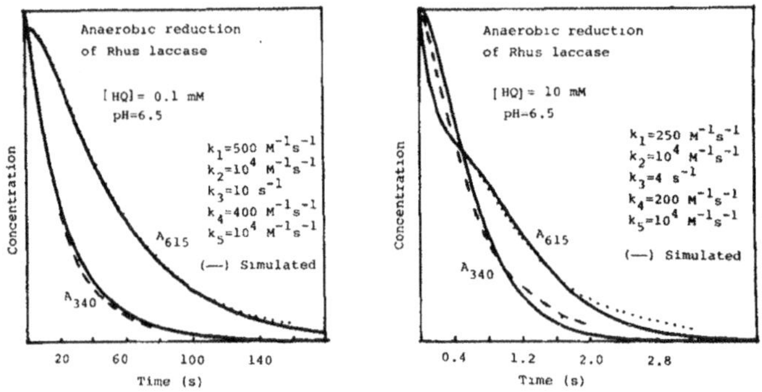

Fig. 7 Experimental (dashed and dotted lines) and simulated (full lines) of types 1 and 3 copper sites in *Rhus* laccase.

Similarly the reduction patterns do not agree with a mechanism where type 2 Cu(II) is reduced by type 1 Cu(I) but suggest that there is an interaction between these metal sites. The first electron will thus reduce type 1 copper. That this metal is the primary electron accepting site has been demonstrated with the type 2 copper-depleted *Rhus* laccase or with F^--inhibited enzymes (59,69). When type 1 copper is reduced it apparently affects the reduction of type 2 Cu(II) which is reduced by another substrate molecule. As discussed in section 2.2. types 1 and 2 copper sites might be coordinated to neighbouring amino acid residues. If this is the case, a reduction of type 1 copper might cause a conformational change at the type 2 copper site which would affect the rate of reduction of this metal. When both types 1 and 2 copper sites are reduced they simultaneously donate one electron each to the two-electron accepting type 3 copper site. This is consistent with the multiphasic reduction patterns of types 1 and 2 sites and with the lag phase in the type 3 copper reduction. The oxidized types 1 and 2 copper ions are then reduced once more resulting in completely reduced enzyme.

The upper part of the reduction scheme in Figure 8 summarizes the reduction reactions at this pH. Types 1 and 2 copper ions are placed at the upper and lower left side, respectively, of the box which represents the enzyme molecule. The two type 3 copper ions are represented by a pair that can accept two electrons. The full lines in Figure 7 are simulated using this scheme and the rate constants given in the figure. The lower values of k_1 and k_4, used in the simulation at higher substrate concentrations, depend on an actual rate saturation of these reactions at 10mM hydroquinone (69).

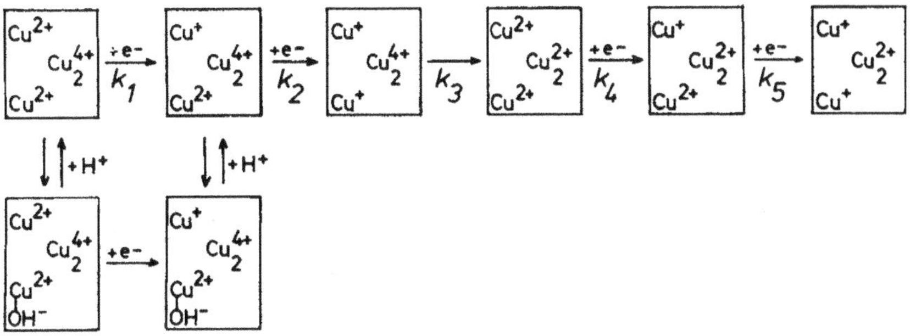

Fig. 8 Anaerobic reduction scheme of *Rhus* laccase at neutral pH.

The reduction patterns of the three metal sites change drastically when pH is raised to 7.4 as shown in Figure 6 (59,69). Type 1 Cu(II) is now reduced about ten times faster than at pH 6.5. Type 2 Cu(II) reduction shows an initial lag phase and a much slower reduction rate. The type 3 copper site is reduced in two phases. About 50 % of this site is reduced in a second order reaction. The remaining absorbance of this site is reduced in a reaction which is independent of hydroquinone concentrations higher than 1 mM. The results are consistent with the following view. About 50% of the molecules are reduced in the same patterns as at pH 6.5. The rest of the molecules are inactive due to a type 2 $Cu(II)-OH^-$ complex which is formed at this pH. Although type 1 Cu(II) is rapidly reduced in all molecules the reduction of types 2 and 3 copper sites are not possible in the inhibited molecules until the OH^- ion, bound to type 2 Cu(II), is eliminated. This reaction occurs in a slow reaction (about 0.5 sec^{-1}) which is independent of the substrate concentration. The reduction of all three metal sites can be simulated using the complete reduction scheme in Figure 8 and the rate constants published in references (59,69).

The anaerobic reduction pattern of the *Polyporus* laccase (15,66) can be explained with a similar model (69). At pH 5.5 all molecules are inhibited as they have OH^- bound to type 2 Cu(II) (59). Although type 1 Cu(II) is rapidly reduced in a second order reaction the type 2 and 3 copper sites are not reduced until the OH^- bound to type 2 Cu(II) is eliminated which occurs in a slow (1 sec^{-1}) reaction.

A number of anionic inhibitors, such as F^-, N_3^- and CN^-, bind strongly to type 2 Cu(II). The inhibition by F^- is best understood. On addition of only one F^- per molecule to the *Polyporus* laccase about 90 % is bound to type 2 copper and the enzyme is almost completely inhibited (71). Since the catalytic activity increases simultaneously with the release of F^- into the solution, in turn-over experiments, this inhibitor apparently affects the reduction of this metal in a similar manner as shown for OH^-. The inhibitory actions by different substances have recently been reviewed (3,5) and is therefore not further considered here.

5.2. The Oxygen Reaction.

In the early studies of the reaction between reduced ceruloplasmin and dioxygen it was observed that type 1 Cu(I) was reoxidized in two phases. Only about 50 % of this metal was reoxidized with a second-order rate constant of $3x10^6$ $M^{-1}sec^{-1}$ (72). The rest of type 1 copper was reoxidized in a much slower reaction which was independent of dioxygen concentration . This result indicates that only one of the type 1 copper ions forms an active unit together with one type 2 and two type 3 copper ions.

In later studies with this enzyme a new transient species, with an absorption maximum at 420 nm, was formed at the same time as the rapid reoxidation of the 610 and 330 nm chromophores. This new absoption species thereafter dacays within 2 seconds and it was suggested to be an important intermediate in the catalytic reaction (73). Similar reaction intermediates have also been observed in laccases (74,75) which shows that they represent a common species in blue oxidases.

This intermediate state was first suggested to represent peroxide formed by a two-electron transfer from the two-electron acceptor to dioxygen. This peroxide would be stabilized by the type 2 Cu(II) until it was reduced to water in another two-electron transfer step from type 3 copper (74). However, studies with the *Rhus* laccase demonstrated that both the 615 and 330 nm chromophores are developed simultaneously with the optical intermediate (75). These reactions are dependent on the dioxygen concentration with a rate constant of $5x10^6$ $M^{-1}sec^{-1}$. The optical intermediate then disappears in a much slower reaction, which is first order with respect to the protein concentration, with a half-time of 1-20 seconds in the different enzymes. The type 2 Cu(I) is reoxidized in a reaction which is simultaneous with the decay of the optical intermediate, indicating that these reactions are coupled.

These reactions suggest that dioxygen rapidly receives three electrons from the types 1 and 3 copper sites and that the optical intermediate is associated with an oxygen species which is bound to the protein. This intermediate then receives a fourth electron from type 2 Cu(I) and the reoxidation reaction is completed. Since only three electrons are rapidly transferred to dioxygen some paramagnetic intermediates would also be formed. No new EPR signals were observed at 77 K but studies at 10 K led to the discovery of new paramagnetic intermediate species in both tree (Figure 9) and fungal laccases and in ceruloplasmin (76,77). This paramagnetic signals show unusual relaxation properties similar to those of $O^{\bar{\cdot}}$. That this reaction intermediate is derived from dioxygen was shown in reoxidation studies with $^{17}O_2$ (78).

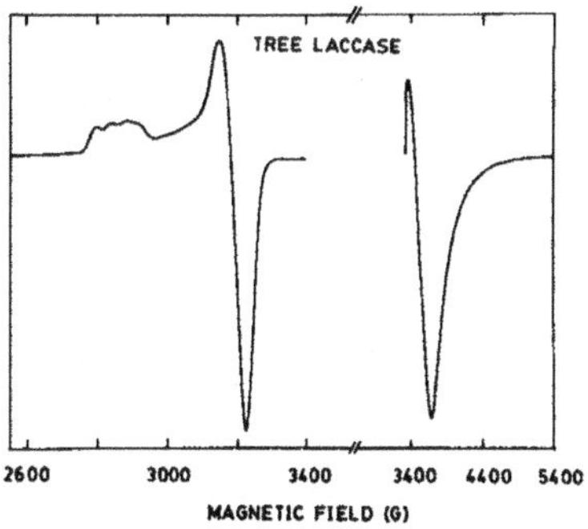

Fig. 9 EPR spectra at 9 GHz of reoxidized *Rhus* laccase showing the type 1 Cu(II) (left) and the oxygen radical signal (right).

The oxygen intermediate state appears to be a normal step in catalysis since it is also formed under turn-over conditions (59).

During reoxidation of type 2 copper-depleted *Rhus* laccase (Figure 4B) the same EPR detectable intermediate is formed as in the native enzyme (60). This demonstrates that type 2 copper is not necessary in the reoxidation reactions and it is not involved in the stabilization of the oxygen intermediate as proposed in several studies (56,79,80).

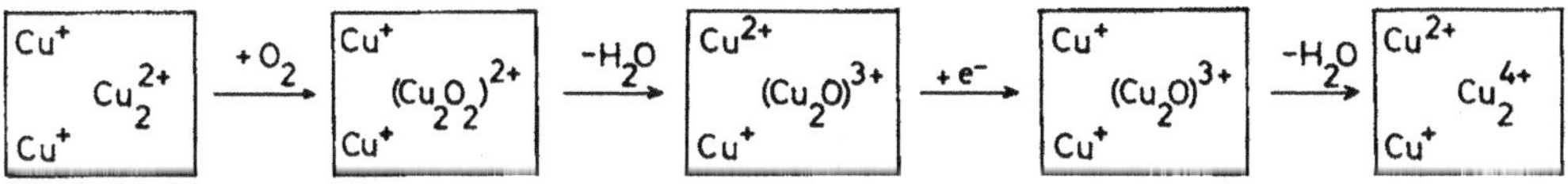

Fig. 10 Scheme showing the reaction between dioxygen and fully reduced laccase

The scheme in Figure 10 summarizes the present knowledge about the reoxidation reactions of laccase which is prior reduced with an excess of electrons. It is not yet established at what reduction state the enzyme reacts with dioxygen. It was suggested that the one-electron reduced tree laccase is rapidly reoxidized by dioxygen (81,82). However, studies with the type 2-depleted enzyme demonstrate that the reduced type 1 copper is very slowly reoxidized in the presence of dioxygen (60). It is also obvious that type 2 Cu(I) cannot be the oxygen binding site since it is not even involved in the reoxidation reactions (60). These results therefore suggest that dioxygen reacts with the reduced type 3 copper pair and forms a peroxide bound to this site. An electron from type 1 Cu(I) is then transferred to this type 3 copper-peroxidase complex and the paramagnetic oxygen intermediate is formed. It is thought that the intermediate is an $O^{\cdot}$ ion which is bound to the copper pair. It is, however not known whether the O-O bond is broken or not in this state. Therefore the oxygen complex could be either $(Cu_2O_2)^+$ or $(Cu_2O)^{3+}+H_2O$. The last step in the oxygen reduction is an electron transfer from reduced type 1 copper in turnover reactions or from type 2 Cu(I) in reoxidation reactions. Since this latter reaction is so slow it cannot be of catalytic importance. During the catalytic cycle type 1 copper is then reduced three times while type 2 is reduced only once.

ACKNOWLEDGMENTS

The author is indebted to Maria Setterbom for typing the manuscript. This work was supported by grants from Statens Naturvetenskapliga Forskningsråd.

REFERENCES

(1) Malmström, B.G., Andréasson, L.-E., and Reinhammar, B.: 1975, in *The Enzymes*, Vol. XII, 3rd ed., Oxidation-Reduction, Part B, Academic Press, Inc., p.507.

(2) Fee, J.A.: 1975 in *Structure and Bonding*, Dunitz, J.D., Hemmerich, P., Holm, R.H., Ibers,J.A., Jørgensen,C.K., Nielands,J.B., Reinen,D., and Williams, R.J.P., Eds., Springer Verlag, Berlin, Heidelberg, New York, Vol. 23, p.1.

(3) Reinhammar, B., Laccase, in *Copper Proteins*, Lontie, R., Ed., CRC Press, Boca Raton, Florida, in press.

(4) Holwerda, R.A., Wherland, A., and Gray, H.B.: 1976, in *Annual Review of Biophysics and Bioengineering*, Mullins, L.J., Hagins, W.A., Stryer, L., and Newton, C., Eds., Annual Review Inc., Palo Alto, California, Vol. 5, p.363.

(5) Reinhammar, B., and Malmström, B.G.: 1981, in *Copper Proteins*, Vol. 3 of *Metals in Biology*, Spiro, T.G., Ed., John Wiley & Sons, Inc., New York, pp.109-149.

(6) Laurie, S.H., and Mohammed, E.S.: 1980, *Coordination Chemistry Reviews*, 33, pp.279-312.

(7) Dawson, C.R.: 1966, in *The Biochemistry of Copper*, Peisach, J., Aisen, P., and Blumberg, W.E., Eds., Academic Press, New York, p.305.

(8) Yoshida, H.: 1883, J. Chem. Soc., 43, p.472.

(9) Fåhraeus, G., Tullander, V., and Ljunggren, H.: 1958, Physiologia Plantarum, 2, p.631.

(10) Esser, K.: 1963, Arch. Microbiol., 46, p.217

(11) Froener, S.C., and Eriksson, K.-E.: 1974, J. Bacteriol., 120, p.458.

(12) Peisach, J., and Levine, W.G.: 1965, J. Biol. Chem., 240, p.2284.

(13) Fåhraeus, G., and Ljunggren, H.: 1961, Biochim. Biophys. Acta, 46, p.22.

(14) Fåhraeus, G.: 1961, Biochim. Biophys. Acta, 54, p.192.

(15) Malmström, B.G., Finazzi Agrò, A., and Antonini, E.: 1969, Eur. J. Biochem., 9, p.383.

(16) Osaki, S., Johnson, D.A., and Frieden, E.: 1966, J. Biol. Chem., 241, p.2746.

(17) Marchesini, A., Kronech, P.M.H.: 1979, Eur. J. Biochem., 101, p.65.

(18) Briving, C.,: 1975, Fungal Laccase B: studies on molecular properties, primary structure and chemical modifications, Ph.D. thesis, University of Göteborg, Sweden.

(19) Rydén, L., and Björk, I.: 1976, Biochemistry, 15, p.3411.

(20) Dwulet, F.E., and Putnam, F.W.: 1981, Proc. Natl. Acad. Sci., USA, 78, p.2805.

(21) Strothkamp, K.G., and Dawson, C.R.: 1974, Biochemistry, 13, p.434.

(22) Colman, P.M., Freeman, H.C., Guss, J.M., Murata, M., Norris, V.A., Ramshaw, J.A.M., and Venkatappa, M.P.: 1978, Nature, 272, p.319.

(23) Adman, E.T., Stenkamp, R.E., Sieker, L.C., and Jensen, L.H.: 1978, J. Mol. Biol., 123, p.35.

(24) Briving,C., Gandvik, E.-K., and Nyman, P.O.: 1980, Biochem. Biophys. Res. Commun., 93, p.454.

(25) Reinhammar, B.: 1980, J. Inorg. Biochem., 18, pp.27-39.

(26) Reinhammar, B., Malkin, R., Jensen,P., Karlsson, B., Andréasson, L.-E., Aasa, R., Vänngård, T., and Malmström, B.G.: 1980, J. Biol. Chem., 255, p.5000.

(27) Fielden, E.M., Roberts, P.B., Bray, R.C., Lowe, D.J., Mautner, G.N., Rotilio, G., and Calabrese, L.: 1974, Biochem. J., 139, p.49.

(28) Bertrand, G.: 1894, Compt. Rend. Acad. Sci., 118, p.1215.

(29) Keilin, D., and Mann, T.: 1939, Nature, 143, p.23.

(30) Nakamura, T.: 1958, Biochim. Biophys. Acta, Short Commun., 30, p.640.

(31) Broman, L., Malmström, B.G., Aasa, R., and Vänngård.: 1962, J. Mol. Biol., 5, p.301.

(32) Blumberg, W.E., Levine, W.G., Margolis, S., and Peisach, J.: 1964, Biochem. Biophys. Res. Commun., 15, p.277.

(33) Malmström, B.G., Reinhammar, B., and Vänngård, T.: 1968, Biochim. Biophys. Acta, 156, p.67.

(34) Malmström, B.G., Reinhammar, B., and Vänngård, T.: 1970, Biochim. Biophys. Acta, 205, 48.

(35) Malmström, B.G., Reinhammar, B., and Vänngård, T.: 1968, Biochim. Biophys. Acta, 156, p.67.

(36) Moltoris, H.P., and Reinhammar, B.: 1975, Biochim. Biophys. Acta, 386, p.493.

(37) Lerch, K., Deinum, J., and Reinhammar, B.: 1978, Biochim, Biophys. Acta, 534, p.7.

(38) Deinum, J., and Vänngård, T.: 1973, Biochim. Biophys. Acta, 310, p.321.

(39) Deinum, J., Reinhammar, B., and Marchesini, A.: 1974, FEBS Lett. 42, pp.241-245.

(40) Ehrenberg, A., Malmström, B.G., Broman, L., Mosbach, B.: 1962, J. Mol. Biol., 5, p.450.

(41) Aisen, T., Koenig, S.H., and Lilienthal, H.R.: 1967, J. Mol. Biol. 28, p.225.

(42) Moss, T.H., and Vänngård, T.: 1974, Biochim. Biophys. Acta, 371, p.39.

(43) Fee, J.A., Malkin, R., Malmström, B.G., and Vänngård, T.: 1969, J. Biol. Chem., 244, p.4200.

(44) Reinhammar, B.: 1972, Biochim. Biophys. Acta, 275, p.245.

(45) Reinhammar, B., and Vänngård, T.: 1971, Eur. J. Biochem., 18, p.463.

(46) Dooley, D.M., Scott, R.A., Ellinghaus, J., Solomon, E.I., and Gray, H.B.: 1978, Proc. Natl. Acad. Sci., USA, 75, p.3019.

(47) Petersson, L., Ångström, J., and Ehrenberg, A.: 1978, Biochim. Biophys. Acta, 526, p.311.

(48) Reinhammar, B.: 1982, J. Inorg. Biochem., submitted.

(49) Farver, O., and Pecht, I.: 1979, FEBS Lett., 108, p.436.

(50) Falk, K.-E., and Reinhammar, B.: 1972, Biochim. Biophys. Acta, 285, p.84.

(51) Blumberg, V.E.: 1966, in *The Biochemistry of Copper*, Peisach, J., Aisen, P., and Blumberg, W.E., Eds., Academic Press, New York, p.576.

(52) Gray, H.B., and Solomon, E.I.: 1981, in *Copper Proteins*, Vol. 3 of *Metal Ions in Biology* series, Spiro, T.G., Ed., John Wiley & Sons, Inc., New York, pp.1-40.

(53) Bergman, C., Gandvik, E.-K., Nyman, P.O., and Strid, L.: 1977, Biochem. Biophys. Res. Commun., 77, p.1052.

(54) Dooley, D.M., Rawlings, J., Dawson, J.H., Stephens, P.J., Andréasson, L.-E., Malmström, B.G., and Gray, H.B.: 1979, J. Am. Chem. Soc., 101 p.5038.

(55) Vänngård, T.: 1972, in *Biological Applications of Electron Spin Resonance*, Swartz, H.M., Bolton, J.R., and Borg, D.C., Eds., John Wiley & Sons, Inc., p.411.

(56) Brändén, R., and Deinum, J.: 1978, Biochim. Biophys. Acta, 524, p.297.

(57) Gunnarsson, P.-O., Nylén, U., and Pettersson, G.: 1973, Eur. J, Biochem., 37, p.47.

(58) Peisach, J., and Blumberg, V.E.: 1974, Arch. Biochem. Biophys., 165, p.691.

(59) Andréasson, L.-E., and Reinhammar, B.: 1979, Biochim. Biophys. Acta, 568, p.145.

(60) Reinhammar, B., and Oda, Y.: 1979, J. Inorg. Biochem., 11, p.115.

(61) Schoot Uiterkamp, A.J.M., Van der Deen, H., Berendsen, H.C.J., and Boas, J.F.: 1974, Biochim. Biophys. Acta, 372, p.407.

(62) Mondovi, B., Graziani, M.T., Mims, W.B., Oltzick, R., and Peisach, J.: 1977, Biochemistry, 16, pp.4198-4202.

(63) Cline, J., Reinhammar, B., and Hoffman, B.: to be published.

(64) Bränden, R., and Reinhammar, B.: 1975, Biochim. Biophys. Acta, 405, p. 236.

(65) Andréasson, L.-E., Bränden, R., Malmström, B.G., Strömberg, C., and Vänngård, T.: 1973, in *Oxidases and Related Redox Systems* (Proceedings

of the Second International Symposium), King, T.E., Mason, H.S., and Morrison, M., eds., University Press, Baltimore.

(66) Andréasson, L.-E., Malmström, B.G., Strömberg, C., and Vänngård, T.: 1973, Eur. J. Biochem., 34, p.434.

(67) Pecht, I., and Faraggi, M.: 1971, Nature New Biology, 233, p.116.

(68) Holverda, R.A., and Gray, H.B.: 1974, J. Am. Chem. Soc., 96, p. 6008

(69) Andréasson, L.-E., and Reinhammar, B.: 1976, Biochim. Biophys. Acta, 445, p.579.

(70) Petersen, L.Chr., and Degn, H.: 1978, Biochim. Biophys. Acta, 526, p.85.

(71) Brändén, R., Malmström, B.G., and Vänngård, T.: 1973, Eur. J. Biochem. 36, p.195.

(72) Carrico, R., Malmström, B.G., and Vänngård, T.: 1971, Eur. J. Biochem. 22, p.127.

(73) Manabe,T., Manabe, N., Hiromi, K., and Hatano, H.: 1972, FEBS Lett., 23, p.268.

(74) Andréasson, L.-E., Brändén, R., Malmström, B.G., and Vänngård, T.: 1973, FEBS Lett., 32, p.187.

(75) Andréasson, L.-E., Brändén, R., and Reinhammar, R.: 1976, Biochim. Biophys. Acta, 438, p.370.

(76) Aasa, R., Brändén, R., Deinum, J., Malmström, B.G., Reinhammar, B., and Vänngård, T.: 1976, FEBS Lett., 61, p.115.

(77) Brändén, R., and Deinum, J.: 1978, Biochim. Biophys. Acta, 524, p.297.

(78) Aasa, R., Brändén, R., Deinum, J., Malmström, B.G., Reinhammar, B., and Vänngård, T.: 1976, Biochem, Biophys. Res. Commun., 70, p.1204.

(79) Brändén, R., Deinum, J., and Coleman, M.: 1978, FEBS Lett., 89,p.180.

(80) Brändén, R., and Deinum, J.: 1977, FEBS Lett., 33, p.144.

(81) Farver, O., Goldberg, M., and Pecht, I.: 1980, Eur. J. Biochem., 104, p.71.

(82) Goldberg, M., and Pecht, I.: 1978, Biophys. Journal, 24, p.371.

METAL REPLACEMENT STUDIES OF CHINESE LACCASE

Margaret C. Morris, Judith A. Blaszak, Helen R. Engeseth and David R. McMillin

Department of Chemistry, Purdue University, West Lafayette, Indiana 47907, U.S.A.

ABSTRACT

After introducing laccase, a blue copper oxidase which is isolated from the oriental lacquer tree we discuss what is known about the copper-binding sites of the enzyme. As an introduction to the technique of metal substitution, we then describe some studies of smaller proteins which contain only a blue, or type 1, copper site, including ^{1}H-NMR experiments. Next we consider the problem of reversibly removing and replacing the native copper ions of laccase and describe a reasonably successful procedure for the enzyme from the Chinese lac tree. Finally, we report preliminary studies of the first mixed-metal derivative of laccase, a hybrid enzyme which appears to contain one Hg(II) and three coppers per molecule with the mercury in the type 1 binding site.

INTRODUCTION

The laccases, along with ceruloplasmin and ascorbate oxidase, belong to the class of blue copper oxidases which have the virtue of being water soluble and therefore conveniently studied in homogeneous solution. In contrast to cytochrome oxidase, the laccases do not appear to be part of an energy-yielding electron transport chain; one function of tree laccase may be to promote the polymerization of urushiols as a protective measure where wounding occurs.(1) Depending on the source, laccases vary in molecular weight (especially carbohydrate content), stability and substrate specificity, but as far as is known, all involve a single polypeptide chain (1,2) which binds four copper atoms in three distinct binding sites.(3)

The type 1 site is a one-electron acceptor and is responsible for the intense visible absorption of the enzyme. The most intense of the visible bands has been firmly established as a ligand-to-metal charge-tranfer transition involving a cysteine sulfur and Cu(II).(4,5) The type 2 copper site is also a one-electron acceptor and is observed to bind small anions.(3,6) It contributes little to the visible absorption,(7)

I. Bertini, R. S. Drago, and C. Luchinat (eds.), The Coordination Chemistry of Metalloenzymes, 201–206.

but gives an EPR signal characteristic of a tetragonal coordination environment. The type 3 site involves two copper atoms and functions as a two-electron acceptor.(3) It does not give an EPR signal in the oxidized state, where it is in fact diamagnetic,(8,9) but it is associated with a shoulder which occurs at 330 nm in the absorption spectrum.

Various kinetic and spectral properties (e.g. EXAFS results) of laccase are difficult to interpret because so many sites are involved. Chemical modification studies can lead to some important simplifications in this regard, and our group has been interested in modifying the metal-binding sites. But before discussing our results with laccase, we will introduce the method of metal-replacement by describing results on simpler systems.

METAL-REPLACEMENT STUDIES OF SMALL BLUE PROTEINS

Azurin, a bacterial electron transferase, has been particularly amenable to metal-replacement studies. Its three dimensional structure is now reasonably well established, the copper center being located in a pseudo-tetrahedral binding site consisting of a cysteine sulfur, a methionine sulfur and imidazole nitrogens from two histidine residues.(10,11) Early on, metal-replacement studies were useful in assigning the ligand-to-metal charge-transfer absorption bands which occur in the visible region. (12) More recently we have determined that the magnetic moment of the nickel center in Ni(II)Az, the nickel(II) derivative of azurin, is 3.2 B.M.,(13) and we have observed d-d transitions from the metal center in the near infra-red region as well.(14) Both observations are consistent with a pseudotetrahedral site symmetry despite the fact that nickel(II) prefers a square planar coordination geometry in the presence of thiolate donors.(14) These results are in keeping with the notion that the protein imposes a rather rigid environment about copper so as to minimize the Franck-Condon barrier to electron-transfer reactions.(15)

Further structural information has come from ^{1}H-NMR studies of Ni(II)Az. (13) They provide direct evidence that a conformational change occurs in the pH regime where histidine 35 is titrated (Figure 1). Just such a structural change has been postulated to explain the kinetics of electron-transfer between azurin and cytochrome C_{551} and envisioned to have a possible regulatory role.(16) More recently, ^{113}Cd-NMR studies of a series of cadmium(II) proteins have shown that the environment about the metal in stellacyanin is very similar to that of azurin despite the fact that stellacyanin has no methionine residues.(17)

In concluding this section, we note that as an alternative to the approach described above where a spectral probe ion has been used, in a polymetallic protein it may be advantageous to insert a spectroscopically silent metal ion so as to enhance the spectral signatures of other sites.

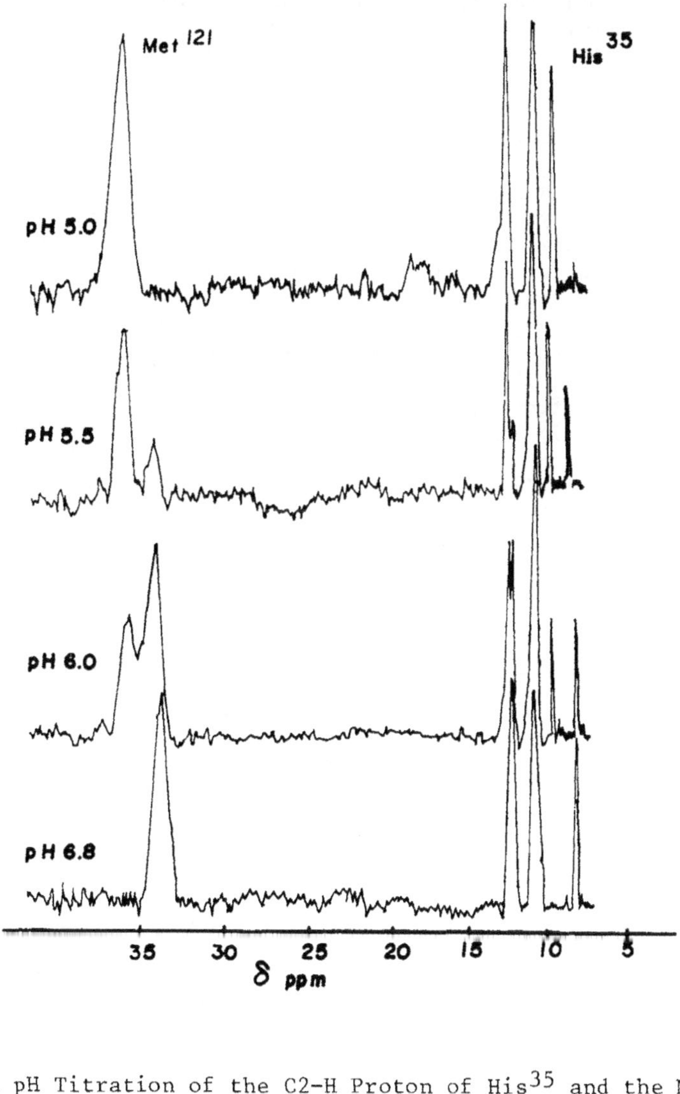

Figure 1. A pH Titration of the C2-H Proton of His[35] and the Methyl Protons of the Copper Ligand Met[83]. Distinct ligand resonances for the protonated and deprotonated proteins establish the existence of two structures which interconvert slowly on the NMR time scale. The spectra are obtained with a Nicolet 85 kG NMR spectrometer which is supported by NIH grant RR01077.

THE REMOVAL AND REPLACEMENT OF COPPER FROM CHINESE LACCASE

The first step in characterizing the metal binding properties of a peptide is to understand the removal and re-incorporation of the native metal ion. As it happens the literature regarding laccase is somewhat confused on this point.(12,18-21) On the other hand, the prevailing view is that metal ions are inserted post-translationally during *in vivo* synthesis, (22-24) and this argues that reconstitution should be possible.

In the case of Chinese laccase we have found that dialysis against a pH 7.2 solution nominally 50 mM in cyanide completely removes copper.(25) Not surprisingly since a type 1 site is known to be present, we can titrate one free sulphydryl group from the apoprotein. If the apoprotein is incubated anaerobically in the presence of a slight excess of copper(I) and then exposed to oxygen, the solution rapidly develops a blue color. Analysis shows that the crude re-metalated protein is generally somewhat low in copper, but exhibits significant absorbance at 614 and 330 nm as well as EPR signals from type 1 and type 2 copper. Using a standard ferrocyanide assay, it exhibits an activity which is comparable to that of native protein which further suggests that substantial restoration of all sites has been achieved. The crude re-metalated protein is a mixture of species which can be partially resolved by column chromatography. In so doing we find that the major portion of protein elutes with the same retention time as a native control, exhibits the correct metal content and gives a native-like EPR spectrum (Figure 2). However, even this material appears to be somewhat heterogeneous; in particular, some of the

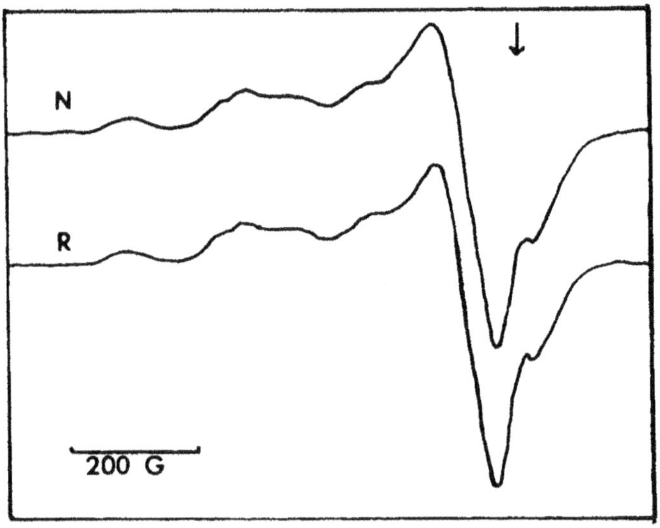

Figure 2. X-Band EPR Spectra of Native(N) and Re-metallated Laccase(R). The marker designates the g=2 position

protein present does not give the spectral properties of a type 1 copper in its native environment.(25) Although we are continuing our efforts to improve the procedure, the current one is judged to be reasonably successful, and we have begun attempts to prepare mixed-metal hybrid derivatives of laccase.

THE PREPARATION OF A MIXED-METAL HYBRID OF LACCASE

From the point of view of coordination chemistry there can be no doubt that other metal ions besides copper will bind to the protein. Indeed, Spiro and co-workers have already shown that cobalt(II) will bind at the type 1 site of tree laccase.(26) However, cobalt(II) has also been found to bind to the binuclear sites of tyrosinase (27) and hemocyanin, (28) sites which probably bear some resemblance to the type 3 site of laccase. If one attempts to prepare mixed-metal derivatives of laccase beginning with the apoprotein, the challenge will probably be to direct particular metal ions toward specific sites. Our first experiments have been designed to take advantage of the unique combination of pseudotetrahedral binding geometry with two sulfur donors which characterizes the type 1 site, properties which might form the basis for an intrinsic selectivity for the heavy metals of group IIB.

If a solution of apolaccase is treated with an excess of one unit of Hg(II) per mole of protein followed by three units of Cu(I) and is then exposed to oxygen, virtually no blue color develops nor does any significant activity according to our standard assay. At the same time, the major portion of the resulting protein binds to a cation-exchange column in the same way as native protein, suggesting that it has a full complement of metal ions; moreover, the analytical data are consistent with the formulation $HgCu_3P$ where P denotes the polypeptide.(29) So far, the redox state of all copper centers has not yet been fully defined, but the type 2 EPR signal is observed as is absorbance at 330 nm. Thus the preliminary results suggest that we have succeeded in preparing a hybrid derivative of laccase which contains Hg(II) in the type 1 site and copper in all others. Work currently underway is aimed at further characterizing this species and exploring its spectral, chemical and kinetic properties.

ACKNOWLEDGEMENTS

Our research has been supported by the National Institutes of Health through grant number GM 22764. The ^{1}H-NMR studies have been carried out in collaboration with E.L. Ulrich and J.L. Markley while the ^{113}Cd-NMR studies have been carried out in collaboration with J.D. Otvos.

REFERENCES

(1) Mayer, A.M. and Harel, E.: 1979, Phytochem. 18, pp. 193-215.

(2) Fee, J.A.: 1975, Struct. and Bond. 23, pp. 1-60.
(3) Malkin, R. and Malmström, B.G.: 1970, Adv. Enzymol. Relat. Subj. Biochem. 33, pp. 177-244.
(4) Tennent, D.L. and Mc Millin, D.R.: 1979, J. Am. Chem. Soc. 101, pp. 2307-2311.
(5) Gray, H.B. and Solomon, E.I.: 1981 in *Copper Proteins* (Spiro, T., Ed.) Wiley: New York, Chapter 1.
(6) Malkin, R.: 1973 in *Inorganic Biochemistry* (Eichhorn G., Ed.) Vol. II, Elsevier: New York, Chapter 21.
(7) Morpurgo, L., Graziani, M.T., Finazzi-Agrò, A., Rotilio, G., and Mondovì, B.: 1980, Biochem. J., 187, pp. 361-366.
(8) Petersson, L., Ångström, J., and Ehrenberg, A.: 1978, Biochim. Biophys. Acta, 526, pp. 311-317.
(9) Dooley, D.M., Scott, R.A., Ellinghaus, J., Solomon, E.I., and Gray, H.B.: 1978, Proc. Natl. Acad. Sci. USA, 75, pp. 3019-3022.
(10) Adman, E.T., Stenkamp, R.E., Seiker, L.C., and Jensen, L.H.: 1978, J. Mol. Biol., 123, pp. 35-47.
(11) Adman, E.T. and Jensen, L.H.: 1981, Israel J. Chem. 21, pp. 8-12.
(12) Hauenstein, B.L., Jr. and Mc Millin, D.R.: 1981 in *Metal Ions in Biological Systems* (Sigel, H., Ed.) Vol. XIII, Marcel Dekker: New York, Chapter 10.
(13) Blaszak, J.A., Ulrich, E.L., Markley, J.L., and Mc Millin, D.R., submitted for publication.
(14) Engeseth, H.R., Mc Millin, D.R., and Ulrich, E.L., submitted for publication.
(15) Holwerda, R.A., Wherland, S., and Gray, H.B.: 1976, Ann. Rev. Biophys. Bioeng. 5, pp. 363-396.
(16) Silvestrini, M.C., Brunori, M., Wilson, M.T., and Darley-Usmar, V.M.: 1981, J. Inorg. Biochem. 14, pp. 327-338.
(17) Engeseth, H.R., Mc Millin, D.R., and Otvos, J.D., to be published.
(18) Tissières, A.: 1948, Nature (London) 162, pp. 340-341.
(19) Omura, T.: 1961, J. Biochem. (Tokyo) 50, pp. 389-391.
(20) Ando, K.: 1970, J. Biochem. (Tokyo) 68, pp. 501-508.
(21) Malmström, B.G., Andreásson, L.-E., and Reinhammar, B.: 1975 in *The Enzymes* (Boyer, P.D., Ed.) vol. XII, Part B, Academic Press: New York, Chapter 8.
(22) Ruis, H.: 1979, Can. J. Biochem. 57, pp. 1122-1130.
(23) Rayton, J.K., Harris, E.D.: 1979, J. Biol. Chem. 254, pp. 621-626.
(24) Keyhani, E. and Keyhani, J.: 1975, Arch. Biochem. Biophys. 167, pp. 596-602.
(25) Morris, M.C., Hauenstein, B.L., Jr., and Mc Millin, D.R., submitted for publication.
(26) Larrabee, J.A., and Spiro, T.G.: 1979, Biochem. Biophys. Res. Commun. 88, pp. 753-760.
(27) Rüegg, Ch., and Lerch, K.: 1981, Biochemistry 20, pp. 1256-1262.
(28) Susuki, S., Kino, J., Kimura, M., Mori, W., and Nakahara, A.: 1982, Inorg. Chim. Acta 66, pp. 41-47.
(29) Morris, M.C. and Mc Millin, D.R., to be published.

REACTIONS OF RHUS VERNICIFERA LACCASE WITH AZIDE AND FLUORIDE. FORMATION OF A "HALF-MET-N_3-TYPE" BINDING.

L. Morpurgo, A. Desideri*, and G. Rotilio

C.N.R. Center of Molecular Biology, Institute of Biological Chemistry, University of Rome; Department of Physics, University of Calabria, Cosenza*, Italy.

ABSTRACT

The reactions of Rhus vernicifera laccase with N_3^- and F^- have shown that native enzyme samples are not homogeneous. A fraction of molecules contain reduced Type 3 Cu, unlike the major portion which is fully oxidized. Molecules with reduced Type 3 Cu bind N_3^- with higher affinity than the fully oxidized ones, giving an adduct with optical absorption bands at 400 and 500 nm and a very low $A_{\parallel}$ value of 100 Gauss in the EPR spectrum. The analogy of these properties with those of the N_3^- adducts of half-met-hemocyanin and half-met-tyrosinase suggests a similar type of binding, with N_3^- bridging a reduced copper of the Type 3 pair and the oxidized Type 2 Cu. This implies that the Type 2 and Type 3 Cu are close to each other at a distance not exceeding 5-6 A. Temperature effects on the conformation of laccase copper centers are discussed in relation to the properties of the fluoride derivatives.

Rhus vernicifera laccase is a multicopper oxidase that contains one "blue" paramagnetic Type 1 Cu(II), one paramagnetic Type 2 Cu(II) and a pair of magnetically coupled cupric ions, the Type 3 Cu(II) (1). The interactions of each type of copper with the solvent are not clear (2), although the Type 2 Cu is the best candidate for binding exogenous molecules (3,4). There are indications that also the Type 3 Cu may be involved in the binding of inhibitors (small anions such as N_3^-) (3,4) and oxygen reduction products such as H_2O_2 (5).

By optical absorption spectroscopy at least two reactions have been shown to occur between the native enzyme and N_3^-, their equilibrium constants differing by about two orders of magnitude (4,6). The reaction occurring at higher N_3^- concentrations produces a band at about 400 nm, assigned to a charge transfer $N_3^- \rightarrow Cu(II)$, quite usual with proteins containing a single copper ion open to solvent access. The reaction occurring at lower N_3^- concentrations produces two absorption bands at 400 and 500 nm, of relatively low intensity, approximately 500 M^{-1} cm^{-1}. Absorption bands at 500 nm have been so far only observed

I. Bertini, R. S. Drago, and C. Luchinat (eds.), The Coordination Chemistry of Metalloenzymes, 207–213.

in N_3^- derivatives of proteins containing binuclear copper centers, such as multicopper oxidases (6,7), half-met-hemocyanins (8), and half-met-tyrosinase (9). In tree laccase the intensity of the 500 nm band was found to increase upon partial reduction of the enzyme, under conditions very likely to involve electron donation to Type 3 Cu(II) (3).

1. Reactions with N_3^- as studied by EPR

A recent re-investigation of these reactions by EPR has shown that stoichiometric concentrations of N_3^- convert a small portion of the Type 2 Cu(II) signal into that of a species with low $A_{\parallel}$ value of 90-100 Gauss, whereas at higher N_3^- concentrations ($\simeq$0.1 M) the predominant signal due to Type 2 Cu(II) has $A_{\parallel}$ close to 135 Gauss (Fig. 1), a value that was confirmed by the 35 GHz spectrum. The EPR spectra recorded at alkaline pH (pH 7.8), where the low affinity N_3^- binding is strongly inhibited (6), unequivocally showed that only a small fraction of laccase Type 2 Cu(II) can be converted to the 90-100 Gauss $A_{\parallel}$ species, since addition of a large N_3^- excess, up to 200 times the enzyme concentration, did not cause additional decrease of the first hyperfine line of native Type 2 Cu(II) over that produced by less than stoichiometric N_3^- to give the 90-100 Gauss species. These results suggest that laccase samples are not homogeneous as far as the reaction with N_3^- is concerned.

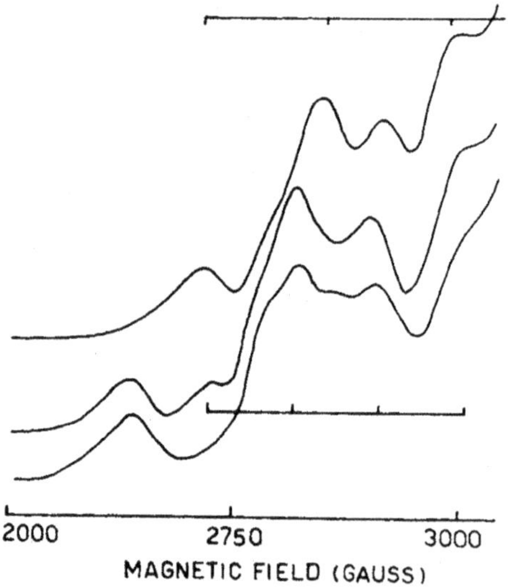

Figure 1. EPR spectra of native laccase (bottom curve); treated with one mole NaN_3 per mole enzyme (middle curve); treated with 0.1 M NaN_3 (upper curve). 0.5 mM enzyme in 0.1 M sodium acetate buffer pH 4.2. EPR conditions: microwave frequency 9.15 GHz, microwave power 20 mW, temperature 100 K. Only the $g_{\parallel}$ region of the spectrum is shown.

2. Reactions with N_3^- in the presence of redox reagents

A homogeneous pattern of reaction between laccase and N_3^- is observed upon treatment of the enzyme with weak oxydizing or reducing agents. Prior addition of 1-3 moles H_2O_2 per mole enzyme prevented the formation of the 400 and 500 nm absorbing species with 90-100 Gauss $A_{\parallel}$, upon addition of stoichiometric N_3^-. On addition of excess N_3^- all Type 2 Cu(II) was converted into the 135 Gauss $A_{\parallel}$ species, absorbing only at 400 nm (Fig. 2). On the other hand, anaerobic treatment in the presence of N_3^- with a weak reducing agent, such as hydroquinone at a stoichiometric ratio to the enzyme, converted all Type 2 Cu(II) EPR signal into the species with $A_{\parallel}$ = 100 Gauss (Fig. 3), irrespective of pH, in the pH range 4.2 - 7.8, and N_3^- concentration, in the range 1.5 - 200 moles per mole enzyme. It should be noted that the dashed curve of Fig. 3 represents a transient state of a slow reduction process, which eventually lead to almost complete reduction of Type 1 Cu and of some Type 2 Cu(II)-N_3 complex.

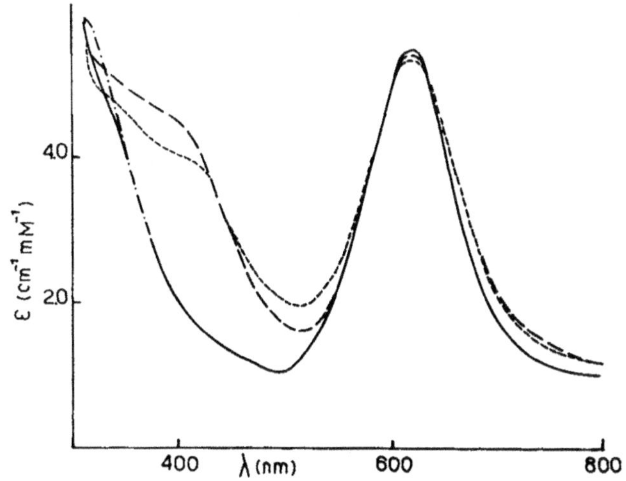

Figure 2. Optical absorption spectra of native laccase (——); treated with 67 mM NaN_3 (---); treated with stoichiometric H_2O_2 (— • —); treated with stoichiometric H_2O_2 and 67 mM NaN_3 (= = —). 0.13 mM enzyme in 0.05 M phosphate buffer pH 6.0.

Since the total integrated intensity of the EPR spectrum of native laccase was neither affected by H_2O_2 treatment nor by hydroquinone at the stage shown in Fig. 3, it seems safe to assume that only the Type 3 Cu was involved in possible redox reactions. This type of copper is EPR-silent and is monitored by the 330 nm band in the optical spectrum (1). In agreement with earlier findings (5), Fig. 2 shows that this band increases upon treatment with H_2O_2. The increase had been assigned to formation of a hydrogen peroxyde adduct of laccase (5), but in the light of the present results it may indicate re-oxidation of some Type 3 Cu that is in the reduced state in the native protein. Therefore it is reasonable to identify the two fractions of native laccase giving different N_3^- adducts, as molecules having the Type 3 Cu in different oxidation states, i.e. oxidized Type 3 Cu(II) in the major fraction and reduced Type 3 Cu(I) in the minor fraction. The former fraction gives rise with excess N_3^- to the 135 Gauss $A_{\parallel}$ species which is not detectable in the presence of reducing agents, while the latter one binds N_3^- at very low concentrations, producing the EPR signal with 100 Gauss $A_{\parallel}$ and the optical absorptions at 400 and 500 nm.

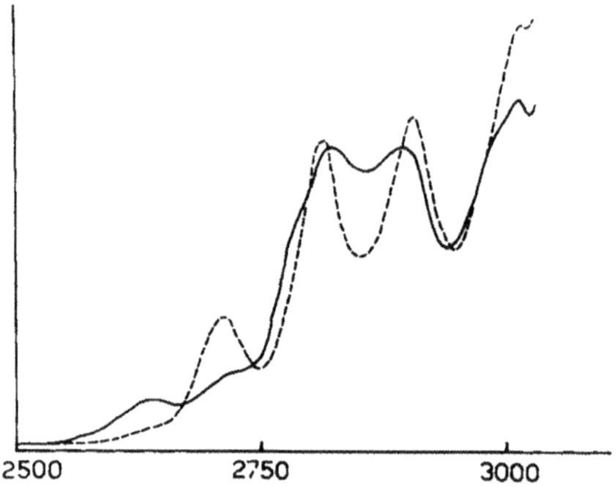

Figure 3. EPR spectra of native laccase treated with 12 mM NaN_3 (full line); one mole hydroquinone per mole enzyme added anaerobically and immediately frozen (dashed line). 0.4 mM enzyme in 0.05 M phosphate buffer pH 6.2. EPR conditions as in Fig. 1.

3. Half-met-N_3-type binding in laccase

The properties of the N_3^- derivative of laccase containing reduced Type 3 Cu, that is a high binding constant, an $A_{\parallel}$ value of about 100 Gauss, and the presence in the optical spectrum of a 500 nm band, are very similar to those exhibited by half-met-N_3-hemocyanins (8) and half-met-N_3-tyrosinase (9). In these compounds the band at 500 nm was shown to be diagnostic of N_3^- bridging one reduced and one oxidized copper ion. If a similar type of binding does also occur in laccase molecules with the Type 3 Cu reduced, the reaction with N_3^- of the two fractions present in the native enzyme can be schematically represented as follows, taking into account the fact that the Type 2 Cu(II) EPR signal is modified in any case:

$Cu_3(I)$ — N_3 — $Cu_2(II)$ $Cu_3(I)$		high affinity absorption bans at 400 and 500 nm $A_{\parallel}$ = 100 Gauss, $g_{\parallel}$ = 2.28
$Cu_3(II)$ $Cu_3(II)$	$Cu_2(II)$ | N_3	low affinity absorption band at 400 nm $A_{\parallel}$ = 135 Gauss, $g_{\parallel}$ = 2.24

where the subscripts refer to Type 3 and Type 2 Cu. The bridging N_3^- implies close proximity of Type 2 and Type 3 Cu in the 5-6 A range (8).

4. Reactions with fluoride

The effect of H_2O_2 on the chemical heterogeneity of native laccase samples can also be inferred by the fact that a well resolved superhyperfine pattern is produced by fluoride in the EPR spectrum of the H_2O_2-treated enzyme. The pattern is indicative of the binding in the equatorial plane of the Type 2 Cu(II) of one or two magnetically equivalent ^{19}F ($I = \frac{1}{2}$) nuclei, depending on F^- concentration (Fig. 4, curves a and b respectively), as already shown to occur in Polyporus laccase (10). The pattern is far less resolved in the native, untreated enzyme.

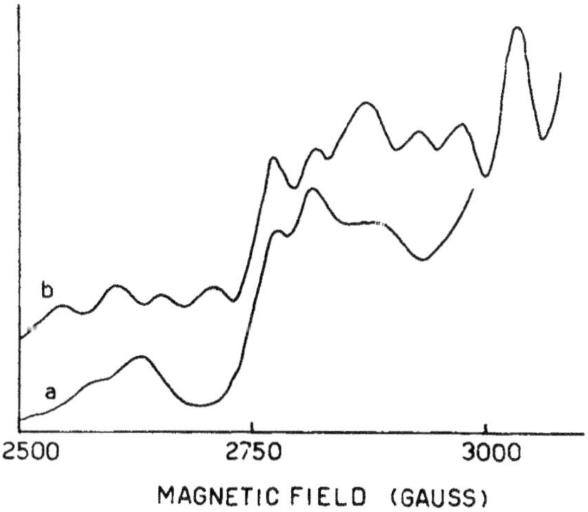

Figure 4. EPR spectra of H_2O_2-treated laccase in the presence of NaF. 0.33 mM enzyme in 0.05 M phosphate buffer pH 6.0 plus 2.3 mM NaF (curve a); plus 330 mM NaF (curve b). EPR conditions as in Fig. 1.

5. Temperature effects

The ^{19}F superhyperfine pattern is lost above -30 °C and the general shape of the EPR spectrum changes in the same fashion as that of the native enzyme (Table 1).

Table 1

Temperature dependence of native Laccase EPR parameters (11)

	Type 1 Cu(II)		Type 2 Cu(II)	
	$g_{\parallel}$	$A_{\parallel}(10^4 cm^{-1})$	$g_{\parallel}$	$A_{\parallel}(10^4 cm^{-1})$
77 °K	2.30	38	2.24	195
Room temp.	2.19	60	2.28	146

In the case of Type 1 Cu the change may be related to a different degree of distortion from regular tetrahedral geometry. In the case of Type 2 Cu the change suggests a shift of the metal coordination geometry from square planar, with possibly one or two more distant axial ligands, to trigonally distorted square pyramidal, a geometry that better accounts (11) for the high redox potential of Type 2 Cu (390 mV), very close to that of the flattened tetrahedral Type 1 Cu (430 mV). Apparently the square planar species can accomodate two fluorides in the equatorial plane, giving rise to the superhyperfine pattern.

At higher temperature the geometry is different, consistently with the changes observed in the native enzyme. Probably a single F^- is bound. At room temperature binding of fluoride is demonstrated by ^{19}F NMR relaxation (12) and other spectral studies (13). However excess fluoride has very small effects on the enhancement of water proton relaxivity produced by laccase (2,14). This may indicate that fluoride either binds without displacing any water molecule as it does in Cu(II) substituted carbonic anhydrase (15), or it displaces equatorial H_2O as in galactose oxidase (16), leaving an axial fast-exchanging H_2O molecule unaffected. In any case these results support the idea that the Type 2 Cu coordination site has a very high degree of flexibility.

REFERENCES

(1) Fee, J.A.: 1975, Struct. Bond., 23, pp.1-60
(2) Rigo, A., Orsega, E.F., Viglino, P., Morpurgo, L., Graziani, M.T., and Rotilio, G.: 1979, in "Metalloproteins. Structure, molecular function and clinical aspects", U. Weser, ed., pp. 25-35
(3) Morpurgo, L., Graziani, M.T., Desideri, A., and Rotilio, G.: 1980, Biochem. J., 187, pp. 367-370

(4) LuBien, C.D., Winkler, M.E., Thamann, T.J., Scott, R.A., Co, M.S., Hodgson, K.D., and Solomon, E.I.: 1981, J. Am. Chem. Soc., 103, pp. 7014-7016
(5) Farver, O. and Pecht, I.: 1981, in "Metal Ions in Biology. 3. Copper proteins", Spiro, T.G., ed., pp. 151-192.
(6) Morpurgo, L., Rotilio, G., Finazzi Agrò, A., and Mondovì, B.: 1974, Biochim. Biophys. Acta, 336, pp. 324-328
(7) Mondovì, B., Avigliano, A., Rotilio, G., Finazzi Agrò, A., Gerosa, P., and Giovagnoli, C.: 1975, Mol. Cell. Biochem., 7, pp. 131-135
(8) Himmelwright, R.S., Eichman, N.C., LuBien, C.D., and Solomon, E.I.: 1980, J. Am. Chem. Soc., 102, pp. 5378-5388
(9) Himmelwright, R.S., Eickman, N.C., LuBien, C.D., Lerch, K., and Solomon, E.I.: 1980, J. Am. Chem. Soc., 102, pp. 7339-7344
(10) Malkin, R., Malmström, B.G., and Vänngård, T.: 1968, FEBS Lett., 1, pp. 50-54
(11) Morpurgo, L., Calabrese, L., Desideri, A., and Rotilio, G.: 1981, Biochem. J., 193, pp. 639-642
(12) Rigo, A., Viglino, P., Argese, E., Terenzi, M.: 1979, J. Biol. Chem., 254, pp. 1756-1758
(13) Bränden, R., Malmström, B.G., and Vänngård, T.: 1973, Eur. J. Biochem., 36, pp. 195-200
(14) Goldberg, M., Vuk-Pavlović, S., and Pecht, I.: 1980, Biochemistry, 19, pp. 5181-5189
(15) Bertini, I., Canti, G., Luchinat, C., and Scozzafava, A.: 1978, J. Chem. Soc. Dalton, pp. 1269-1273
(16) Marwedel, B.J., Kosman, D.J., Bereman, R.D., and Kurland, R.J.: 1981, J. Am. Chem. Soc., 103, pp. 2842-2847

STRUCTURAL MAGNETIC CORRELATIONS IN EXCHANGE COUPLED SYSTEMS

Dante Gatteschi

Istituto di Chimica Generale e Inorganica, Facoltà di Farmacia, University of Florence, Florence, Italy

ABSTRACT

The formalism relevant to exchange coupled pairs is reviewed. The relations connecting the spin hamiltonian parameters, g, A, D, ..., to those of the individual ions are provided, and their validity is shown through some examples. The isotropic exchange coupling constant is discussed for series of complexes as a function of the electronic structure of the metal ions and of the nature of the bridge ligands.

1. SPIN HAMILTONIAN APPROACH TO EXCHANGE COUPLING

When two metal atoms are interacting to the extent that they are no longer to be considered individually, but not yet fully bound by a strong covalent bond, it is said that they are exchange coupled. In other terms exchange coupling is a form of weak covalent bonding which can be determined either by direct interaction (direct exchange) or by the overlap of the metal orbitals with the orbitals of intervening ligands (indirect exchange or super-exchange). It is the latter type of interaction which will be the matter of this article.

There are several approaches possible to calculate the energy levels of the exchange coupled pairs, but the simplest one is the spin hamiltonian which allows one to interpret the spectral and magnetic properties of interacting metal ions. This simple approach is valid when the ground states of the two metal orbitals are orbitally non degenerate (1).

For most purposes the spin hamiltonian can be written in the form:

$$\hat{H} = \underline{\hat{S}}_1 \cdot \underline{\underline{J}} \cdot \underline{\hat{S}}_2 \qquad |1|$$

where $\underline{\underline{J}}$ is a matrix connecting the two spin operators $\hat{S}_1$ and $\hat{S}_2$ which refer to metal atom 1 and 2 respectively. It can be easily shown that the hamiltonian |1| can be rewritten in an equivalent way (2), according to:

I. Bertini, R. S. Drago, and C. Luchinat (eds.), The Coordination Chemistry of Metalloenzymes, 215–228.

$$\hat{H} = J\,\hat{\underline{S}}_1 \cdot \hat{\underline{S}}_2 + \hat{\underline{S}}_1 \cdot \underline{\underline{D}} \cdot \hat{\underline{S}}_2 + \underline{d} \cdot \hat{\underline{S}}_1 \times \underline{S}_2 \qquad |2|$$

The first term in |2| is often referred to as the isotropic or Heisenberg exchange. When it is the leading term of the hamiltonian |2| , i.e. when $J \gg D, d$ then J reflects the weak bonding interaction between the two metal ions. It will have different expressions according to the different models within which it is evaluated (see below). In the following we will always assume implicitly valid the assumption that J is the leading term in |2|.

The effect of $J\hat{\underline{S}}_1 \cdot \hat{\underline{S}}_2$ on the energy levels of the dinuclear species will be that of grouping them according to the angular momentum addition rules:

$$|S_1 - S_2| \leq S \leq S_1 + S_2 \qquad |3|$$

the energies of the S states are given by:

$$E(S) = \frac{J}{2}|S(S+1) - S_1(S_1+1) - S_2(S_2+1)| \qquad |4|$$

The second term of the spin hamiltonian |2| , $\hat{\underline{S}}_1 \cdot \underline{\underline{D}} \cdot \hat{\underline{S}}_2$, is often referred to as anisotropic exchange and is given by the sum of two contributions

$$\underline{\underline{D}} = \underline{\underline{D}}^{ex} + \underline{\underline{D}}^{dip} \qquad |5|$$

$\underline{\underline{D}}^{dip}$ is determined by the through-space magnetic interaction between the dipole moments centered on metal 1 and 2 respectively, while $\underline{\underline{D}}^{ex}$ is brought about by spin orbit coupling and is physically determined by the exchange interaction between one ion in its ground state and the other one in an excited state (3). Since this exchange contribution depends on spin orbit coupling mixing it is expected to be in any case much smaller than the isotropic contribution. The anisotropic term is expected to yield a zero field splitting of the S-states, for $S>1$.

The third term, $\underline{d} \cdot \hat{\underline{S}}_1 \times \hat{\underline{S}}_2$, often referred to as either antisymmetric or Dzialonshinsky-Moriya exchange, (2,4) is also determined by spin orbit coupling admixture of excited states into the ground state. It is as yet the less well characterized term (5), also because it vanishes by symmetry in the case of two identical metal ions.
When $J \gg D$, S will be a good quantum number and all the S states will be characterized by spin hamiltonian parameters which can be calculated by projecting according to standard procedures (6) the correspinding spin hamiltonian parameters of the individual ions onto the S manifold;

$$\begin{aligned} \underline{\underline{g}}_c &= c_1 \underline{\underline{g}}_1 + c_2 \underline{\underline{g}}_2 \\ \underline{\underline{A}}_{c1} &= c_1 \underline{\underline{A}}_1 \\ \underline{\underline{A}}_{c2} &= c_2 \underline{\underline{A}}_2 \\ \underline{\underline{D}}_c &= c_3 D_1 + c_4 \underline{\underline{D}}_2 + c_5 \underline{\underline{D}} \end{aligned} \qquad |6|$$

where 1 and 2 refer to the two metal ions, g_i, A_i, D_i are the Zeeman, nuclear hyperfine, and zero field splitting tensors of ion i, D is defined in |1| . The c_1, c_2, c_3, c_4, c_5 coefficients are defined as:

$$c_1 = |S(S+1) + S_1(S_1+1) - S_2(S_2+1)|/2S(S+1)$$

$$c_2 = |S(S+1) + S_2(S_2+1) - S_1(S_1+1)|/2S(S+1).$$

$$c_3 = |1/S(2S-1)|\{|3|S(S+1)+S_1(S_1+1)-S_2(S_2+1)|^2 - 4(S+1)^2 S_1 (S_1+1)|/4(S+1)^2 + 3|(S+1)^2 - (S_1-S_2)||(S_1+S_2+1)^2 - (S+1)^2|/(4S+1)^2(2S+3)\} \qquad |7|$$

$$c_4 = |1/S(2S-1)|\{|3|S(S+1) + S_2(S_2+1) - S_1(S_1+1)|^2 - 4(S+1)^2 S_2 (S_2+1)|/4(S+1)^2 + 3|(S+1)^2 - (S_1-S_2)^2||(S_1+S_2+1)^2 - (S+1)^2|/(4S+1)^2(2S+3)\}$$

$$c_5 = (1-c_3-c_4)/2$$

The values of the coefficients in the case of $S_1 = 1/2$, S_2 ranging from 1/2 to 5/2 are given in Table I.

TABLE I. The Coupling Coefficients for Pairs Containing One S = 1/2 Ion.

S_1	1/2	1/2	1/2	1/2	1/2	1/2	1/2	1/2	1/2
S_2	1/2	1	1	3/2	3/2	2	2	5/2	5/2
S	1	3/2	1/2	2	1	5/2	3/2	3	2
c_1	1/2	1/3	-1/3	1/4	-1/4	1/5	-1/5	1/6	-1/6
c_2	1/2	2/3	4/3	3/4	5/4	4/5	6/5	5/6	7/6
c_3	--	--	--	--	--	--	--	--	--
c_4	--	1/3	--	1/2	3/2	3/5	7/5	4/6	8/6
c_5	1/2	1/3	--	1/4	1/4	1/5	-1/5	1/6	-1/6

A perusal of Table I shows that in the case of two identical metal ions the g values of the pair are identical to the values of the single ion. This has been verified experimentally in several cases. A representative example is that of $Cu_2(pyO)_2Cl_4(H_2O)_2$ where pyO is pyridine-N-oxide, whose structure (7) is shown below

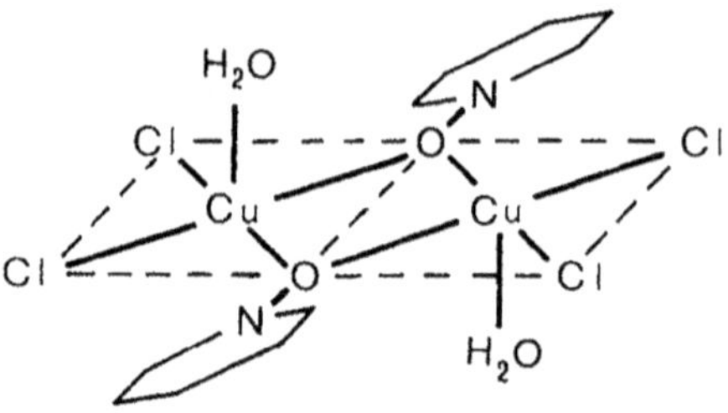

The coupling was found to be fairly strong and antiferromagnetic. The EPR spectra of the triplet excited state yielded (8) g_z=2.32, g_y=2.08, g_x= 2.06; D_{zz}=-0.102, D_{yy}=0.022, D_{xx}=0.081 cm^{-1}. If in the preparation of the copper compound some zinc chloride is added, in the solid some copper-zinc pairs are formed. Since the copper-copper species are antiferromagnetically coupled, at low temperature the host lattice becomes diamagnetic and only the copper-zinc heterodinuclear species will be detected by EPR spectroscopy. The spectra are those of a S=1/2 system, with g values which are experimentally indistinguishable from those of the Cu-Cu species.

It is also possible to substitute some manganese for the zinc, forming copper-manganese pairs (9,10). Since Mn has S_2=5/2, the resulting states are S=2 and S=3. The single crystal EPR spectra show four transitions as expected for an S=2 state split in zero field. The hyperfine splitting is determined by both the Mn and Cu nuclei. The principal g and A values are given in Table II.

Table II. Spin Hamiltonian Parameters for $(CuM)(pyO)_2Cl_4(H_2O)_2$ Complexes.

	Cu-Cu	Cu-Mn (S=2) Exp.	Cu-Mn Calc.	Cu-Zn
g_z	2.323	1.954	1.95	2.32
g_y	2.076	1.989	1.99	2.08
g_x	2.063	1.993	1.99	2.06
A_z,Cu	-----	23	23	139
A_y,Cu	-----	<7		
A_x,Cu	-----	<3		
A_z,Mn	-----	81		
A_y,Mn	-----	89		
A_x,nn	-----	89		

Using Table I, the g values for the copper ion, and assuming g=2 for the manganese ion, the experimental values are perfectly reproduced.

In the $M(prp)_2en\ Mn(hfa)_2$ complex, (11) ($(prp)_2en$ is the Schiff base formed by 2-hydroxypropiophenone and 1,2-diaminepropane, hfa is hexafluoroacetylacetonate, M=Ni,Cu) it was possible to measure the zero field splitting parameters for the Ni-Mn complex, (12) yielding the single ion D value, $D_{5/2}=0.049\ cm^{-1}$, since the nickel ion is diamagnetic. For the Cu-Mn complex D was measured for the S=2 and S=3 states. $D_2=0.034\ cm^{-1}$, $D_3=0.047\ cm^{-1}$. Using Table I we find that the following relations should hold:

$$D_2 + D_3 = 2\, D_{5/2}$$

The equation is not verified by the experimental values. The origin of the difference between calculated and expected values is not yet explained.

A possible application of the relations | 6 | is that of obtaining the spin hamiltonian parameters, and therefore information on the electronic structure, of metal ions which do not yield directly EPR spectra. For instance it is well known that nickel(II) ions, in symmetry lower than octahedral, do not give any EPR signal due to large zero-field splittings and/or unfavourable spin lattice relaxation. However from the spectra of Cu-Ni pairs, knowing the values of g for the copper ion, it was possible to calculate the g values of the nickel ions, as shown in Table III.

Table III. g Values for Nickel(II) Complexes Calculated from the Spectra of Copper-Nickel Pairs.

	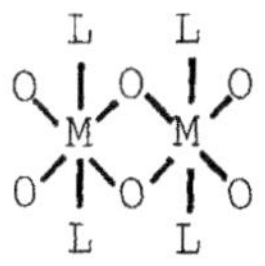	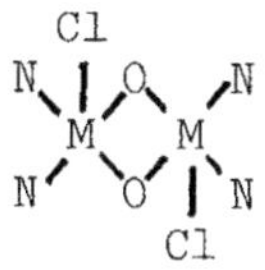	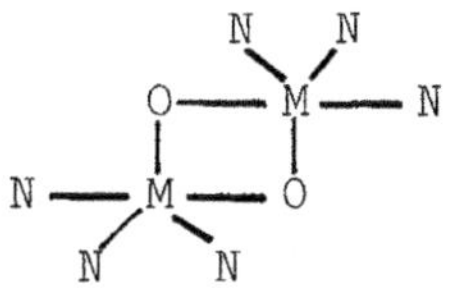
Ref.	13	14	15
Cu-Zn	2.34	2.24	2.25
	2.08	2.08	2.16
	2.08	2.05	2.02
Cu-Ni	2.15	2.09	2.75
	2.21	2.41	2.02
	2.25	2.49	2.02
Ni	2.18	2.12	2.61
	2.20	2.32	2.09
	2.21	2.38	2.06

It is apparent that the calculated values reflect the symmetry of the nickel complex, with quasi isotropic g values for hexacoordinated, more anisotropic values for square pyramidal complexes, and very highly anisotropic values for trigonal bipyramidal complexes.

2. MOLECULAR ORBITAL APPROACH TO EXCHANGE COUPLING.

In the assumptions we mentioned the most relevant parameter of the spin hamiltonian is J, which must be related to the electronic structures of the M and M' complexes, to the nature of the bridge ligands to the topology of the bridge. This can be done through the comparison of fundamental theories with the energy levels calculated through the spin hamiltonian approach.

The first successfull approach to the calculation of the energy levels of coupled metal pairs is due to Anderson (16) who developed his superexchange model within a Valence Bond formalism. The description of the model is beyond the scope of the present article and I will only give an extremely schematic and qualitative outline. The unpaired electrons on each metal center are assigned to an orbital, which may be named magnetic orbital. Goodenough (17) and Kanamori (18) gave some simple symmetry rules for predicting the sign of the exchange integrals: the overlap of the magnetic orbitals couples the spins antiparallel in the low energy state, while orthogonality of the magnetic orbitals couples the spins parallel in the low energy state. Overlap means that the two magnetic orbitals can overlap effectively to the same ligand orbital while orthogonality means that one magnetic orbital overlaps one ligand orbital which is orthogonal to the second metal magnetic overlap, as shown below

The mechanism for ferromagnetic coupling is most effective if the two magnetic orbitals have large areas of overlap.

In the last years there have been also several attempts to translate the superexchange formalism into the Molecular Orbital approach (19, 20), essentially because in the latter model it is much simpler to perform calculations. It must be mentioned however that the Molecular Orbital approach suffers of many theoretical limitations for the description of the electronic structure of weakly interacting centers, the principal being that it understimates the electron correlation, in the sense that at the lowest level of approximation it considers as equally probable covalently bound states, M-M', and ionic states in which one electron has been transferred from one metal to another. Although in principle it is possible to overcome this inconvenience, this can be done only with configuration interaction calculations, which for the molecules of interest requires computer time beyond any tolerable limit. It is possible however to use perturbation treatments which allow to reduce considerably the computer time needed.

Recently an ab initio molecular orbital calculation has been performed on copper acetate hydrate (21). The J value was calculated as the sum of two contributions J_F and J_{AF}. The ferromagnetic contribution, J_F, which is given by the exchange integrals between two localized molecular orbitals, one mainly centered on one metal and the second on the other, has a numerical value of $-234\ cm^{-1}$. The antiferromagnetic contribution, J_{AF} depends essentially on the difference of energy, Δ, between the two molecular orbitals which are formed by the two magnetic orbitals, and has a numerical value of $478\ cm^{-1}$. The calculated J value

$$J = J_{AF} + J_F$$

is $244\ cm^{-1}$ in fair agreement with the experimental value of $286\ cm^{-1}$.

A computationally simpler procedure (22) has been used to calculate the extent of coupling between two iron centers in the compound $Fe_2S_2(SH)_4{}^{n-}$

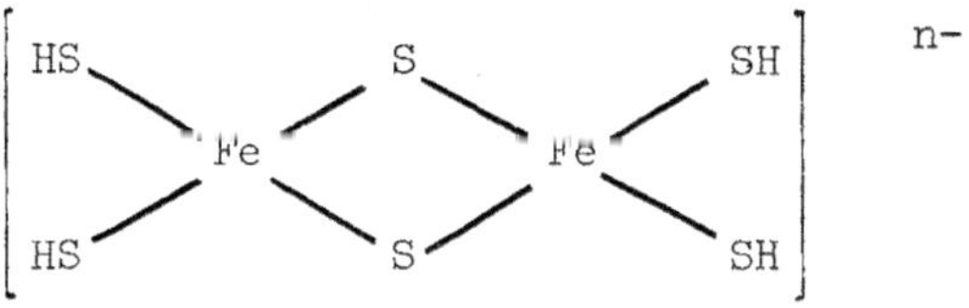

where n=2,3 , which can be considered as a model of 2-Fe ferrodoxins. The used model was called Xα-Valence Bond, which through the use of the SCF Xα molecular orbital model allows the fast calculation of the energy levels. The calculated J values compare satisfactorily with those experimentally observed in model compounds and in the real proteins.

Although these recent developments show that the possibility of calculating the exchange coupling integrals begins to become real, it is still necessary to use simplified models in a semiquantitative way in order to relate the experimental J values to the structural and electronic characteristic of the metal complexes. Extended Hückel calculations have been performed in order to show that much can be learnt about the extent of antiferromagnetic coupling in series of similar complexes just looking at the difference in energy, Δ, between the two molecular orbitals which are formed by the two magnetic orbitals (20). Similar results can be reached also using an Angular Overlap Model (23).

As an example of the possibility of relating the J values to structural features, di-μ-oxo-bridged copper(II) complexes appear particularly well suited due to the large number of examples where both the X-ray crystal structure and the magnetic data are available (24). In many cases the structures can be referred to the simple scheme

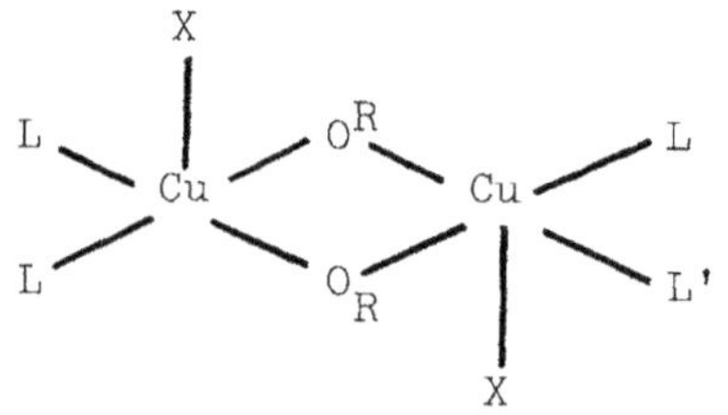

i.e. the two copper ions are either in a square planar or in a square pyramidal coordination environment such that the magnetic orbitals are x^2-y^2 in nature, as shown below

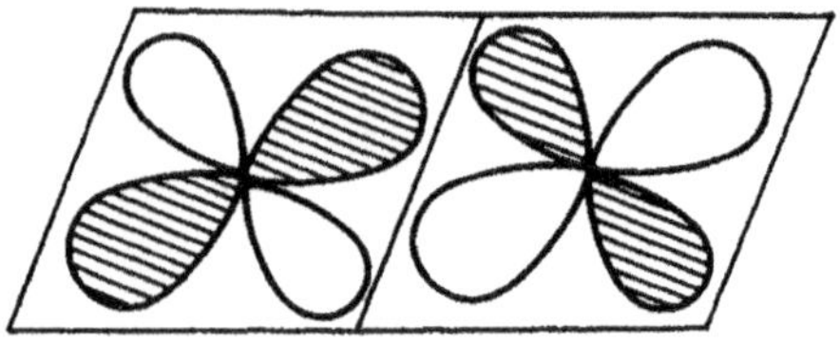

The extent of the exchange coupling is expected to depend on the Cu-O-Cu angle, ϕ and on the nature of the bridge ligand. In Figure 1 the experimentally observed J values are plotted versus ϕ. It is apparent that the di-hydroxo bridged complexes follow a linear relationship the coupling being ferromagnetic for $\phi < 96°$. When R is different from H there is no longer a linear relationship, but at any rate it seems safe to state that an increase in ϕ tends to make the coupling more antiferromagnetic. This is to be expected since as ϕ deviates from 90° the two magnetic orbitals are no longer orthogonal, the extent of the overlap increasing as ϕ becomes larger. From Figure 1 we also learn that the alkoxo are more effective than the hydroxo ligands to transmit the exchange interaction and the

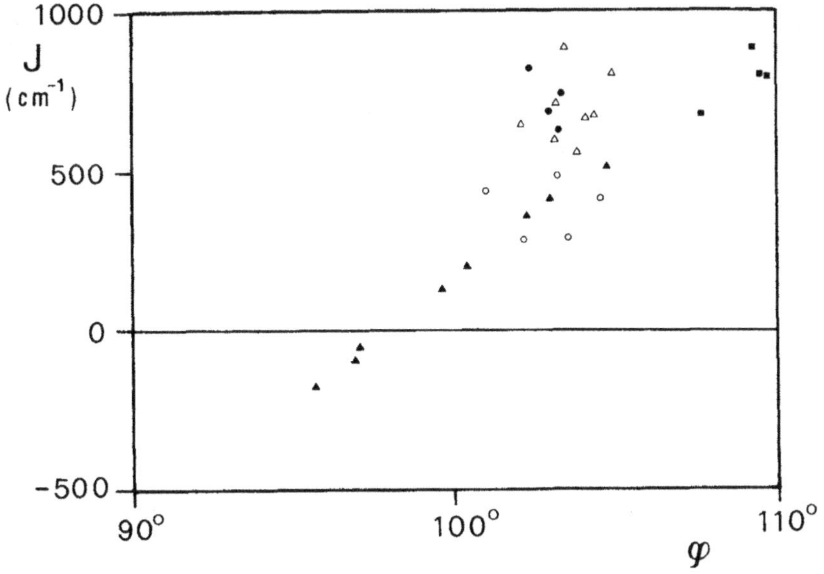

Figure 1. The dependence of J on the Cu-O-Cu angle, ϕ, for di-μ-oxo bridged copper(II) complexes. ▲ hydroxo, △ alkoxo, ○ phenoxo, ● keto, ■ N-oxo bridging ligands.

keto ligands seem to be similar to the former. Phenoxo and N-oxo ligands on the other hand seem to be similar to the hydroxo even if the N-oxo give very strongly coupled systems, presumably due to the large ϕ angles which were observed in their complexes.

Another interesting series to illustrate the dependence of the coupling on the electronic structure of the complexes is provided by CuM $(fsa)_2en \cdot XY$, where $(fsa)_2en$ is N,N'-(2-hydroxy-3-carboxybenzylidene)-1,2-diaminoethane, X and Y are additional ligands coordinated as shown below

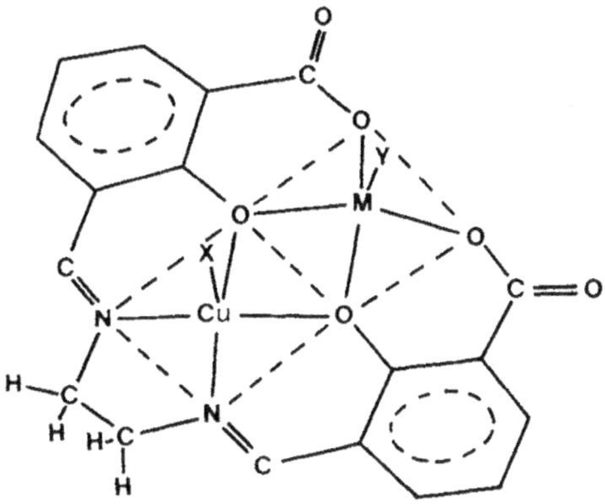

The copper ion in any case occupies the N_2O_2 site. When M=Cu the coupling is fairly large, 660 cm^{-1}, as expected due to the large Cu-O-Cu angle (30). When a nickel(II) ion is substituted into the O_4 site the J value decreases to 150 cm^{-1} (31). Since the two unpaired electrons on the nickel center are to a good approximation in x^2-y^2 and z^2 orbitals the experimental J value can be decomposed as

$$J = \frac{1}{2} (J_{x^2-y^2,x^2-y^2} + J_{x^2-y^2,z^2}) \qquad |8|$$

|8| is obtained from the general expression

$$J = \frac{1}{4 S_1 S_2} \Sigma_{a,b} J_{a,b} \qquad |9|$$

where the sum is over all the magnetic orbitals of the ground configuration on the two centers 1 and 2. a refers to center 1 and b to 2. $J_{x^2-y^2,x^2-y^2}$ is expected to be antiferromagnetic, while since x^2-y^2 and z^2 are orthogonal, $J_{x^2-y^2,z^2}$ can be either weakly ferromagnetic or non magnetic. Thus the J value much smaller in the Cu-Ni pair as compared to Cu-Cu appears to be justified.

When more unpaired electrons are present on the metal M in general only the x^2-y^2, x^2-y^2 pathway remains effective to yield antiferromagnetic coupling, while all the others are either ferromagnetic or non magnetic. The ferromagnetic coupling becomes dominant with M=VO (IV) (32) or Cr (III), (33), since in these cases one unpaired electron is an xy orbital which is orthogonal to x^2-y^2. The coupling between the two is expected to be fairly effective since the two orbitals have large overlap areas.

This relatively simple framework becomes definitely much more complicated when at least one of the two ions has an orbitally degenerate ground state. This can occur for instance with octahedral ions which have T ground levels, such as high spin cobalt(II), low spin iron(III) etc. It does not occur generally with ions having E ground levels since in this case the Jahn-Teller effect is expected to be strong enough to remove the degeneracy. The assumptions of the spin hamiltonian formalism break down and the exchange interaction can by no means be expressed with only one parameter. Although several theoretical models are available for the interpretation of the coupling, the main difficulty at present is that of collecting enough experimental data to obtain meaningful values for the parameters.

We have found (34,35) that the EPR spectra can be a powerful tool for measuring exchange interactions between an orbitally degenerate and an orbitally non degenerate ground state. We studied Ni-Co pairs in octahedral complexes, where the nickel(II) ion has a ground $^3A_{2g}$ level and the cobalt ion has $^3T_{1g}$. The EPR spectra of these pairs, recorded at 4.2 K show that a Kramers doublet is the ground level. The principal g values for the two complexes we studied are given in Table IV. In the trik complex the two metal ions are connected through monoatomic oxygen bridges, while in the dhph the bridge is biatomic.

TABLE IV. Experimental and Calculated g Values of Nickel(II)-Cobalt(II) Pairs in the trik and dhph Complexes.[a]

	trik		dhph	
	exp.	calcd.	exp.	calcd.
g_z	2.1	2.1	0.6	0.6
g_x	1.2	1.2	0.9	1.0
g_y	0.3	0.5	2.1	2.1
J_{A1}		4 cm^{-1}		30 cm^{-1}
J_{A2}		45		30
J_{B2}		45		30

[a] trik = dimetal(II)bis(1,5-diphenyl-1,3,5-pentanetinate)tetrakis pyridine;dhph = diaquo-1,4-dihydrazinephtalazine metal(II) chloride hydrate

while Figure 2 shows the principal directions. Although the two sets of

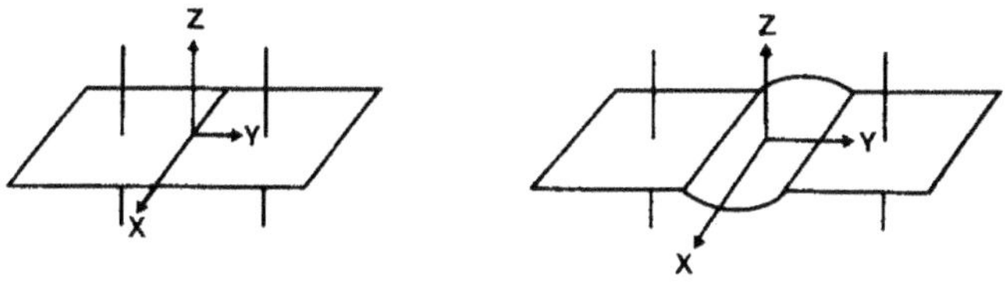

Figure 2. Schematic orientation of the principal g axes in the trik (left) and dhph (right) complexes.

g values are not too dissimilar there is not correspondence along the principal directions: the largest g value of 2.1 is found close to z for the trik and close to y for the dhph complex.

The spectra were analyzed through the hamiltonian

$$\hat{H} = \hat{H}_{Co} + \hat{H}_{Ni} + \hat{H}_{Co-Ni} \qquad |10|$$

$\hat{H}_{Co}$ and $\hat{H}_{Ni}$ are the single ion hamiltonians and $\hat{H}_{Co-Ni}$ has the form (13)

$$\hat{H}_{Co-Ni} = \underline{\hat{J}}\, \underline{\hat{S}}_1 \cdot \underline{\hat{S}}_2 \qquad |11|$$

where $\hat{J}$ is now an orbital operator, which has three components, corresponding to the orbital interaction of the ground A_{2g} state of nickel with the three symmetry split levels of cobalt T_{1g}. According to the approximate C_{2v} symmetry of the complex the three J parameters are labelled as J_{A1}, J_{A2} and J_{B2}. The g values of the pair were calculated by diagonalizing the hamiltonian matrix within 36x36 matrix of $^3A_{2g}X^4T_{1g}$ and applying the appropriate Zeeman operator within the lowest Kramers doublet. The calculated J values are given in Table 5. It can be seen that the J values are always antiferromagnetic. In the dhph complex they are very similar to each other while in the trik complex one value is markedly smaller than the others. An analysis of the exchange pathways shows that in J_{A1} is present the $J_{x^2-y^2,xy}$ integral which according to established results (32) is ferromagnetic. It can therefore reduce J values in the trik complex. In the dhph complex the two-atom bridge must reduce the extent of the overlap between the x^2-y^2 and xy orbital, thus making the ferromagnetic pathway less effective.

REFERENCES

(1) Dirac, P.A.M.: 1926, Proc. Roy. Soc., A112, p. 661; 1929, A123, p. 714; Heisenberg, A.: 1926, Z. Physik, 38, p. 441; Van Vleck, J.H.: 1934, Phys. Rev., 45, p. 405

(2) Moryia, T. in Magnetism, ed. by Rado, G.T.; Suhl, H., Vol. 1, Academic Press, New York, 1963

(3) Kanamori, J. in Magnetism, ed. by Rado, G.T.; Suhl, H., Vol. 1, Academic Press, New York, 1963, p. 161

(4) Dzialoshinsky : 1958, J. Phys. Chem. Solids, 4, p. 241

(5) Bencini, A.; Gatteschi, D.: Mol. Phys., in press

(6) Scaringe, J.; Hodgson, D.J.; Hatfield, W.E.; 1978, Mol. Phys., 35, p. 701

(7) Estes, E.D.; Hodgson, D.J.: 1976, Inorg. Chem., 15, p. 348

(8) Kokoszka, G.F.; Allen, H.C., Jr.; Gordon, G.: 1967, J. Chem. Phys., 46, p. 3020

(9) Paulson, J.A.; Krost, D.A.; McPherson, G.L.; Rogers, R.D.; Atwood, J.L.: 1980, Inorg. Chem., 19, p. 2519

(10) Bulluggiu, E.: 1980, J. Phys. Chem. Solids, 41, p. 1175

(11) O'Connor, C.J.; Freybey, D.P.; Sinn, E.: 1979 , Inorg. Chem., 18, p. 1077

(12) Banci, L.; Bencini, A.; Gatteschi, D.: 1981, Inorg. Chem., 20, p. 2734

(13) Banci, L.; Bencini, A.; Benelli, C.; Dei, A.; Gatteschi, D.: 1981, Inorg. Chem., 20, p. 1399

(14) Banci, L.; Bencini, A.; Gatteschi, D.; Dei, A.: 1979, Inorg. Chim. Acta, 36, p. L419
(15) Banci, L.; Bencini, A.; Dei, A.; Gatteschi, D.: 1981, Inorg. Chem., 20, p. 393
(16) Anderson, P.W. in Solid State Physics, ed. by Seitz, F.; Tumbull, D., 1963, Academic Press, New York, p. 99
(17) Goodenough, J.B., Magnetism and Chemical Bond, Interscience Publishers Inc., New York, 1963
(18) Kanamori, J.: 1959, Phys. Chem. Solids, 10, p. 87
(19) Kahn, O.; Briat, B.: 1976, JCS Faraday II, 72, p. 268
(20) Hay, P.J.; Thibeault, J.C.; Hoffman, R.: 1975, J. Am. Chem. Soc., 97, p. 4884
(21) de Loth, P.; Cassoux, P.; Daudey, J.P.; Malrien, J.P.: 1981, J. Am. Chem. Soc., 103, p. 4007
(22) Norman, J.G., Jr.; Ryan, P.B.; Noodleman : 1980, J. Am. Chem. Soc., 102, p. 4279
(23) Bencini, A.; Gatteschi, D.: 1978, Inorg. Chim. Acta, 31, p. 11
(24) Hodgson, D.J.: 1975, Progr. Inorg. Chem., 19, p. 173
(25) Lewis, D.L.; McGregor, K.T.; Hatfield, W.E.; Hodgson, D.J.: 1974, Inorg. Chem., 13, p; 1013; Lewis, D.L.; Hatfield, A.E.; Hodgson, D.J.: 1972, Inorg. Chem., 11, p. 2216; Casey, A.T.; Hoskins, B.F.; Whillans, F.D.: 1970, JCS Chem. Comm., p. 904; Estes, E.D.; Hatfield, W.E.; Hodgson, D.J.: 1974, Inorg. Chem., 13, p. 1654; Mitchell, T.P.; Bernard, W.H.; Wasson, J.R.: 1070, Acta Cryst., B26, p. 2096; Mayeste, R.J.; Mayers, E.A.: 1970, J. Phys. Chem., 74, p. 3497; Toofan, M.; Boushehri, A.; Ul-Haque, M.: 1976, JCS Dalton, p. 217; Crawford, V. M.; Richardson, H.W.; Wasson, J.R.; Hodgson, D.J.; Hatfield, W.E.: 1976, Inorg. Chem., 15, p. 2107
(26) Bertrand, J.A.; Kirkwood, C.E.: 1972, Inorg. Chim. Acta, 6, p. 248; Willett, R.D.; Breneman, G.L.: Inorg. Chem., in press; Timmons, J. H.; Martin, J.W.L.; Martell, A.E.; Rudolf, P.; Clearfield, A.; Loeb, S.J.; Willis, C.J.: 1981, Inorg. Chem., 20, p. 181; Le May, H.S., Jr.; Hodgson, D.J.; Pruettiang Kura, P.; Theriot, L.J.: 1979, JCS Dalton, p. 781; Nassif, P.J.; Boyko, E.R.; Thompson, L.D.: 1974, Bull. Chem. Soc. Japan, 47, p. 2321; Merz, L.; Haase, W.: 1978, Acta Cryst., B34, p. 2128; Haase, W.: 1973, Chem. Ber., 106, p. 3132
(27) Blake, A.B.; Fraser, L.R.: 1974, JCS Dalton, p. 2554; Lintvedt, R. L.; Glick, M.D.; Tomlonovic, B.K.; Gavel, D.P.; Kuszaj, J.M.: 1976, Inorg. Chem., 15, p. 1633; Heeg, M.J.; Mack, J.L.; Glick, M.D.; Lintvedt, R.L.: 1981, Inorg. Chem., 20, p. 833; Guthrie, J.W.; Lintvedt, R.L.; Glick, M.D.: 1980, Inorg. Chem., 19, p. 2949
(28) Sager, R.S.; Williams, R.J.; Watson, W.H.: 1967, Inorg. Chem., 6, p. 951; Johnson, D.R.; Watson, W.H.: 1971, Inorg. Chem., 10, p. 1281
(29) Gluvichinsky, P.; Mockler, G.M., Healy, P.C.; Sinn, E.: 1974, JCS Dalton, p. 1156; Countryman, R.M.; Robinson, W.T.; Sinn, E.: 1974, Inorg. Chem., 13, p. 2013; Butcher, R.J.; Sinn, E.: 1976, Inorg. Chem., 15, p. 604
(30) Galy, J.; Jand, J.; Kahn, O.; Tola, P.: 1979, Inorg. Chim. Acta, 36, p. 229
(31) Morgenstern-Baradau, I.; Rerat, M.; Kahn, O.; Jand, J.; Galy, J.: Inorg. Chem., in press

(32) Kahn, O.; Galy, J.; Journaux, Y.; Jand, J.; Morgenstern-Baradau, I.: 1982, J. Am. Chem. Soc., 104, p. 2165
(33) Journaux, Y.; Kahn, O.; Coudanne, H.: Angew. Chem., Inst. Ed. Engl., in press
(34) Banci, L.; Bencini, A.; Benelli, C.; Gatteschi, D.: Inorg. Chem., in press

STRUCTURAL AND MAGNETIC INVESTIGATIONS ON MODEL Cu_4O_4 CUBANE - LIKE CLUSTERS

W.Haase and L.Walz
Institut für Physikalische Chemie der
Technischen Hochschule Darmstadt
D-6100 Darmstadt, Germany

and

F.Nepveu
Laboratoire de Biospectroscopie
CTB SC 13 INSERM
31400 Toulouse, France

INTRODUCTION

It is well known that the binuclear copper centers play an important role in the catalytic processes of the multicopper containing proteins (1). In simple molecules the electronic interaction between the two copper centers is currently described, but many questions remain unsolved. In order to gain more understanding of the relationships between exchange interactions and geometric and electronic factors, a detailed description of structures and magnetic properties of the Cu_4O_4 cubane-like molecules is helpful.

STRUCTURAL DESCRIPTION

Molecules with a cubane-like Cu_4O_4 core are becoming increasingly well investigated (2) (3). Most of them are derived from aminoalcohols (A) and iminoalcohols (B).

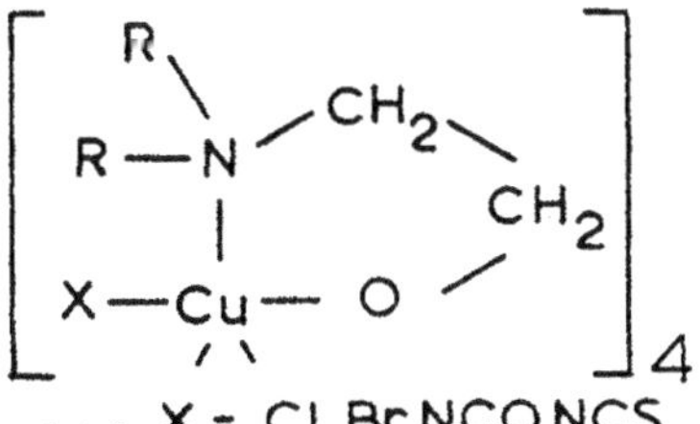

(A) X = Cl, Br, NCO, NCS
R = ALKYL

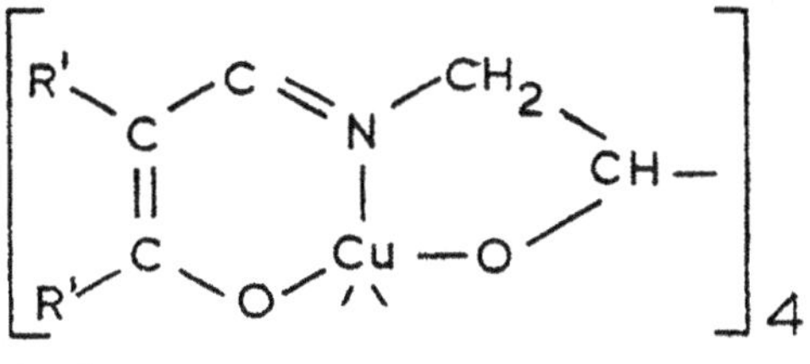

(B)

I. Bertini, R. S. Drago, and C. Luchinat (eds.), The Coordination Chemistry of Metalloenzymes, 229–234.

At first, considering the different possible polynuclear species in the crystalline state with both ligands (A) and (B), a particular aspect must be emphasized.
The complexes derived from 2-aminoalcohols exist as dimeric, tetrameric or polymeric species (2). With Schiff base ligands, including an imino-alcohol five-membered ring, (B), tetrameric structures of the cubane type are currently observed (4-7). The first structural determinations derived from (A) and (B) were reported respectively for chloro (2-diethylamino-ethanolato) copper (II) by Haase (8) and for the Schiff base copper (II) complex with 7-hydroxy-4-methyl-5-aza-hept-4-en-2-on by Bertrand et al. (4).
According to Mergehenn (9), the various forms of cubane-like Cu_4O_4 molecules can be understood as derived from an idealized Cu_4O_4 core (Fig. 1) by stretching the four Cu-O bonds (Cu(1) - O(3), Cu(2) - O(4), Cu(3) - O(2), Cu(4) - O(1)) (Type I) or by stretching the four Cu - O bonds within the pseudodimeric units (Cu(1) - O(2), Cu(2) - O(1), Cu(3) - O(4), Cu(4) - O(3)) (Type II). Between these types (I and II) a continuous series of molecular forms exists. In all cases, the copper atoms are five coordinated: three oxygen, one nitrogen atoms and one additional X ligand (A); four oxygen and one nitrogen atoms in case of (B). The coordination geometries can be described as square pyramides or as trigonal bipyramides with various distortions.

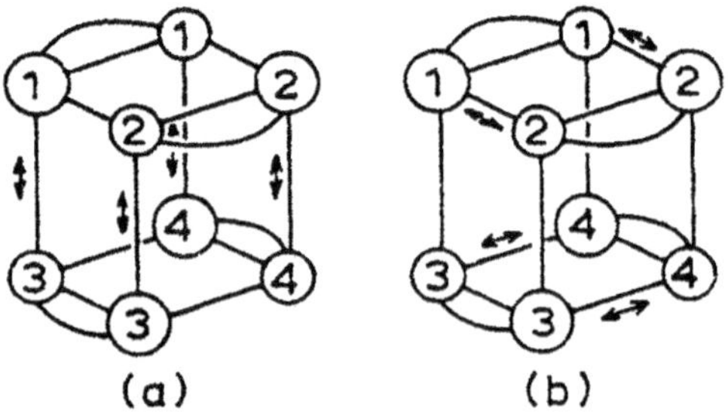

Fig. 1 Cu_4O_4 core. (a) Type I, (b) Type II

MAGNETIC PROPERTIES AND STRUCTURAL DATA

For molecules under investigation with a Cu_4O_4-core the highest possible symmetry is S_4 (Fig.2). The exchange Hamiltonian , $\hat{H}_{ex}$, in its simplest form, is:

$$\hat{H}_{ex} = -2 \sum_{i<j} J_{ij} S_i \cdot S_j$$

and for a Cu_4O_4 core with S_4 symmetry follows:

$$\hat{H}_{ex} = -2 J_{12} (S_1S_2 + S_3S_4) - 2J_{13} (S_1S_3 + S_1S_4 + S_2S_3 + S_2S_4).$$

The energy values corresponding to this Hamiltonian are:
$E_1 = -J_{12}-2J_{13}$, S' = 2; $E_2 = -J_{12}+2J_{13}$, S'=1; $E_3 = J_{12}$, S' = 1; $E_4 = 3J_{12}$, S' = 0; $E_5 = -J_{12} + 4J_{13}$, S' = 0 (with $S_i = S_j = 1/2$ and S' = total spin).

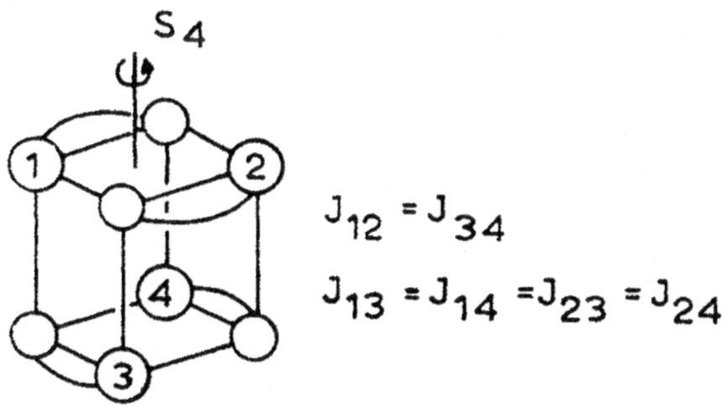

Fig.2 Cu_4O_4 core with S_4 symmetry

It should be noted that for the lower symmetries (C_{2v}, C_2, C_1) the energy value expressions are more complicate.
The type I form has a singlet ground state (S' = 0) while the type II form has a quintet ground state (S' = 2). In both cases, the coupling within the pseudodimeric units is antiferromagnetic ($-J_{12} \gg 0$ cm^{-1}) whereas the coupling between the pseudodimeric units is ferromagnetic ($J_{13} \gg 0 cm^{-1}$). Related values for these alkoxo-bridged complexes are presented in Table (1).

Within the pseudodimer		Between the pseudodimer	
Cu(1)-Cu(2)	: 2.90-3.52 Å	Cu(1)-Cu(3)	: 3.16-3.46 Å
Cu(1)- O(2)	: 1.95-2.52 Å	Cu(1)- O(3)	: 2.50-1.98 Å
Cu(1)- O(1)-Cu(2)	: 95- 104 °	Cu(1)- O(3)-Cu(3)	: 102- 109 °
		Cu(1)- O(3)-Cu(4)	: 101 - 89 °
$-J_{12}$	: 93 - 0cm^{-1}	J_{13}	: 47 - 0cm^{-1}

Table 1: Mean values of the geometrical factors and exchange integrals for tetrameric compounds, (A) after (2).

The values of the exchange integrals are obtained by fitting procedure of the experimental susceptibilities to a theoretical $\chi(T)$ equation based on the isotropic Heisenberg-Dirac-van Vleck model (HDvV model). The character of the $\chi(T)$ curve is the result of the temperature-dependent population of the spin states($E_1, \ldots, E_5$) given by the Boltzmann distribution. The exchange interaction between the four paramagnetic copper ions is interpreted by superexchange mechanism via oxygen-bridges. Correlations between the values of the exchange integrals, J_{ij}, and the bridging angle Cu-O-Cu have been reported (10). From the theory it is known that J_{ij} is dependent from the distances between the paramagnetic centers. In a recent work (11,12), a first combined relation, in the form $2J_{12}(\Phi, r) = f(\Phi) \cdot f(r)$, was proposed, with Φ= Cu(1)-O(1)-Cu(2) (bridging angle) and r = Cu(1)-O(2) (distance), for the antiferromagnetic contribution. The crossover angle (Φ =95.8°), for these alkoxo-bridged compounds, is slightly smaller than the same one ($\Phi \sim 98°$) for the hydroxy-bridged dimers. The equation:

$$\ln 2J_{12}/\mathrm{cm}^{-1}/(\Phi - 95.8^{o}) = 17.97 - 6.83(r/\text{Å})$$

leads to a good agreement with the fitted values. For the ferromagnetic contribution, a relationship between $\ln 2J_{13}$ and Cu(1)-O(3) distance was only observed. Due to the lack of data for compounds derived from an iminoalcohol (B), these two relations cannot be directly used.

EXTENSION OF THE CORRELATIONS BETWEEN MAGNETISM AND STRUCTURES

The magnetic properties and the molecular structural data can be connected. In tetrameric molecules with a Cu_4O_4 core, if the crystal structure is unknown, the principal molecular form can be suggested as type I, type II, or intermediate forms, from the magnetic behaviour. This proceeding allows to propose, with various reservations, structural aspects in other molecular aggregations (e.g. tetrameric step structures, dimeric or polymeric species).
The results as described above were obtained with well defined crystalline samples. In some cases, several molecular forms are observed in different crystal structures. For instance bromo(2-diethylaminoethanolato) copper (II) exhibits three different species (dimeric, polymeric, tetrameric type II (with CCl_4) forms) which can be isolated by different crystallization processes (13). A fourth intermediate form (type I) can be observed by loss of the tetrachloromethane. For $(\text{Cu-EIA})_4$ transitions between different forms were also observed (5,14):

$$\alpha\text{-(Cu-EIA)}_4\text{(type I)} \underset{}{\overset{C_6H_6}{\rightleftharpoons}} \text{Cu-EIA/solvent} \underset{CH_3OH}{\rightleftharpoons} \beta\text{-(Cu-EIA)}_4\text{(type II)}$$

In solution, the situation is more complicate since no definite molecular forms exist. It is apparent that the structural and magnetic properties from static systems, in the solid state, cannot be directly extented to the related dynamic ones in solution and in sites of metalloenzymes. However comprehensive studies, in model systems, can procure informations, for instance the proximity of the metal centers.

EXAMPLE

The reaction of the pyridoxal with the biogenic amine, 2-amino-1-(3',4'-dichlorophenyl)ethanol and copper (II) acetate monohydrate leads to a tetrameric molecule in methanol.

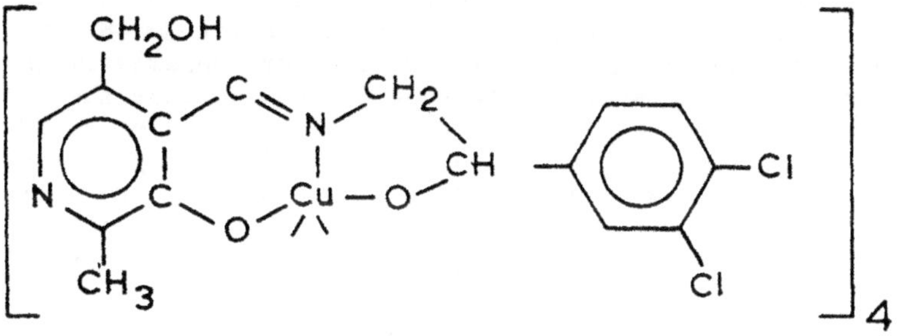

This compound crystallizes with 9 molecules of methanol ($C_{64}H_{56}Cu_4N_8O_{12}Cl_8 \cdot 9CH_3OH$, I 2/c, with the cell parameters (7)

a = 27.062 (5), b = 25.062 (5), c = 26.390 (5) Å, ß = 92.39°. V = 17781 $Å^3$). These tetrameric molecules show ferromagnetic coupling in the ground state (S' = 2). The temperature dependence of the magnetic susceptibilities is given in figure 3 (7).

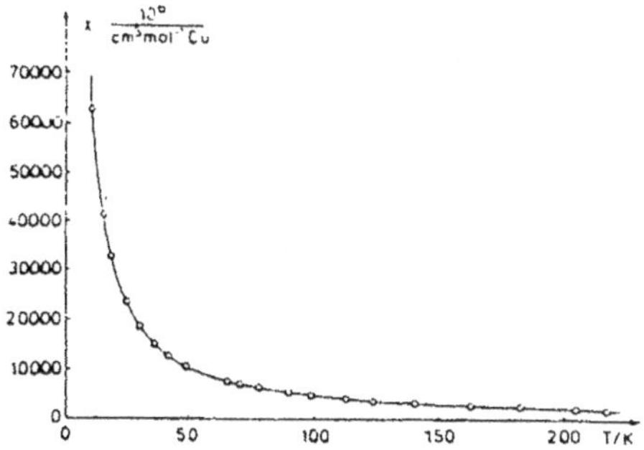

Fig. 3 Temperature dependence of the magnetic susceptibilities (full circles: experimental values, full line : calculated values). Fitted data: g = 2,08, $J_{12} = -9,5cm^{-1}$, $J_{13} = +20,5cm^{-1}$

The tetrameric molecules in the crystalline state are associated by hydrogen bonds in the form:

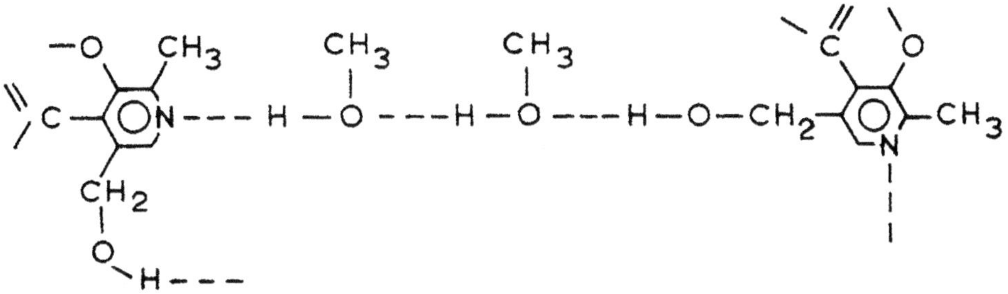

The molecular weight of this tetrameric compound, including the molecules of methanol, $9CH_3OH$, is 1943 or equivalent to 486/1Cu. In tyrosinase (Neurospora) the molecular weight per copper atom is 20000 (1) and in bovine erithrocyte superoxide dismutase the weight is 15600/Cu (15). In our compound this ratio is smaller by 30 - 40 times. The sensibility of the method for determining magnetic susceptibilities supports such a divisor factor. Thus magnetic properties can procure helpful supplementary data on the active site of multi-center metalloenzymes for which structural determinations are not yet possible.

REFERENCES

(1) Solomon,E.I.:1981, Copper Proteins, Spiro,T.G., Ed, Wiley - Interscience,pp.41-108.
(2) Merz,L.,and Haase,W.:1980,J.Chem.Soc.Dalton,pp.875-879 and references therein.
(3) Nieminen,K.:1979,Acta Chem.Scand. A33,pp.375-381;Muhonen,H.:1981,Finn.Chem.Lett. pp.94-96;Smolander,K.:1982,Acta Chem.Scand. A36,pp.189-194.
(4) Bertrand,J.A.,and Kelley,J.A.:1970,Inorg.Chim.Acta 4,pp.203-209.
(5) Mergehenn,R.,Merz,L.,Haase,W., and Allmann,R.:1976, Acta Cryst. B32,pp.505-510.
(6) Nepveu,F.,Laurent,J.P.,Bonnet,J.J.,Walz,L.,and Haase,W.:J.Chem. Soc.Dalton,in press.
(7) Walz,L.,Paulus,H.,Haase,W.,Langhof,H.,and Nepveu,F.:to be published.
(8) Haase,W.:1973,Chem.Ber. 106,pp.3132-3148.
(9) Mergehenn,R.,and Haase,W.:1977,Acta Cryst.B33,pp.1877-1882.
(10) Hatfield,W.E.:1974,Extended Interactions Between Metals Ions, Interrante L.V.,Ed.,ACS Symposium Series,10,pp.108-141.
(11) Merz,L.:Dissertation,1980,Technische Hochschule Darmstadt.
(12) Merz,L.,and Haase,W.: to be published
(13) Mergehenn,R.,Merz,L.,and Haase,W.:1980,J.Chem.Soc.Dalton,pp.1703-1709 and references therein.
(14) Merz,L.,and Haase,W.:1976,Z.Naturforsch.31a,pp.177-182.
(15) see chapter 11 in this book.

NEUROSPORA TYROSINASE: INTRINSIC AND EXTRINSIC FLUORESCENCE

M. Beltramini & K. Lerch

Biochemisches Institut der Universität Zürich, 8028 Zürich, Switzerland

Abstract The comparison of the tryptophan emission in different derivatives of *Neurospora* tyrosinase indicates the presence of tryptophan residue(s) near the copper active site. For the tryptophan-copper site pair, interaction distances of 27.5 Å and 11.9 Å are calculated in oxy- and (carbon-monoxy)-tyrosinase. The fluorescent probe 8-anilino-1-naphtalene sulphonate binds to apotyrosinase when the metal is removed and it is displaced by addition of Cu^{2+} ions indicating binding of the active site. The observed changes in the emissive properties of the probe suggest the presence of a hydrophobic environment for the tyrosinase active site.

INTRODUCTION

Tyrosinase is a copper-containing monoxygenase widely distributed in micro-organisms, plants and animals. It catalyzes the o-hydroxylation of monophenols and the oxidation of o-diphenols by dioxygen and its biological role is directed towards the formation of melanin pigments (1,2). The enzyme isolated from *Neurospora crassa* is a monomeric protein containing a pair of copper ions bound to a polypeptide chain of 46'000 M_r (2). The metal, essential for the enzymatic activity (3) is present in the resting enzyme in the divalent state but is E.P.R. undetectable (4), suggesting antiferromagnetic spin coupling between the two metal ions (5,6). This "type 3" copper unit is an active site feature common to other copper proteins (hemocyanin, laccase, ceruloplasmin and ascorbic acid oxidase) involved in different biological functions. Several derivatives of tyrosinase are present during the catalytic cycle of the enzyme: oxy-($Cu_2^{2+}O_2^{2-}$), desoxy-(Cu_2^{1+}) and met-(Cu_2^{2+}) tyrosinase (7). The oxygen, in oxytyrosinase, is coordinated as peroxide forming a bridge between the two divalent Cu ions (6). Oxytyrosinase displays spectral features very similar to those of oxyhemocyanin (6): an intense band at ~345 nm (ε~18'000 $M^{-1}cm^{-1}$) and a second one at 600 nm (ε~ 1000 $M^{-1}cm^{-1}$). Apotyrosinase specifically binds cobalt at the copper active site with a stoichiometry of 2 g-atoms Co^{2+} per mole of protein, but this derivative is devoid of enzymatic activity and oxygen binding properties (8).

I. Bertini, R. S. Drago, and C. Luchinat (eds.), The Coordination Chemistry of Metalloenzymes, 235–239.

INTRINSIC FLUORESCENCE

The tryptophan fluorescence in *Neurospora* tyrosinase has a maximum of emission at ∿330 nm (λ_{exc} = 294 nm). The comparison of emissive properties between apo-, met-, desoxy, Co^{2+}- and oxytyrosinase shows that, although the position of emission maximum is virtually the same in all derivatives, binding of metal and peroxide to the active site brings about a strong decrease of the quantum yield (Table 1, Fig. 1A).

Table 1: Emissive properties of different derivatives of *Neurospora* tyrosinase. (F = fluorescence intensity, P = phosphorescence intensity)

Derivative	λem (nm)	Q quantum yield	F/F_{apo}	P/P_{apo}
oxy-	331	0.026	0.20	0.19
met-	332	0.051	0.40	0.32
desoxy-	332	0.046	0.36	--
Co^{2+}	332	0.050	0.39	--
apo-	330	0.129	1	1

Similar effects are observed for the tryptophan triplet emission measured at 77 K (Table 1, Fig. 1B).

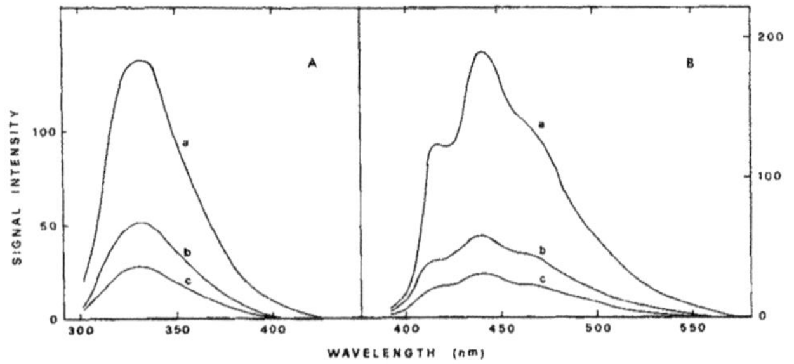

Fig. 1. Fluorescence (A) and phosphorescence (B) spectra of apo-(a), met-(b), and oxytyrosinase (c).

In addition, the lifetime of the tryptophan triplet state decreases upon metal binding (τ_{apo} = 5.7 sec; τ_{met} = 4.3 sec) but does not change appreciably upon binding of O_2 (τ_{oxy} = 3.9 sec). Quenching of tryptophan in proteins by interaction with a transition metal has been observed in several metalloproteins (9-13). In the case of tyrosinase, where binding of metal ions does not produce absorption bands in the region of tryptophan emission, the observed quenching can be attributed to a direct interaction of the tryptophan(s) dipole moment and the metal centre. This fact strongly suggests the presence of tryptophan(s) residues near the active site. Introduction of peroxide brings about a further emission quenching and the presence of an intense absorption band overlapping the tryptophan emission strongly suggests a non-radiative energy transfer mechanism of quenching (Fig. 2).

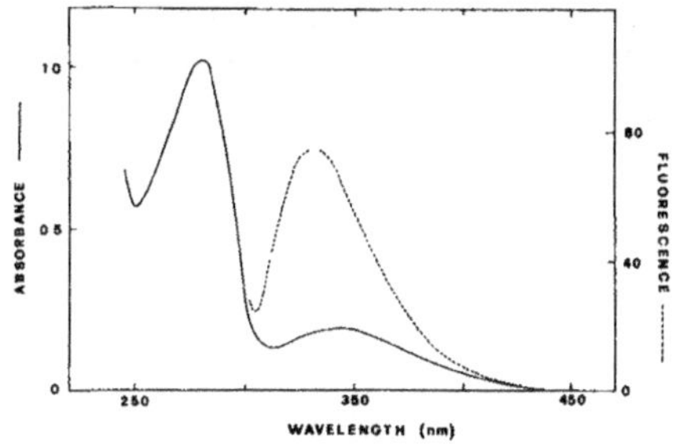

Fig. 2. Absorption (solid line) and emission (dotted line) spectra of oxytyrosinase.

This hypothesis is supported by the fact that the degree of quenching is directly proportional to the degree of oxygenation (13). Furthermore, in spite of a decrease of phosphorescence intensity the lifetime of tryptophan triplet state does not change, in agreement with a decrease in the population of the precursor singlet, as expected in the case of dipole-dipole interaction. In the case of the energy-transfer process it is possible, according to Förster's theory, to calculate the distance R between the tryptophan (donor) and the $Cu^{2+}{}_2O_2{}^{2-}$ chromophore (acceptor). The parameter R_o (critical distance) is defined as the distance at which the probability of transfer of excitation energy is equal to that of deactivation by other processes. The distance R_o (Å) is given by the following equation (14)

$$R_o{}^6 = \frac{9 \ln 10\ K^2\ q\ J}{128\ \pi^5\ n^4\ N} \qquad (1)$$

where K^2 is a mutual transition dipole geometrical factor, n the refractive index of medium, N the Avogadro's number and q the quantum yield of donor in absence of transfer.
The J value is the overlap integral between the donor emission and the acceptor absorption spectra and can be calculated as follows:

$$J = \int F(\lambda)\cdot\varepsilon(\lambda)\cdot\lambda^4\ d\lambda$$

where $\varepsilon(\lambda)$ is the molar extinction coefficient of acceptor and $F(\lambda)$ is the fluorescence intensity of donor normalized so that $\int F(\lambda)\cdot d\lambda = 1$. Using the values $K^2 = 2/3$ (assuming freely rotating tryptophan), $q = 0.046$, $J_{oxy-} = 24.23 \times 10^{-12}$ cm^6 $mole^{-1}$ and $n = 1.33$ (water), $R_o = 26.3$ Å. As tyrosinase contains 12 tryptophyl residues (2) it is not possible to relate the quenching effect upon oxygenation to a single donor group. Under these conditions one cannot calculate an interchromophore distance from R_o. Nevertheless, using the "equivalent oscillators" model (13), it is possible to calculate a distance value (R) for a system consisting of a single donor-acceptor pair with parallel orientations:

$$\frac{F_o}{F} - 1 = \left(\frac{R_o}{R}\right)^6 \qquad (2)$$

F and F_o are the quantum yields of oxytyrosinase (with energy transfer) and desoxytyrosinase (without energy transfer), respectively. According to eq.2 R is 27.5 Å and in the protein the distance value corresponds to an upper limit of interaction distance within which a tryptophan residue is expected to undergo quenching. The resolution of the interaction distance in case of oxytyrosinase is not very high because of the large overlap integral J_{oxy}.

In a previous study (15), desoxytyrosinase was shown to bind CO with a stoichiometry of 1CO/2Cu. The formation of a CO-tyrosinase complex brings about a quenching of tryptophan fluorescence similar to that observed upon oxygenation. As for oxytyrosinase, binding of CO leads to an absorption band overlapping the tryptophan emission. Therefore, also in this case a non-radiative energy-transfer process can be proposed as quenching mechanism. The position of maximum ($\lambda \sim$ 305 nm) and the low intensity ($\varepsilon \sim$ 1400 M^{-1} cm^{-1}) of the Cu-CO absorption band makes the overlap integral with the tryptophan emission spectrum about two orders of magnitude smaller than the one calculated for oxytyrosinase (J_{CO} = 0.254 x 10^{-12} cm^6 $mole^{-1}$) resulting in a better resolution of the interchromophoric distance. According to the formalism previously described, an interaction distance of 11.9 Å is calculated.

EXTRINSIC FLUORESCENCE

Fluorescence probes are small molecules which change their emissive properties upon binding to specific sites in proteins. The fluorescence behaviour of 8-anilino-1-naphtalenesulphonate (ANS) depends strongly on the environment. It shows a weak emission with a maximum at $\sim$ 520 nm. However, in a non-polar medium the maximum of emission is blue-shifted and the quantum yield drastically increases (16). Therefore, it is possible to follow the binding of the probe to a hydrophobic site of a protein. In the presence of oxy-, met- or Co^{2+}-tyrosinase, ANS shows a broad emission band (λ_{exc} = 390 nm) with a maximum at 515 nm. However, the emission spectrum of ANS drastically changes in the presence of apotyrosinase showing higher intensity and blue shift of the emissiom maximum to 490 nm (Fig. 3), indicating binding of the probe to the protein.

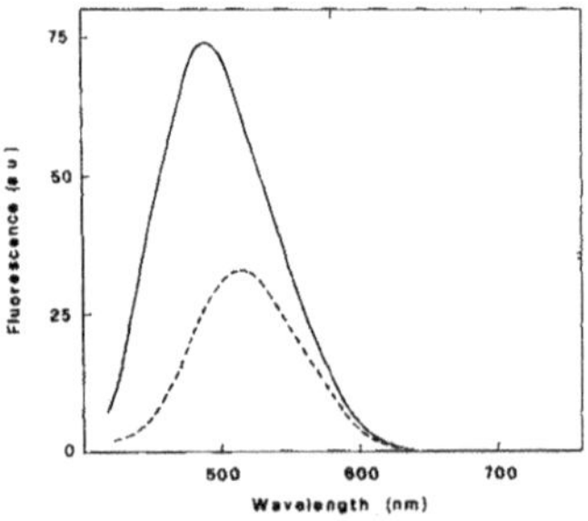

Fig. 3. Emission spectra of ANS (λ_{exc}=390 nm) in buffer (dotted line) and in the presence of apotyrosinase (solid line). The dotted line describes also the spectrum observed in the presence of met-, Co^{2+}-, oxytyrosinase and apotyrosinase + Cu^{2+} ions.

Upon addition of Cu^{2+} ions to the apotyrosinase-ANS complex the emissive properties of the probe are completely reversed, strongly indicating that ANS binds to the active site of apotyrosinase. It is interesting that despite the fact that ANS is displaced almost immediately, the recovery of the enzymatic activity is a rather slow process (10-15 hrs). This finding is in line with the observation of Kertesz et al. (17) reporting a discrepancy between the fast binding of Cu^{2+} ions to apotyrosinase ($k_{on} > 5 \times 10^6$ M^{-1} sec^{-1}) and the slow recovery of enzymatic activity. The observed changes in the emissive properties of ANS upon binding to apotyrosinase suggest a hydrophobic character for the protein active site as previously proposed by a resonance Raman study on oxytyrosinase (18).

References

(1) Mason, H.S.: 1975, Ann. Rev. Biochem. 34, pp. 595-634
(2) Lerch, K.: 1981, in Metal Ions in Biological Systems (Siegel, H. ed.) Vol. 13; Marcel Dekker, New York, pp. 143-186
(3) Kubowitz, F.: 1938, Biochem Z. 299, pp. 32-97
(4) Deinum, J., Lerch, K. & Reinhammar, B.: 1976, FEBS Lett. 69, pp. 161-164
(5) Schoot Uiterkamp, A.J.M. & Mason, H.S.: 1973, Proc. Natl. Acad. Sci. U.S.A., 70, pp. 993-996
(6) Himmelwright, R.S., Eickman, N.C., LuBien, C.D., Lerch, K. & Solomon, E.I.: 1980, J. Am. Chem. Soc. 102, pp. 7339-7344
(7) Winkler, M.E., Lerch, K., Solomon, E.I.: 1981, J. Am. Chem. Soc. 103, pp. 7001-7003
(8) Rüegg, C. & Lerch, K.: 1981, Biochemistry 20, pp. 1256-1262
(9) Bannister, W.H. & Wood, E.J.: 1971, Comp. Biochem. Physiol. 40 B, pp. 7-18
(10) Finazzi-Agrò, A., Rotilio, G., Avigliano, L., Guerrieri, P., Boffi, V. & Mondovi, B.: 1970, Biochemistry 9, pp. 2009-2014
(11) Morpurgo, L., Graziani, M.T., Finazzi-Agrò, A., Rotilio, G., Mondovi, B.: 1980, Biochem. J. 187, pp. 361-366
(12) McMillin, D.R., Rosenberg, R.C. & Gray, H.B.: 1974, Proc. Natl. Acad. Sci. U.S.A. 71, pp. 4760-4762
(13) Shaklai, N. & Daniel, E.: 1970, Biochemistry 9, pp. 564-568
(14) Lehrer, S.S.: 1969, J. Biol. Chem. 244, pp. 3613-3617
(15) Kuiper, H.A., Lerch, K., Brunori, M. & Finazzi-Agrò, A.: 1980, FEBS Lett. 111, pp. 232-234
(16) Stryer, L.: 1968, Science 162, pp. 526-533
(17) Kertesz, D., Rotilio, G., Brunori, M., Zito, R. & Antonini, E.: 1972, Biochem. Biophys. Res. Comm. 49, pp. 1208-1215
(18) Eickman, N.C., Solomon, E.I., Larrabee, J.A., Spiro, T.G. & Lerch, K.: 1978, J. Am. Chem. Soc. 100, pp. 6529-6531

COMPARISON OF TWO FUNGAL TYROSINASES

D.A. Robb

Department of Bioscience and Biotechnology, University of Strathclyde, Taylor Street, Glasgow G4 ONR, U.K.

ABSTRACT

Although the molecular properties of tyrosinases prepared from mushroom (Agaricus bisporus) and Neurospora crassa are dissimilar it is suggested that the active site of both comprises a binuclear copper complex.

INTRODUCTION

Tyrosinase is the sole enzyme in the metabolic pathway leading to the black pigment melanin. Both the pigment and the enzyme are found at most phylogenetic levels but they are not of universal occurrence. The enzyme catalyses the first two steps in the pathway (Fig. 1a) the hydroxylation of tyrosine (tyr) to 3,4-dihydroxyphenylalanine (dopa) and the subsequent oxidation of dopa to a quinone (dpq). Thus the enzyme is both an oxidase (Fig. 1b) and a hydroxylase (Fig. 1c) and as such is given two entries in the IUB classification of enzymes, namely E.C. 1.10.3.1 and E.C. 1.14.18.1. Oxygen which is required for both activities, is reduced to water without the liberation of hydrogen peroxide. The specificity of the enzyme is broad, both L- and D-isomers and simpler phenols being oxidised. An interesting feature of the hydroxylase reaction is that an o-diphenol is required for efficient hydroxylation of a phenol i.e. to convert tyrosine to dopa efficiently, requires the product (dopa) to be already present.

Consideration of tyrosinase is relevant to the theme of this conference because it is the simplest of the five oxidases able to convert oxygen to water (laccase, cytochrome oxidase, ascorbate

I. Bertini, R. S. Drago, and C. Luchinat (eds.), The Coordination Chemistry of Metalloenzymes, 241–246.

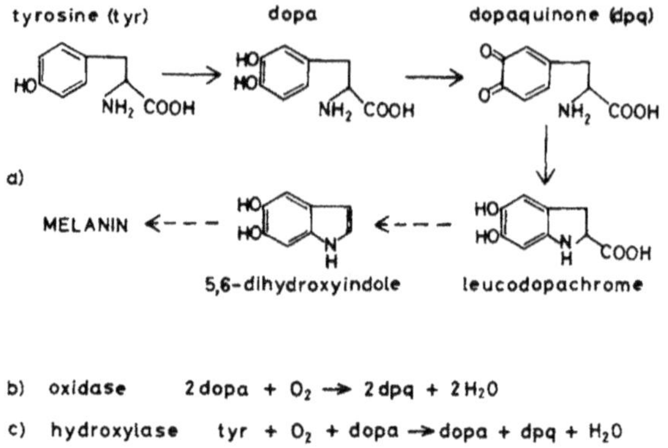

Figure 1 Reactions of tyrosinase

oxidase and ceruloplasmin all contain additional redox centres). It is also a naturally occurring example of an antiferromagnetically coupled Cu-Cu dimer; thus this article is a logical extension of the contributions of Drs. Gatteschi and Haase on magnetic exchange coupling in model compounds.

MOLECULAR PROPERTIES

The best sources of tyrosinase are mushroom (Agaricus bisporus) and bread mould (Neurospora crassa) and the mushroom enzyme is commercially available with a low degree of purity (~10%). Table I summarises our work on the molecular properties of both enzymes.

Table 1 Physico-chemical Properties of Mushroom and Neurospora Tyrosinases (1,2)

		Mushroom	Neurospora
Molecular Weight		110,000	47,000-170,000
Subunit MW	H^a	48,000	46,000
	H^b	45,000	
	L	13,000	
Isoelectric Point		4.70-4.95	7.8
s_{20w} (sedimentation velocity)		6.48S	4.50S
(active enzyme centrifugation)		6.53S	4.30S

It shows that the mushroom enzyme has a complex quaternary structure, the major species in a purified preparation most probably being a tetramer comprising two large polypeptides (H) and two small ones (L). Two types of larger polypeptides (H^a and H^b) are currently recognised (2).

The Neurospora enzyme is characterised as an aggregating system based on a single polypeptide the size of the H subunit. Active enzyme centrifugation is a sensitive technique used to determine if the presence of substrates alters the molecular properties (3). In this case no change was detected. Similar molecular properties have been reported by others (4-6) and it is now apparent that the dimeric form of the mushroom enzyme (HL) and the monomeric Neurospora enzyme each contain two atoms of copper per mole.

The purified enzyme from either source is straw coloured and has a relatively strong absorption band in the 280-290 nm region indicative of a high tyrosine and tryptophan content. A small absorption at 345nm is apparent and this is increased on the addition of hydrogen peroxide to yield a derivative termed oxytyrosinase. This derivative is characterised by two peaks, at 345 nm (ε=9,000 $M_{Cu}^{-1}cm^{-1}$) and 600nm (ε=600 $M_{Cu}^{-1}cm^{-1}$) and these absorption bands are dependent on the presence of Cu and oxygen (7). Hydrogen peroxide is thought to combine with the Cu (II) form (mettyrosinase) of the isolated enzymes to form oxytyrosinase, a form which is only present to a small extent in the isolated preparation. Removal of oxygen by evacuation produces a colourless derivative, deoxytyrosinase, which has the property of reforming oxytyrosinase on admission of oxygen. None of these derivatives gives an EPR spectrum.

PROPERTIES OF THE COPPER SITE

The following observations have led to the proposal that the tyrosinase active site comprises an exchange coupled Cu-Cu dimer.

(i) Mettyrosinase acts as a two electron acceptor and accepts one electron per copper, $E_o' = 0.36V$, n = 2 (8).

(ii) Resonance Raman spectroscopy shows oxytyrosinase to be a peroxy complex since $^{16}O_2$ and $^{18}O_2$ produce peaks at 755 cm^{-1} and 714 cm^{-1} respectively (9), indicating a two electron transfer from deoxytyrosinase upon binding oxygen.

(iii) Binding studies show that one mole of the substrate analogue, benzoate, binds per two moles of copper (10) and similar binding ratios have been determined for carbon monoxide (an oxygen analogue) and H_2O_2 (5,7,11).

(iv) Magnetic susceptibility measurements show mettyrosinase to be diamagnetic (8), requiring an exchange integral (J) $>$625 cm^{-1}.

(v) Addition of nitric oxygen and a trace of oxygen to deoxytyrosinase produces an EPR signal at g = 4 (Fig. 2a). The change in quantum number (ΔM) of two is characteristic of a magnetic dipole . . dipole coupled copper pair, about 0.6nm apart with J$<$30 cm^{-1} (12-14).

The observations may be represented as follows:-

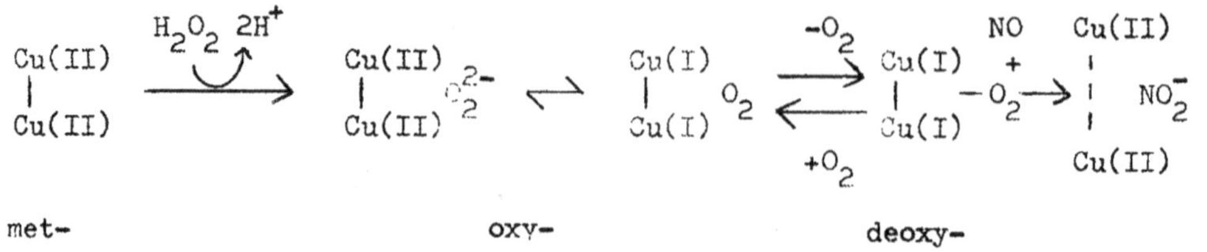

The two copper atoms in the pair do not always show equivalent behaviour. Thus it is possible to prepare a series of derivatives in which one of the copper atoms is Cu(II) and the other is Cu(I). Such semi-met derivatives yield EPR spectra typical of mononuclear copper as indicated in Fig. 2b.

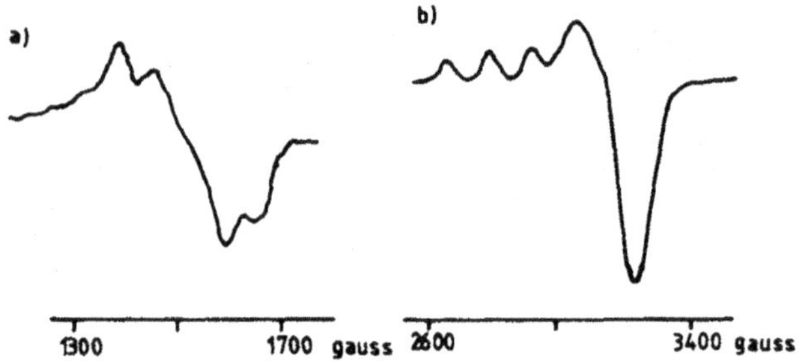

Figure 2 Low temperature EPR spectra of tyrosinase in a) $g=4$ region of a met-nitrite derivative (12) and b) $g=2$ region for the semi-met-nitrite form; $g_{\parallel}=2.296$ $g_{\perp}=2.078$ (Reprinted with permission from Ref. (14). Copyright 1982 Amer. Chem. Soc.)

IDENTIFICATION OF ENDOGENOUS LIGANDS

In the absence of an X-ray structural determination we must rely on spectroscopic and covalent modification studies. The latter have benefited greatly from the determination of the amino acid sequence of the *Neurospora* enzyme (15). Both the mushroom and *Neurospora* enzymes are susceptible to photochemical oxidation in the presence of a suitable photosensitizer and cyanide (16); if the apoenzyme is oxidised, cyanide may be omitted (17). When three histidine residues in *Neurospora* apotyrosinase are photooxidised (residues 188, 193, 289) the enzyme cannot recombine with copper to form an active enzyme. Also a further histidine residue (residue 306) is destroyed during catalytic oxidation of catechol (15). This modification leads to inactivation of the enzyme and has been observed previously with model complexes and H_2O_2 when the imidazole nitrogen is a ligand (18). An additional histidine residue (305) is adjacent to the one which is inactivated and may be expected to have some role; since free cysteine is absent in the *Neurospora* enzyme (15), SH cannot be a ligand. If the possibility that the copper pair interact directly through a metal-metal bond is discounted, then a bridging ligand is required to explain the diamagnetic nature of mettyrosinase. The identity of this ligand is uncertain but from spectroscopic evidence an oxygen atom donated by a carboxylate or phenolate has been suggested (14).

INHIBITION BY ANIONS

Halide ($F^- > Cl^- > Br^-$), azide and cyanide ions bind more strongly to tyrosinase than they do to superoxide dismutase. Values for the dissociation constants with F^-, N_3^-, and the substrate analogue p-nitrocatechol are given in Table II which also shows them to be markedly dependent on pH.

Table II Inhibition of Mushroom Tyrosinase (19)

	Experimental pKi				
pH	4.1	5.2	6.1	7.0	8.0
Azide	4.7	4.7	3.8	2.85	-
Fluoride	3.6	2.85	1.9	1.45	-
p-Nitrocatechol	-	-	3.7	3.5	2.85
	Calculated pKi*				
Hydrazoic acid	4.80	5.25	5.15	5.05	-
Hydrofluoric acid	4.35	4.60	4.55	4.90	-
p-Nitrocatechol	-	-	3.85	3.90	4.10

* assuming unionised form is responsible for inhibition.

This dependancy is largely removed if the data are corrected on the assumption that only the unionised ligand binds. It appears therefore that a proton acceptor such as a histidine residue or a liganded H_2O or OH^- is present at the active site and that binding of the undissociated acid and of the anion may be represented as

$$Enz + HA \longrightarrow H^+ \; EnzA^-$$
$$Enz + H^+ + A^- \longrightarrow H^+ \; EnzA^-$$

In conclusion a scheme for the binuclear copper site of oxytyrosinase may be proposed (Fig. 3) which is consistent with the spectroscopic observations. The two copper atoms have tetragonal symmetry being about 0.35 nm apart and bridged by oxygen and an unidentified ligand R. Other ligands are thought to be provided by histidine residues and water molecules which are displaced by the phenolic substrate. It may be noted that a very similar model is also proposed for oxyhemocyanin which shows a very weak tyrosinase activity. Upon displacement of oxygen, treatment with nitric oxide and a trace of oxygen leads to loss of the bridging ligand R and separation of the copper atoms to 0.6nm.

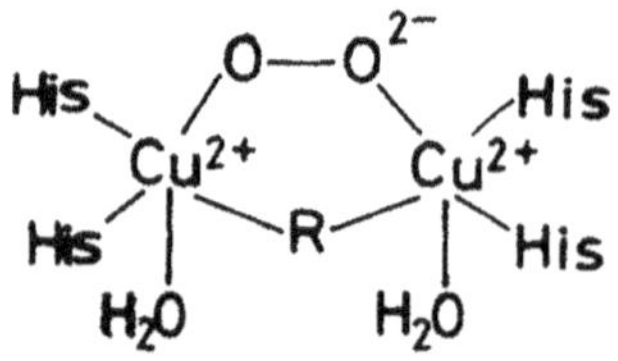

Figure 3 Proposed structure for the active site of oxytyrosinase

REFERENCES

(1) Robb, D.A. and Gutteridge, S. : 1981, Phytochemistry 20, pp 1481-1485.
(2) Robb, D.A. : 1979, Biochem. Soc. Transac. 7, pp 131-132.
(3) Cohen, R. and Mire, M. : 1971, Eur. J. Biochem 23, pp 267-275.
(4) Strothkamp, K.G., Jolley, R.L. and Mason, H.S.: 1976, Biochem. Biophys. Res. Commun. 70, pp 519-524.
(5) Lerch, K. : 1976, FEBS Lett. 69, pp 157-160.
(6) Gutteridge, S. and Mason, H.S. : 1980, in "Biochemical and Clinical Aspects of Oxygen" (Caughey, W.S. ed.) Academic Press New York. pp 589-602.
(7) Jolley, R.L., Evans, L.H., Makino, N. and Mason, H.S. : 1974, J. Biol. Chem. 249, pp 335-345.
(8) Makino, N., McMahill, P., Mason, H.S. and Moss, T.H. : 1974, J. Biol. Chem. 249, pp 6062-6066.
(9) Eickman, N.C., Solomon, E.I., Larrabee, J.A., Spiro, T.G. and Lerch, K. : 1979, J. Am. Chem. Soc. 100, pp 6529-6531.
(10) Duckworth, H. and Coleman, J.E. : 1970, J. Biol. Chem. 245, pp 1613-1625.
(11) Kubowitz, F. : 1938, Biochem. Zeit. 299, pp 32-57.
(12) Schoot Uiterkamp,A.J.M. and Mason, H.S. : 1973, Proc. Nat. Acad. Sci. U.S.A. 70, pp 993-996.
(13) Schoot Uiterkamp, A.J.M., van der Deen, H., Berendsen, H.C.J. and Boas, J.F. : 1974, Biochem. Biophys. Acta 372, pp 407-425.
(14) Himmelwright, R.S.,Eickman, N., Lu Bien, C.D. and Lerch, K. : 1980, J. Am. Chem. Soc. 102, pp 7339-7344.
(15) Lerch, K. : 1978, Proc. Nat. Acad. Sci. U.S.A. 75, pp 3635-3639.
(16) Gutteridge, S., Dickson, G. and Robb, D. : 1977, Phytochemistry 16, pp 517-519.
(17) Pfiffner, E. and Lerch, K. : 1981, Biochemistry 20, pp 6024-6035.
(18) Sigel, H. : 1969, Angew. Chemie. Internat. Ed. 8, pp 167-177.
(19) Robb, D.A., Swain, T. and Mapson, L.W. : 1966, Phytochemistry 5, pp 665-675.

ACTIVATION OF MOLECULAR OXYGEN

Russell S. Drago

Department of Chemistry
University of Florida
Gainesville, Florida

This article summarizes the various ways that transition metal ions can function to activate O_2. Included are discussions of autoxidation reactions, Wacker oxidations and oxidations of phenols by metal bound O_2. Metal activation to form H_2O_2 and its use in oxidations are discussed along with some stoichiometric oxygen atom transfer reactions.

INTRODUCTION

The reactivity of molecular oxygen is a two-edged sword to chemists. On the one hand, oxidations are key steps in the industrial synthesis of over 50% of the compounds made(1), while on the other hand, great precautions are taken in many reactions to exclude oxygen from reactants and products. The electronic structure of O_2 is unusual. The ground state is a stable triplet most readily described by the molecular orbital sequence:

$$KK<\sigma_{2s}^{2}<\sigma_{2s}^{*2}<(\pi_{2p_x}=\pi_{2p_y})^{4}<\sigma_{2p_z}^{2}<(\pi_{2p_x}^{*}=\pi_{2p_2}^{*})^{2} \quad (1)$$

The bond dissociation energy of the O_2 molecule is 118 kcal mole^{-1}, yet in spite of this stability, the reaction of O_2 with organic compounds to produce CO_2 and H_2O is thermodynamically favorable. Thus, the very existence of life on this planet is attributed to the kinetic unreactivity of O_2 with most organic compounds. The unreactive nature of O_2 in some systems has been attributed to the spin restrictions associated with the formation of diamagnetic products from paramagnetic O_2. An important factor contributing to the unreactivity of O_2 is the relative reluctance of O_2 to accept one electron to form the superoxide radical in an electron transfer reaction. Reduction potentials for the reduction of O_2 in aqueous solution are summarized in Table 1.

I. Bertini, R. S. Drago, and C. Luchinat (eds.), The Coordination Chemistry of Metalloenzymes, 247–257.

Table 1. Reduction Potentials for Molecular Oxygen and Related Species

Reaction		E^o
$O_2 + H^+ + e^- \rightarrow HO_2$	(1)	-0.32v
$HO_2 + H^+ + e^- \rightarrow H_2O_2$	(2)	+1.68v
$H_2O_2 + e^- \rightarrow HO\cdot + OH^-$	(3)	+0.8v
$\cdot OH + H^+ + e^- \rightarrow H_2O$	(4)	+2.74v

The species HO_2, H_2O_2 and $\cdot OH$ all readily accept one electron in an electron transfer reaction and are very reactive toward organic substrates. In order to survive, life forms are critically dependent upon superoxide dismutase to scavenge HO_2 and catalases as well as peroxidases to decompose H_2O_2. Hydroxyl radicals are so reactive that any of the reducing agents present in natural systems will react with them. The pK's of several of the species in Table 1 have been determined and are relevant to the form of these substances in aqueous solutions. The pK of HO_2 is 4.8, HO_2^- is 11.8 and H_2O_2 is 7.4. The electron affinity(2) of O_2 is 0.87 ev compared to 3.04 for HO_2 and 2.6 for Br_2.

Different products result from the direct reaction of O_2 with alkali metals. Lithium forms Li_2O, while both Na_2O and Na_2O_2 are formed from sodium. The products of the reaction with potassium, rubidium and cesium are superoxides, MO_2. The reason for these differences is not understood but in all instances the thermodynamically stable product is the metal oxide, M_2O.

AUTOXIDATION REACTIONS

The reactivity of O_2 with organic radicals

$$R\cdot + O_2 \rightarrow R - O_2\cdot \qquad (6)$$

is a facile reaction involved in the autoxidation of organic compounds. (3)(4) Initiation of the reaction with an initiator, In·, (equation 7) leads to a free radical chain process:

$$In\cdot + RH \rightarrow InH + R\cdot \qquad (7)$$

$$R\cdot + O_2 \rightarrow RO_2\cdot \qquad (8)$$

$$RO_2\cdot + RH \rightarrow RO_2H + R\cdot \text{ (often slow)} \qquad (9)$$

($RO_2\cdot$ is often relatively stable and abstracts the most weakly bound hydrogen)

$$RO_2H \xrightarrow{M^{n+}} ROO\cdot + RO\cdot + H_2O \quad (10)$$

$$RO\cdot + RH \rightarrow ROH + R\cdot \quad (11)$$

$$ROO\cdot \rightarrow \text{ketones} \quad (12)$$

Substrates containing allylic hydrogens readily form the allyl radical in equation 7 and undergo the subsequent reactions(5) leading to $RC(=O)-CH=CH_2$ and $\overset{H}{RC}(-OH)-CH=CH_2$. The ketone may form in equation 12 by the reaction

$$R-\underset{H}{\overset{O-O\cdot}{C}}-CH=CH_2 \rightarrow \overset{O}{\overset{\|}{RC}}-CH=CH_2 + OH\cdot \quad (12a)$$

$$OH\cdot + RH \rightarrow R\cdot + H_2O \quad (12b)$$

Cyclohexene readily undergoes allylic oxidation to produce(4)

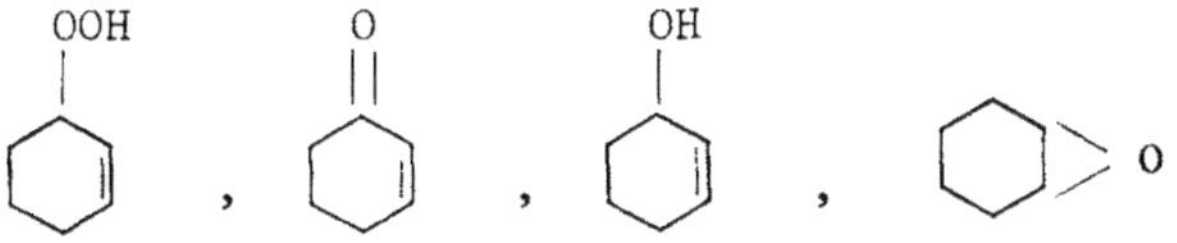

and polymers.

For an illustration of the complex radical chemistry that can occur in these systems the reader is referred to an article by R.K. Jensen et al. and the references therein.(6)

The metal ion chemistry associated with the catalytic hydroperoxide decomposition shown in equation 10, the Haber-Weiss mechanism, is well understood(3):

$$M^{+n} + ROOH \rightarrow M^{+(n+1)} + RO\cdot + OH^- \quad (13a)$$

$$M^{+(n+1)} + ROOH \rightarrow M^{+n} + ROO\cdot + H^+ \quad (13b)$$

Ideally one would like characteristic features that signify this mechanism when new reactions are investigated. As long as purified reagents are used a long induction period is observed for this oxidation. The addition of free radical traps to these systems is often found to increase the induction period for reaction. When cyclohexene is used as the substrate the product distribution shown above is expected though in some instances only the ketone is obtained in radical reactions (7)(8). By proper selection of the metal, selective oxidations result from these free radical processes. If equation 9 is rate determining, a deuterium isotope effect is expected for the RD analogue.

WACKER OXIDATIONS

The Wacker process provides a selective method for oxidizing ethylene to acetaldehyde (9). The net reaction is

$$H_2C = CH_2 + \frac{1}{2}O_2 \rightarrow CH_3CHO$$

The following mechanism is proposed:

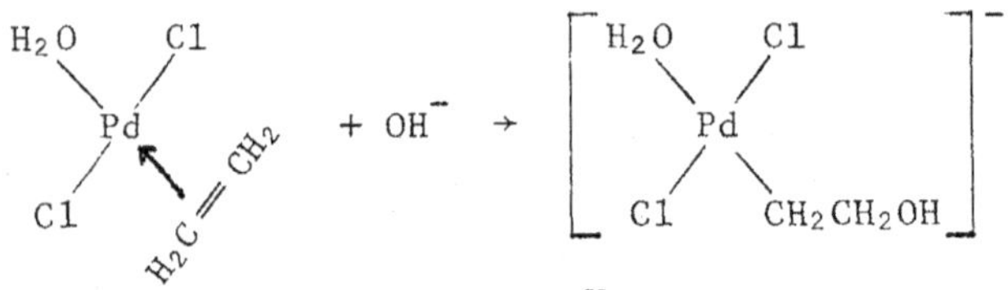

(14)

$$\rightarrow Pd^o + 2Cl^- + H_2O + [CH_3\text{-}CH(H)\text{-}OH]^+ \longrightarrow H^+ + CH_3C(H) = O \qquad (15)$$

$$Pd^o + 2Cu^{2+} \rightarrow 2Cu^+ + Pd^{2+} \qquad (16)$$

$$2Cu^+ + 2H^+ + \frac{1}{2}O_2 \rightarrow 2Cu^{2+} + H_2O \qquad (17)$$

In equation 14, nucleophilic attack on ethylene, made electrophillic by coordination to palladium, is proposed. In equation 15 a reductive elimination occurs in which Pd^o is formed. The ion in brackets is a protonated acetaldehyde which transfers a proton to solvent. Oxygen is involved in the Cu(I) oxidation completing the catalytic cycle.

OXIDATIONS BY METAL BOUND O_2

The binding of oxygen to transition metals involves the pairing up of one or two unpaired electrons in the antibonding orbitals of O_2 with unpaired electrons of the metal. This spin pairing model has been described(10)(11) and is briefly summarized for an end on bonded dioxygen adduct of cobalt (II) by the molecular orbital diagram in Figure 1.

A pairing up of electrons in Ψ_I accounts for the bonding of the oxygen to the metal. The charge on the bound oxygen is determined by the relative magnitude of the cobalt and oxygen coefficients of the Ψ_I molecular orbital. Factors that give rise to a wide range in the charge of the bound O_2 are reported (11). The unpaired electron resides essentially on O_2 regardless of the charge on O_2. The model explains factors influencing the metal-O_2 bond strength which in turn has led to a potential energy storing model to describe cooperativitiy in hemoglobin (11).

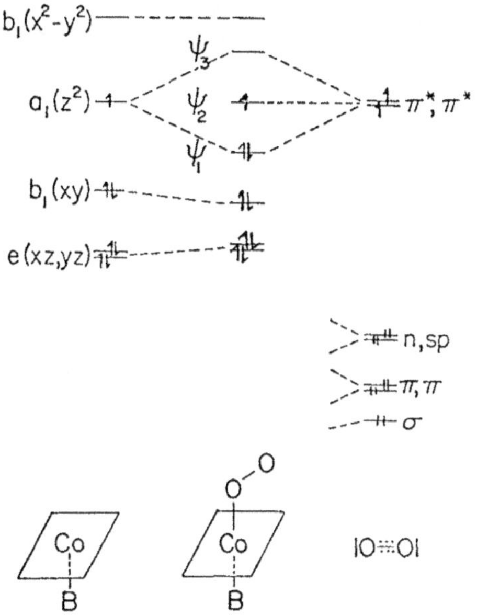

Figure 1

The enhanced basicity of O_2 that arises by virtue of its being bound to a metal ion is illustrated by hydrogen bonding of CF_3CH_2OH to the bound oxygen. In a detailed study of the oxidation of substituted phenols by O_2 it was shown that a metal bound dioxygen initiates the oxidation. The results are consistent with the following mechanism(12):

$$Co(II) + O_2 \leftrightharpoons Co\text{-}O_2$$

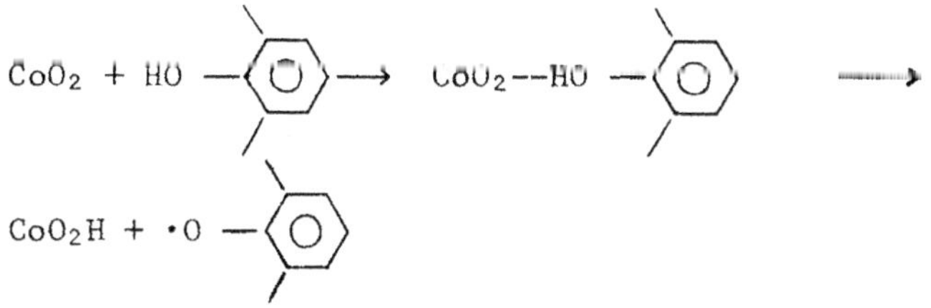

(18)

(19)

$$(CoO_2H \rightarrow Co(II) + HO_2 \rightarrow Co(II) + \frac{1}{2}H_2O_2 + \frac{1}{2}O_2;$$

$$2HO_2 \rightarrow H_2O_2 + O_2 \quad k \sim 10^{10}(Msec^{-1})(13)$$

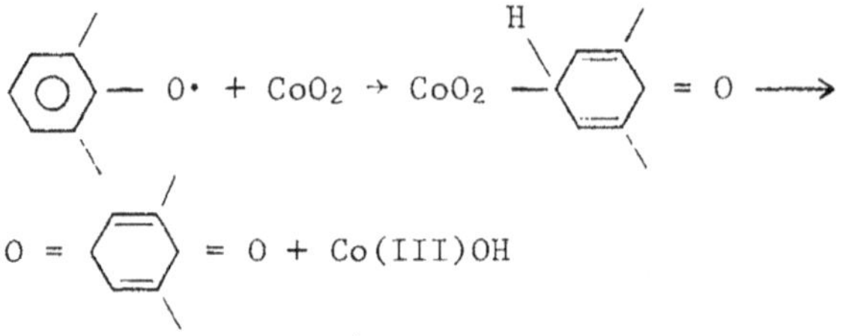

(20)

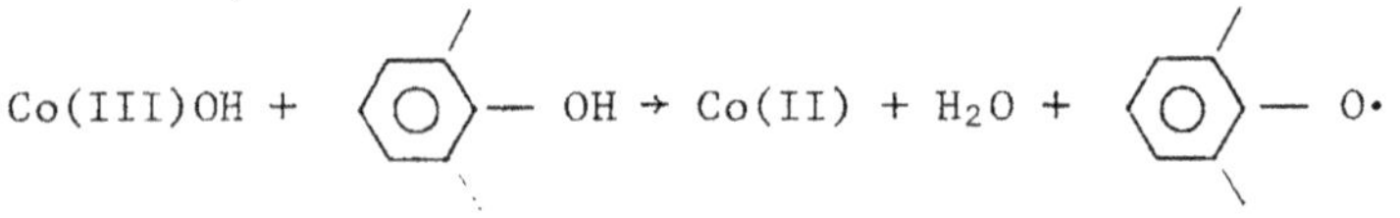

(21)

If the cobalt concentration is reduced, the rate of step 20 is reduced and the phenoxy radicals couple leading to diphenoquinone,

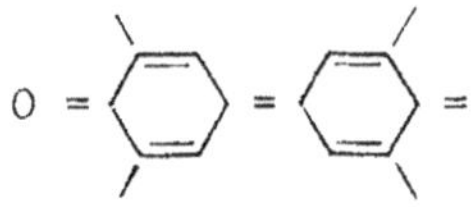

, upon subsequent oxidation. This result shows that attack by O_2 is not competitive with coupling of the phenoxy radicals. Thus, we note that coordination of O_2 to the cobalt ion has increased its tendency to undergo the free radical pairing reactions in equation 20.

Characteristic features of this reaction type include inhibition by hydrogen bonding acids that are stronger than the phenol. Since the reaction stops immediately upon addition of acid steps 19 and 20 are both inhibited. Cobalt(II) complexes that do not form stable O_2 adducts do not catalyze the formation of benzoquinone. No deuterium isotope effect is noted and the reaction is first order in O_2, cobalt complex and phenol (12).

OXIDATIONS BY PRODUCTS DERIVED FROM METAL-O_2 SPECIES

Dudley, Read and Walker(14) have reported that $Rh(P(C_6H_5)_3)_3Cl$ catalyzes co-oxidation of olefins and $(C_6H_5)_3P$ to ketones and phosphine oxides. The mechanism is speculative. More recently the Rh(III)/Cu(II) co-catalyzed oxidation of terminal olefins to 2-ketones occurs with $\geq$98% specificity (15). Simultaneous coordination of both alkene and O_2 to rhodium(I) is proposed. A rearrangement to a peroxymetallocycle and decomposition to a ketone and rhodium(III) is proposed to account for the specificity. The rhodium(III) reportedly undergoes Wacker chemistry to form ketone and regenerate rhodium(I). Work in our laboratory(16) on this system shows that the active catalyst in this system is a rhodium(III) complex and that in the absence of Cu(II) the oxidation mechanism appears to proceed via attack by H_2O_2 formed in the reaction of O_2 with alcohols.

There are several systems in which metal bound O_2 is displaced to produce reduced dioxygen species. The O_2 adduct of N,N′-ethylenebis-(acetylacetoniminato)cobalt(II)pyridine reacts(17) with acid in organic solvents to give H_2O_2. The kinetic order suggested that the binuclear cobalt O_2 complex is involved in the reaction. It has been shown(18) that O_2^{2-} can be displaced from $Pt[P(C_6H_5)_3]O_2$ upon addition of $P(CH_3)_2$-(C_6H_5) to an ethanol solution of the O_2 adduct. At low temperatures H_2O_2 is detected and upon warming this reacts with excess phosphine to produce phosphine oxide.

Bulky phosphine Pt(O) complexes that form O_2 adducts have been used in the stoichiometric conversion of $SO_2 \rightarrow SO_4^{2-}$ and $CS_2 \rightarrow CS_2O^{2-}$. The results of these studies suggest that prior coordination of O_2 to the metal is a requisite step in the oxidations (19). Previous labelling studies(20) are also consistent with oxidation by metal bound O_2.

In view of the possibility of obtaining H_2O_2 from O_2 and metal complex catalysts, metal catalyzed oxidations of organic substrates by H_2O_2 will be mentioned briefly. Iron(II) complexes and H_2O_2 lead to the Fenton reagent which hydroxylates a wide variety of organic substrates. Extensive work by Walling(21) shows that this chemistry usually involves hydroxyl radicals

$$Fe^{2+} + H_2O_2 \rightarrow Fe^{3+} + OH^- + \cdot OH$$

$$\cdot OH + RH \rightarrow R\cdot + H_2O \qquad (22)$$

The organic product obtained depends on whether R· reacts with Fe^{2+} to form Fe^{3+}, or with Fe^{3+} to form Fe^{2+}, or with other radicals to give the coupled product (22).

The hydroxylation of aromatic compounds by O_2 and ascorbic acid at a pH of 7 with iron(II)ethylenediaminetetraacetic acid is reported by Udenfriend _et al._(23). It was claimed that this system is a modified Fenton reagent (24). More recent work(25) suggested that this chemistry may involve oxygen atom transfer from an iron(IV) species because the product distribution is different than that from the reaction of hydroxyl radicals.

OXYGEN ATOM TRANSFER REACTIONS

Using py-Co(TPP)NO_2, phosphines can be oxidized to phosphine oxides with the formation of a cobalt nitrosyl. Molecular oxygen reoxidizes the nitrosyl to the nitro compound to complete the catalytic cycle (26). When Lewis acids are added to the system organic sulfides, alcohols and 1,3-cyclohexadiene are oxidized but monoolefins are not. When $PdCl_2C_2H_4$ is used as a co-catalyst with Co(TPP)NO_2 (in THF or dichloroethane) ethylene is oxidized catalytically to acetaldehyde and propylene to acetone.

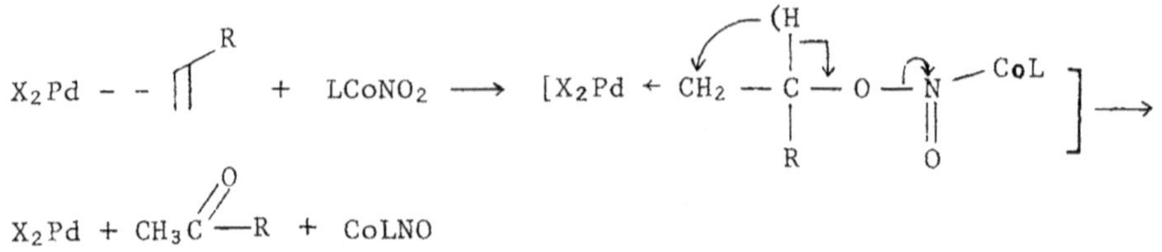

The intermediate is proposed to decompose by a β-hydride elimination followed by a hydride shift. The curved arrows account for the eventual disposition of the electrons.

When bis(acetonitrile)chloronitropalladium(II) is used as a catalyst, O_2 will oxidize(27) norbornene to exo-epoxynorbornene in toluene solution with > 90% yield and 99% selectivity. Reaction of the nitro complex with one equivalent of norbornene in acetone leads to nearly instantaneous and quantitative formation of:

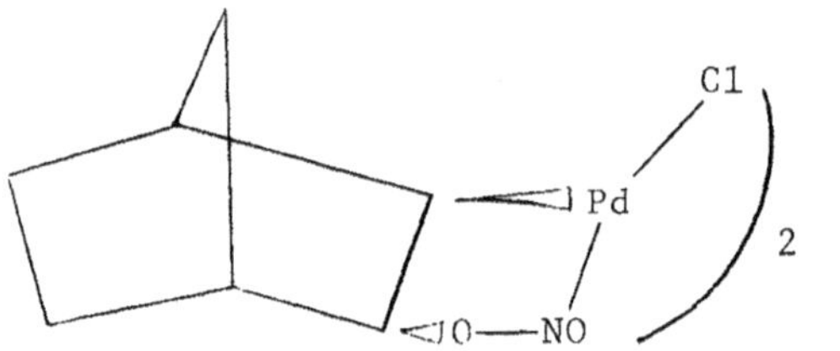

This material decomposes slowly in solution under N_2 to give exo-epoxynorbornene and $[PdClNO]_x$. The nitrosyl can be reoxidized.

The reaction of $Ni(NO_2)_2(PEt_3)_2$ with CO(28) gives CO_2 and $Ni(NO_2)NOL_2$. Tracer and kinetic studies(24)(30) support a mechanism in which an oxygen atom is transferred from the coordinated nitro group to a bound CO.

Olefins are selectively oxidized to epoxides and alcohols by hydroperoxides in the presence of molybdenum and vanadium catalysts. The following mechanism is proposed(31)

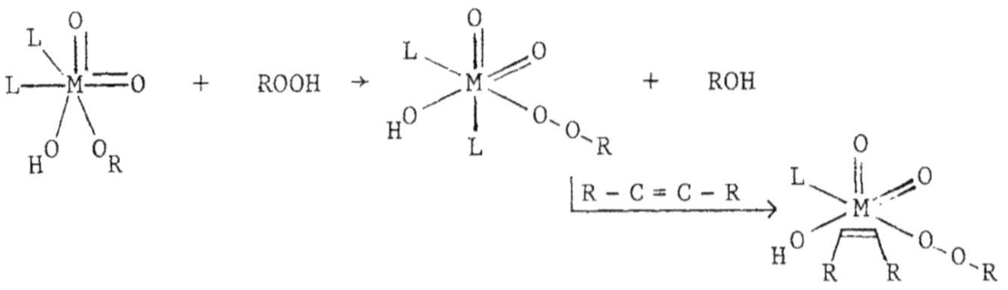

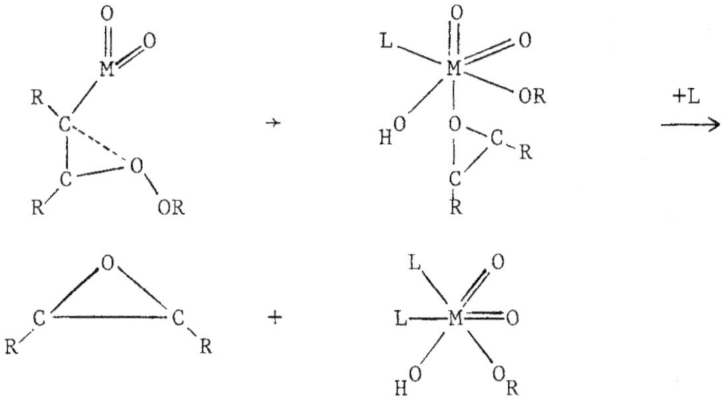

Reactions of chromyl chloride with olefins yields products expected from cis addition. For example RCH=CHR yields

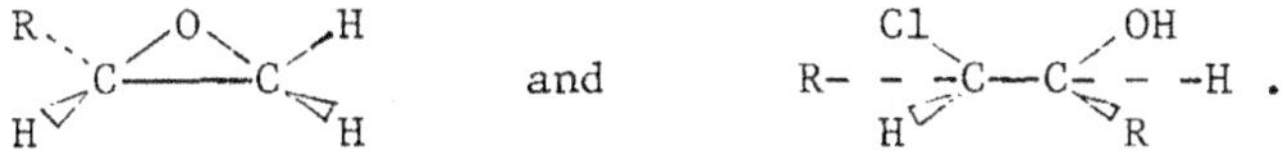

The following mechanism is suggested for epoxide formation:

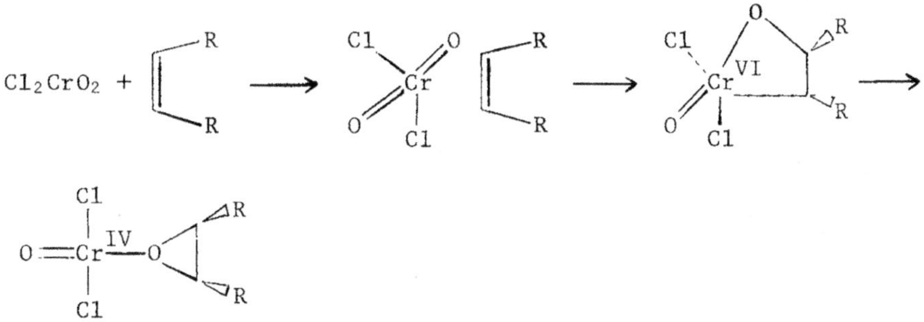

The well characterized reaction involving oxygen transfer from iodosylbenzene (33) should be mentioned for the sake of completeness and will be covered in more detail by Groves.

CONCLUSION

In this article I have attempted to summarize a vast body of literature by selecting those articles of special value to my understanding of how O_2 can be activated. Not included are many interesting reactions for which only minimal mechanistic information is available. Furthermore, I have left topics involving biological systems to others in the school. As in any study in which the chemistry is dominated by kinetic considerations instead of thermodynamic ones, it is difficult to come up with generalizations that can be universally applied. Instead I have tried to give an appreciation for ways that the oxidizing potential of O_2 can be taken advantage of to do selective oxidations. Hopefully these ideas will be of use to chemists in extending this chemistry and in discovering new reaction types for O_2.

Acknowledgement - The support of this work by the National Science Foundation and the Office of Naval Research is appreciated.

REFERENCES

(1) "Selective Catalytic Oxidation of Hydrocarbons: A Critical Analysis". Catalytica Associates Inc., Santa Clara, CA, Multiclient Study Mo 1077 October 1979.

(2) "Bond Energies, Ionization Potentials and Electron Affinities", Vedeneyev, V. I., et al., E. Arnold Publishers, London (1966).

(3) Sheldon, R. A.; Kochi, J. K., Adv. Catal., 1976, 25, 272.

(4) Lyons, J. E. Adv. Chem. Ser. 1974, 132, 64.

(5) Mayo, L. Acc. Chem. Res. 1968, 1, 193.

(6) Jensen, R. K.; Korcek, S.; Mahoney, L. R.; Zinbo, M. J. Am. Chem. Soc. 1981, 103, 1742.

(7) Kurkov, V. R., Pasky, J. Z. and Lavinge, J. B. J. Am. Chem. Soc. 1968, 90, 4743.

(8) Fusi, A.; Ugo, R.; Fox, F.; Pasini, A. and Cenini, S. J. Organomet. Chem. 1971, 26, 417.

(9) Hartley, F. R. J. Chem. Educ. 1973, 50, 263 and references therein.

(10) Tovrog, B. S.; Kitko, D. J.; Drago, R. S. J. Am. Chem. Soc. 1976, 98, 5144.

(11) Drago, R. S.; Corden, B. B. Acc. Chem. Res. 1980, 13, 353 and references therein.

(12) Zombeck, A.; Drago, R. S.; Corden, B. B. and Gaul, J. H. J. Am. Chem. Soc. 1981, 103, 7580.

(13) Howard, J. A.; Ingold, K. V. Can. J. Chem. 1967, 45, 785.

(14) Dudley, C. W.; Read, G.; Walker, P. J. C. J. Chem. Soc. Dalton 1974, 1926.

(15) Mimoun, H.; Machirant, M. M.; de Roch, I. S. J. Am. Chem. Soc. 1978 100, 5437· Mimoun, H.; Igersheim, F. Nouv. J. Chim. 1980, 4, 161.

(16) Nyberg, E. D.; Drago, R. S.; Pribich, D. submitted.

(17) Pignatello, J. J.; Jensen, F. R. J. Am. Chem. Soc. 1979, 101, 5929.

(18) Sen, A.; Halpern, J. J. Am. Chem. Soc. 1977, 95, 8399.

(19) Moody, D. C. private communication.

(20) Ryan, R. R.; Kubos, G. S.; Moody, D. C.; Eller, P. C. Structure and Bonding 1981, 46, 48.

(21) Walling, C.; Johnson, R. A. J. Am. Chem. Soc. 1975, 97, 363 and references therein.

(22) Walling, C.; El-Taliawi, G. M. J. Am. Chem. Soc., 1973, 95, 844 and references therein.

(23) Brodie, B. B.; Axelrod, J.; Shore, P. A.; Udenfriend, S. J. Biol. Chem. 1954, 208, 741.

(24) Breslow, R.; Lukens, L. N. J. Biol. Chem. 1960, 235, 292.

(25) Hamilton, G. A. J. Am. Chem. Soc. 1964, 86, 3391.

(26) Tovrog, B. S.; Mares, F.; Diamond, S. E. J. Am. Chem. Soc. 1980, 102, 6618 and references therein.

(27) Andrews, M. private communication.

(28) Booth, G.; Chatt, J. J. J. Chem. Soc. 1962, 2009.

(29) Doughty, D. T.; Gordon, G.; Stewart, R. P. J. Am. Chem. Soc. 1979, 101, 2645.

(30) Kriege-Simondsen, J. et al. Inorg. Chem. 1982, 21, 230.

(31) Sharpless, K. B.; Verhoeven, T. R. Aldrichimica Acta 1979, 12, 63.

(32) Sharpless, K. B.; Teranishi, A. Y.; Böckvall, J.-E. J. Am. Chem. Soc. 1977, 99, 3120.

(33) Groves, J. T.; Nemo, T. E.; Myers, R. S. J. Am. Chem. Soc. 1979, 101, 1032.

MÖSSBAUER STUDIES ON PUTIDAMONOOXIN - A NEW TYPE OF [2Fe-2S] CONTAINING OXYGENASE COMPONENT WITH A MONONUCLEAR NON-HEME IRON ION AS COFACTOR

E. Bill *, F.-H. Bernhardt **, A.X. Trautwein *

* Universität des Saarlandes, Saarbrücken, FRG and
Medizinische Hochschule Lübeck, Lübeck, FRG.
** RWTH, Aachen, FRG.

Abstract: We report Mössbauer studies on the active site of putidamonooxin, the terminal component of a monooxygenase which turns out to be specific in the sense that three iron atoms are involved in its activation of molecular oxygen, two of them belonging to a [2Fe-2S] ferredoxin-type chromophore, and the third being identified as mononuclear non-heme iron.

I. INTRODUCTION

The conjugated iron-sulfur protein putidamonooxin, PMO, (1-8) is, in the presence of the NADH-putidamonooxin oxidoreductase, a conjugated iron-sulfur protein (2,3,8,9) - and in the presence of NADH and O_2, the oxygen activating component of the 4-methoxybenzoate monooxygenase from *Pseudomonas putida*. The enzyme system catalyzes the O-demethylation of its physiological substrate 4-methoxybenzoate by a monooxygenase reaction of PMO resulting in the formation of 4-hydroxybenzoate and formaldehyde. PMO is not highly specific because different derivatives of benzoic acid can be bound, as indicated either by distinct changes in the visible absorption spectra of the oxidized PMO or by following O_2 consumption and NADH oxidation of the enzyme system (10-14).

II. MÖSSBAUER SPECTROSCOPIC INVESTIGATION

In this communication we present Mössbauer results obtained from investigating PMO under various conditions, i.e. in oxidized, reduced and aerobically reoxidized form, at different temperatures and magnetic fields, and with either the cluster- or the cofactor iron or both

I. Bertini, R. S. Drago, and C. Luchinat (eds.), The Coordination Chemistry of Metalloenzymes, 259–263.

being ^{57}Fe. The detailed experimental procedure of preparing the enzyme samples is described elsewhere (15).

1. [2Fe-2S] Clusters

Comparing our measured Mössbauer spectra of reduced PMO (Fig. 1 a) with the published Mössbauer spectra of other iron-sulfur proteins (16-23) we conclude that putidamonooxin contains [2Fe-2S] clusters. In its reduced form these clusters yield a typical ferric high-spin spectrum (A) and a typical ferrous high-spin spectrum (B). The corresponding spectrum of a [4Fe-4S] cluster would consist of one single, relatively broad doublet only.

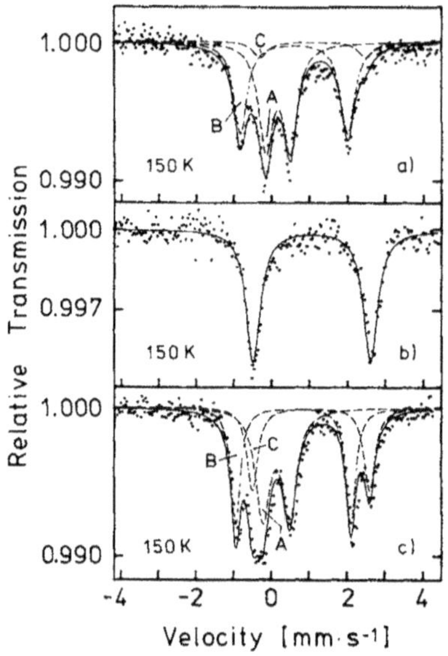

FIGURE 1. Mössbauer spectra of reduced PMO with 4-methoxybenzoate as substrate, enzyme (a) as obtained after gel filtration with only traces of cofactor iron, (b) with ^{57}Fe...[$2^{56}Fe$-2S], and (c) with ^{57}Fe... [$2^{57}Fe$-2S]. (A, B and C are defined in the text).

2. Mononuclear Non-Heme Iron

From Fig. 1 it is obvious that in additon to the quadrupole doublets A and B of the reduced [2Fe-2S] cluster, a line doublet C contributes to the total spectrum. This doublet could be identified as that of a cofactor iron, the role of which is described in the following.
Both, isomer shift and quadrupole splitting of doublet C indicate that the mononuclear non-heme iron in its reduced form is in the ferrous

high-spin state and is penta- or hexa- rather than tetra-coordinated.

The relatively large hyperfine field of 50 Tesla, which is derived from the spectrum of oxidized PMO with ^{57}Fe.../2^{56}Fe-2S/ taken at low temperature in an externally applied field (Fig. 2), indicates that also the oxidized mononuclear non-heme iron is in the high-spin state and presumably penta- or hexa-coordinated (15).

Adding molecular oxygen to the reduced enzyme and measuring the Mössbauer spectra of ^{57}Fe.../2^{56}Fe-2S/, yields an absorption pattern at low temperature in an externally applied field as shown in Fig. 2. At 150 K this magnetic pattern collapses into a line doublet with very small intensity and remains therefore unresolved when oxidized ^{57}Fe.../2^{57}Fe-2S/ is measured at 150 K (Fig. 3). The spectrum of Fig. 3 was obtained from the sample which corresponds to Fig. 1c, after molecular oxygen had been added. The line analysis of Fig. 1c shows that the ratio of reduced

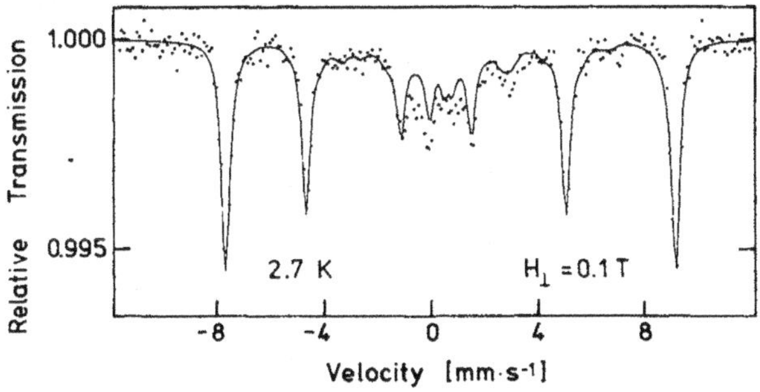

FIGURE 2. Mössbauer spectrum of oxidized PMO with 4-methoxybenzoate as substrate, i.e. ^{57}Fe.../2^{56}Fe-2S/ in a field of 0.1 Tesla applied perpendicular to the direction of γ -rays.

/2^{57}Fe-2S/ clusters to reduced mononuclear non-heme ^{57}Fe is 1:0.65. From a corresponding line analysis of Fig. 3 we learn that only 65 % of the /2^{57}Fe-2S/ clusters have been oxidized, and 35 % of them remained in the reduced form. From this finding it is obvious that the ratio of oxidized /2Fe-2S/ clusters to oxidized mononuclear non-heme iron is 1:1. Thus, during aerobic oxidation of reduced PMO, molecular oxygen is activated by the transfer of one electron from the reduced /2Fe-2S/ cluster and one electron from the reduced mononuclear non-heme iron (15).

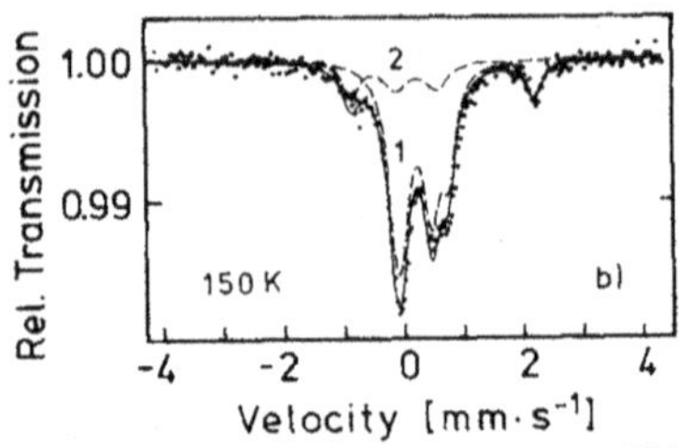

FIGURE 3. Mössbauer spectrum of PMO after adding molecular oxygen to the sample corresponding to Fig. 1c. (Subspectrum 1 represents oxidized [2^{57}Fe-2S] clusters and subspectrum 2 represents reduced [2^{57}Fe-2S] clusters).

III. CONCLUSION

The present results together with former biochemical findings (5,7,10) lead to the general reaction sequence of 4-methoxybenzoate monooxygenase as symbolized by Fig. 4.

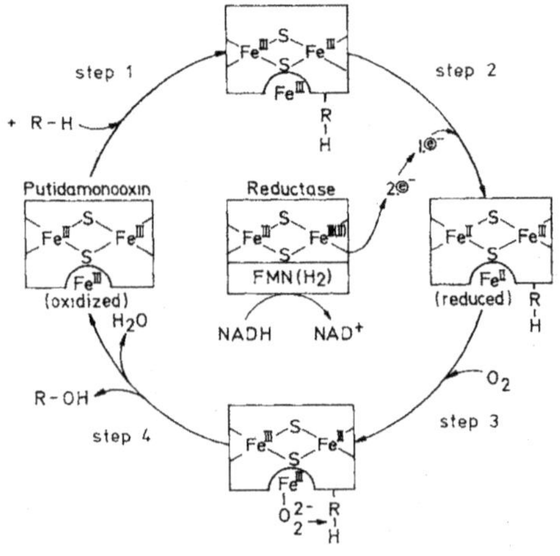

FIGURE 4. Proposed reaction cycle of putidamonooxin for the hydroxylation of 4-methoxybenzoate. R-H=substrate, R-OH=hydroxylated reaction product.

This work was supported by the Deutsche Forschungsgemeinschaft.

REFERENCES

(1) Bernhardt, F.-H., Ruf, H.-H., Staudinger, H.J. & Ulrich, V. (1971) Hoppe-Seyler's Z. Physiol. Chem. 352, pp. 1091-1099.
(2) Bernhardt, F.-H. & Staudinger, H.J. (1973) Hoppe-Seyler's Z. Physiol. Chem. 354, p. 217.
(3) Bernhardt, F.-H., Pachowsky, H. & Staudinger, H.J. (1975) Eur. J. Biochem. 57, pp. 241-256.
(4) Bill, E., Bernhardt, F.-H., Marathe, V.R. & Trautwein, A. (1980) J. Phys. Colloq. 41, pp. C 1-485-486.
(5) Adrian, W., Bernhardt, F.-H., Bill, E., Gersonde, K., Heymann, E., Trautwein, A. & Twilfer, H. (1980) Hoppe-Seyler's Z. Physiol. Chem. 361, p. 211.
(6) Adrian, W., Bill, E., Bernhardt, F.-H. & Trautwein, A.X. (1980) Biophys. Struct. Mech. (suppl.) 6, p. 23.
(7) Bernhardt, F.-H. & Meisch, H.M. (1980) Biochem. Biophys. Res. Commun. 93, pp. 1247-1253.
(8) Twilfer, H., Bernhardt, F.-H. & Gersonde, K. (1981) Eur. J. Biochem. 119, pp. 595-602.
(9) Bernhardt, F.-H. & Pachowsky, H. (1974) IRCS (Int. Res. Commun. Syst.) Med. Sci.-Libr. Compend. 2, p. 1267.
(10) Bernhardt, F.-H., Erdin, N. & Staudinger, H.J. (1973) Eur. J. Biochem. 35, pp. 126-134.
(11) Bernhardt, F.-H., Ruf, H.-H. & Ehrig, H. (1974) FEBS Lett. 43, pp. 53-55.
(12) Bernhardt, F.-H. & Seydewitz, V. (1975) IRCS (Int. Res. Commun. Syst.) Med. Sci.-Libr. Compend. 3, p. 113.
(13) Bernhardt, F.-H. & Ruf, H.-H. (1975) Biochem. Soc. Trans. 3, pp. 878-881.
(14) Bernhardt, F.-H., Heymann, E. & Traylor, P.S. (1978) Eur. J. Biochem. 92, pp. 209-223.
(15) Bill, E., Bernhardt, F.-H., Trautwein, A.X. (1981) Eur. J. Biochem., 121, pp. 39-46.
(16) Sands, R.H. & Dunham, W.R. (1975) Qu. Rev. Biophys. 7, pp. 443-504.
(17) Cammack, R. (1976) J. Phys. Colloq. 37, pp. C 6-137-151.
(18) Münck, E. & Zimmermann, R. (1976) in Mössbauer Effect Methodology (Gruverman, I.J. & Seidel, C.W., eds) vol. 10, 1st. edn, pp. 119-154, Plenum Press, New York and London.
(19) Debrunner, P.G. (1976) in Applications of Mössbauer Spectroscopy (Cohen, R.L., ed.) vol. 1, 1st edn, pp. 171-196, Academic Press, New York, San Francisco, London.
(20) Johnson, C.E. (1975) in Topics in Applied Physics (Gonser, U., ed.) vol. 5, 1st. edn, pp. 139-166, Springer-Verlag, New York, Heidelberg, Berlin.
(21) Cammack, R., Dickson, D.P.E. & Johnson, C.E. (1977) in Iron Sulfur Proteins (Lovenberg, W., ed.) vol. III, pp. 283-330, Academic Press, New York.
(22) Debrunner, P.G., Münck, E., Que, L. & Schulz, C.E. (1977) in Iron Sulfur Proteins (Lovenberg, W., ed.) vol. III, pp. 381-417, Academic Press, New York.
(23) Münck, E. (1978) Methods Enzymol. 54, pp. 346-379.

THE ENZYME-SUBSTRATE INTERACTION IN THE CATECHOL DIOXYGENASES

Lawrence Que, Jr.*
Randall B. Lauffer
Robert H. Heistand, II

Department of Chemistry - Baker Laboratory
Cornell University, Ithaca, New York 14853 U.S.A.

Abstract

Investigations into the enzyme-substrate complex of the catechol dioxygenases have raised the possibility that the catechol may bind to the metal center in a monodentate configuration. Based on NMR studies of model monodentate and chelated catecholate complexes, the substrate is shown to be coordinated to the ferric center through only one oxygen in the catechol 1,2-dioxygenase-4-methylcatechol complex. Rationalizations for this preferred configuration are suggested by the oxygen reactivity of various model compounds.

The catechol dioxygenases are bacterial enzymes which catalyze the oxidative cleavage of o-dihydroxybenzenes to yield aliphatic products (1,2); two of these have been studied in some detail, namely protocatechuate 3,4-dioxygenase (PCD) and catechol 1,2-dioxygenase (CTD). The active sites of these enzymes consist of mononuclear high-spin ferric centers (3,4) coordinated to at least two distinct tyrosinates (5,6). Water has also been implicated as a ligand in the PCD from Brevibacterium fuscum on the basis of the broadening of the high-spin ferric EPR signals in $H_2{}^{17}O$ (7). Other ligands to the metal are currently unknown.

The native enzymes exhibit a brownish-pink color as a result of tyrosinate-to-iron(III) charge transfer interactions (Figure 1, 8-12). Anaerobic addition of substrate generates a purple or grayish-blue complex with increased absorbance at long wavelength (Figure 1). Resonance Raman studies of both PCD and CTD enzyme-substrate complexes show enhanced Raman vibrations due to substrate, indicating catecholate

I. Bertini, R. S. Drago, and C. Luchinat (eds.), The Coordination Chemistry of Metalloenzymes, 265–271.

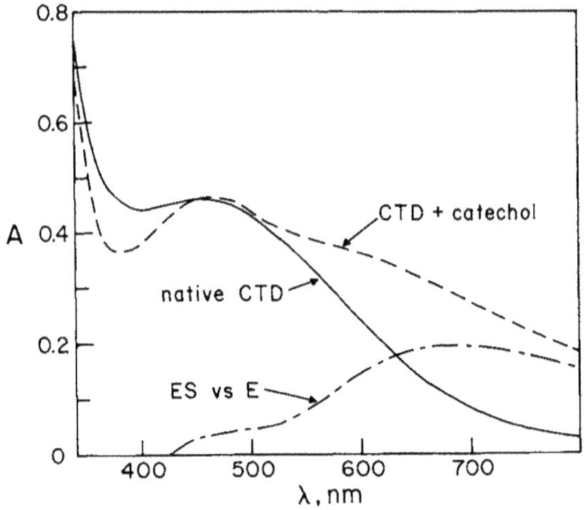

Figure 1. Visible spectra of native CTD and its complex with catechol.

coordination to the metal center (9,11). Furthermore, Mössbauer investigations unequivocally demonstrate that the metal center remains in the high-spin ferric state upon substrate binding (3,4). Two possible coordination modes for the catechol are immediately apparent - one, a chelated configuration with many examples in the literature (13) and the other, where only one oxygen is coordinated, with only two such structures known (14,15).

Suggestions that the monodentate configuration may be important come from steady state inhibition kinetic studies on PCD with isomeric hydroxybenzoates showing that the para hydroxybenzoates are much better inhibitors than the meta isomers (16,17). A comparison of the resonance Raman spectra of PCD complexes with p- and m-hydroxybenzoates (Figure 2) shows the presence of a resonance enhanced ν_{CO} feature due to the coordination of the phenolate oxygen in the p-hydroxybenzoate complex (assigned by deuteration of the inhibitor ring protons) and the absence of this feature in the m-hydroxybenzoate complex. These observations suggest that the substrate may bind to the metal center through only one oxygen (6).

For our spectroscopic studies, we have synthesized (18) the complexes, Fe(salen)catH and $[Fe(salen)cat]^-$, as examples of complexes with monodentate and chelated catecholate configurations (salen = ethylenebis(salicylidenimine) dianion, $catH_2$ = catechol). Salen serves as an analogue for the active site, providing two phenolate functionalities. Fe(salen)catH has been demonstrated to be five-coordinate

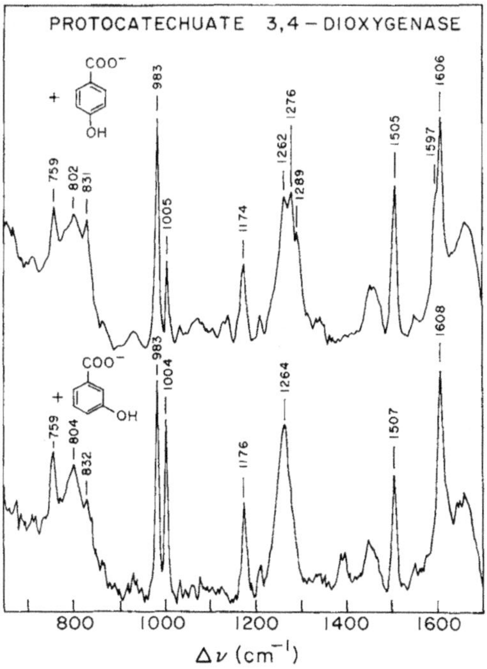

Figure 2. Resonance Raman spectra of PCD complexes with *p*-hydroxybenzoate and *m*-hydroxybenzoate.

in CH_2Cl_2 solution (18), while Fe(saloph)catH has been shown by x-ray crystallography to have a square pyramidal geometry with the tetradentate ligand occupying the basal plane and the catecholate clearly monodentate at the apical site (14, saloph = *o*-phenylenebis(salicylidenimine) dianion). Upon treatment with potassium *t*-butoxide, Fe(salen)catH loses its proton, generating [Fe(salen)cat]$^-$, the potassium salt crystallized in the presence of 18-crown-6. X-ray crystallography on these crystals reveals a complex with distorted octahedral geometry, the catecholate chelated to the iron (19).

The ^{1}H-NMR spectra of these complexes indicate that the mode of catecholate coordination may be deduced directly from the isotropic shifts exhibited by the catecholate protons. Figure 3 shows the 300 MHz NMR spectra of Fe(salen)(OC_6H_4-4-Me), Fe(salen)(4-Me-catH), and [Fe(salen)(4-Me-cat)]$^-$. These spectra have been previously assigned, the observed shifts consistent with a π delocalization mechanism (18). The methyl resonance in Fe(salen)(OC_6H_4-4-Me) is observed at ca. 110 ppm downfield of Me_4Si, while methyl groups meta to the phenolate oxygen, as in Fe(salen)(OC_6H_3-3,5-Me_2), exhibit upfield shifts of

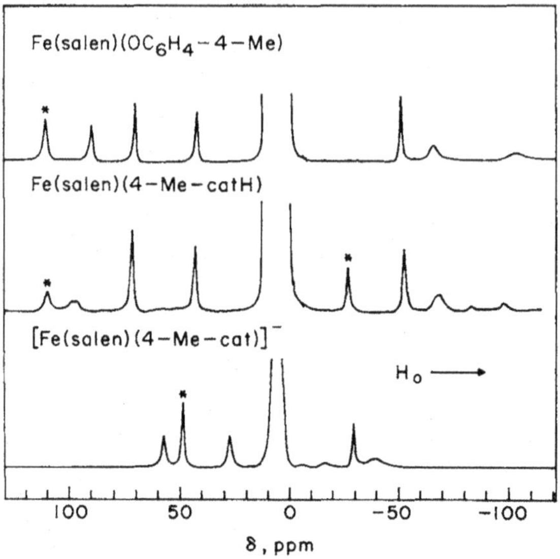

Figure 3. 300MHz ^{1}H NMR spectra of Fe(salen)X complexes - * denotes resonances arising from methyl groups.

ca. 30 ppm. The isotropic shifts of the monodentate catecholate complexes closely resemble those of the phenolate complexes. The methyl resonance in Fe(salen)(4-Me-catH) is found at ca. 110 ppm downfield and ca. 30 ppm upfield, indicating the presence of two isomers in solution, with the shift of the methyl resonance dependent on which catecholate oxygen is coordinated to the iron. In contrast, the methyl resonance in the chelated $[Fe(salen)(4\text{-Me-cat})]^-$ is observed at ca. 50 ppm downfield. This can be readily understood with the realization that in the chelated configuration there are two competing pathways for spin delocalization. The methyl group is para to one oxygen, resulting in a large downfield shift, and at the same time meta to the other oxygen, giving rise to a smaller upfield shift. The partial cancellation of these two effects results in the somewhat smaller downfield shift. These observations thus provide the basis for the NMR determination of the substrate binding configuration in the ES complex of CTD.

CTD is a protein of molecular weight 63,000 with an αβFe composition (20). Catechol is the natural substrate, though 4-methylcatechol is almost as easily cleaved by the enzyme (21). Figure 4 shows the 300MHz ^{1}H NMR spectra of CTD complexed with catechol, 4-methylcatechol, and 4-methyl-d_3-catechol. Reasonably well-resolved peaks in the down-

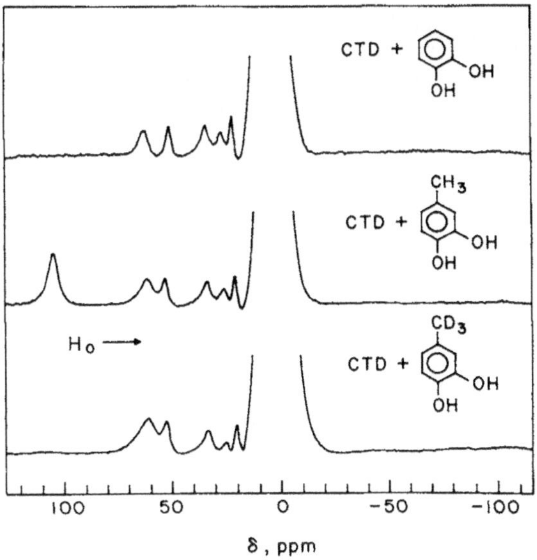

Figure 4. 300MHz ^{1}H NMR spectra of CTD-substrate complexes.

field region are observed, most of which arise from protein residues coordinated to the iron. The assignments of these resonances are currently being pursued. A comparison of the three spectra in Figure 4 clearly reveals the resonance at ca. 100 ppm downfield as arising from the methyl group of 4-methylcatechol. This large downfield shift indicates the coordination of the catechol to the ferric center through only the oxygen para to the methyl group, thus providing direct evidence for the monodentate coordination of substrate in CTD.

Why the requirement for a monodentate configuration? A possible explanation is suggested by studies of the oxygen reactivity of two model complexes (22). Fe(salen)DBcatH reacts readily with oxygen to yield Fe(salen)DBSQ as monitored by NMR spectroscopy (DBcatH_2 = 3,5-di-*tert*-butylcatechol, $DBSQ^-$ = 3,5-di-*tert*-butyl-*o*-benzosemiquinone anion). By contrast, $[Fe(salen)DBcat]^-$ is stable in oxygenated THF solution over a period of 24 hours. Furthermore, Fe(salen)DBSQ reacts with superoxide to yield $[Fe(salen)DBcat]^-$ and oxygen, demonstrating the thermodynamic preference for the chelated catecholate state. Cyclic voltammetric studies on both Fe(salen)DBSQ and $[Fe(salen)DBcat]^-$ show the redox potential of the coordinated semiquinone-catecholate couple to be -180 mV vs. SCE; this is to be compared with the reduction of free $DBSQ^-$ at -1340 mV vs. SCE (23). Thus chelation stabilizes the catecholate oxidation state by more than a volt, rendering the chelated complex unreactive to oxygen. Such shifts in potential have also been observed for other chelated catecholate complexes (24-26).

In conclusion, the data presented above demonstrate that the substrate catechol coordinates to the active site iron in a monodentate configuration and that this configuration appears to be necessary for reaction with oxygen.

Acknowledgments. This work has been supported by the United States Public Health Service Grant GM-25422. The Bruker WM-300 NMR spectrometer was obtained in part with a grant from the National Science Foundation CHE-7904825. L.Q. is an Alfred P. Sloan Foundation fellow (1982-84) and the recipient of an NIH Research Career Development Award 1 K04 AM-00974.

References

(1) Nozaki, M.: 1979, Top. Curr. Chem., 78, pp. 145-186.
(2) Que, L., Jr.: 1980, Struct. Bonding (Berlin), 40, pp. 39-72.
(3) Que, L., Jr., Lipscomb, J.D., Zimmermann, R., Münck, E., Orme-Johnson, N.R., Orme-Johnson, W.H.: 1976, Biochim. Biophys. Acta, 452, pp. 320-334.
(4) Kent, T., Münck, E., Que, L., Jr., Widom, J.: unpublished results.
(5) Que, L., Jr., Heistand, R.H., II, Mayer, R., Roe, A.L.: 1980, Biochemistry, 19, pp. 2588-2593.
(6) Que, L., Jr., Epstein, R.M.: 1981, Biochemistry, 20, pp. 2545-2549.
(7) Lipscomb, J.D., Whittaker, J.D., Arciero, D.M. in "*Oxygenases and Oxygen Metabolism*" (Nozaki, M., Yamamoto, S., Ishimura, Y., Coon, M.J., Ernster, L., Estabrook, R.W., eds.) Academic Press, Tokyo, in press.
(8) Tatsuno, Y., Saeki, Y., Iwaki, M., Yagi, T., Nozaki, M., Kitagawa, T., Otsuka, S.: 1978, J. Am. Chem. Soc., 100, pp. 4614-4615.
(9) Keyes, W.E., Loehr, T.M., Taylor, M.L.: 1978, Biochem. Biophys. Res. Comm., 83, pp. 941-945.
(10) Felton, R.H., Cheung, L.D., Phillips, R.S., May, S.W.: 1978, Biochem. Biophys. Res. Comm., 85, pp. 844-850.
(11) Que, L., Jr., Heistand, R.H., II: 1979, J. Am. Chem. Soc., 101, pp. 2219-2221.
(12) Bull, C., Ballou, D.P., Salmeen, I.: 1979, Biochem. Biophys. Res. Comm., 87, pp. 836-841.
(13) Pierpont, C.G., Buchanan, R.M.: 1981, Coord. Chem. Rev., 38, pp. 44-87.
(14) Heistand, R.H., II, Roe, A.L., Que, L., Jr.: 1982, Inorg. Chem., 21, pp. 676-681.
(15) Sacconi, L., Orioli, P.L., Vaira, M.: 1967, J. Chem. Soc. Chem. Comm., pp. 849-850.
(16) May, S.W., Phillips, R.S., Oldham, C.D.: 1978, Biochemistry, 17, pp. 1853-1860.

(17) Que, L., Jr., Lipscomb, J.D., Münck, E., Wood, J.M.: 1977, Biochim. Biophys. Acta, 485, pp. 60-74.

(18) Heistand, R.H., Lauffer, R.B., Fikrig, E., Que, L., Jr.: 1982, J. Am. Chem. Soc., 104, pp. 2789-2796.

(19) Lauffer, R.B., Heistand, R.H., II, Que, L., Jr.: submitted for publication.

(20) Nakai, C., Kagamiyama, H., Saeki, Y., Nozaki, M.: 1979, Arch. Biochem. Biophys., 195, pp. 12-22.

(21) Fujiwara, M., Golovleva, L.A., Saeki, Y., Nozaki, M., Hayaishi, O.: 1975, J. Biol. Chem., 250, pp. 4848-4855.

(22) Lauffer, R.B., Heistand, R.H., II, Que, L., Jr.: 1981, J. Am. Chem. Soc., 103, pp. 3947-3949.

(23) Nanni, E.J., Jr., Stallings, M.D., Sawyer, D.T.: 1980, J. Am. Chem. Soc., 102, pp. 4481-4485.

(24) Rohrscheid, F., Balch, A.L., Holm, R.H.: 1966, Inorg. Chem., 5, pp. 1542-1551.

(25) Wicklund, P.A., Brown, D.G.: 1976, Inorg. Chem., 15, pp. 396-400.

(26) Balch, A.L.: 1973, J. Am. Chem. Soc., 95, pp. 2723-2724.

FERRIC NITRILOTRIACETATE: AN ACTIVE CENTRE ANALOGUE OF PYROCATECHASE

Michael G. Weller

Inst. Physiolog. Chemie, Anorg. Biochemie, Univ. Tübingen
Hoppe-Seyler-Str. 1, 7400 Tübingen, FRG

Abstract: Ferric nitrilotriacetate Fe(NTA) binds 3,5-di-t-butyl-catechol $DBcatH_2$ to form a ternary complex, $[Fe(NTA)(DBcat)]^{2-}$. Upon exposure to O_2, dioxygenation leads to ring cleavage of the catechol, yielding muconic acid derivatives. Binding and subsequent intradiol ring cleavage of catechol are characteristics of pyrocatechase, a non heme ferric dioxygenase. Thus, Fe(NTA) mimics some properties of the active centre of this enzyme.

Pyrocatechase (catechol 1.2-dioxygenase, EC 1.13.11.1), a non heme ferric dioxygenase (1), catalyzes the intradiol ring cleavage of catechol to cis,cis-muconic acid (scheme 1a). The iron(III) in the active centre of the enzyme is coordinated to tyrosinate ligands of the protein. The substrate catechol is known to bind to the iron first, and a $EFeSO_2$-complex is formed subsequently, befor ring cleavage occurs (2,3). (see preceding contribution for details!) The dioxygenation mechanism remains to be discovered.

Strikingly, Fe^{3+} ions in aqueous solution are found to catalyze catechol oxidation, yielding quinone instead of muconic acid (4) (scheme 1b). This reaction can be demonstrated to proceed via a ferric catecholate 1, ferric semiquinone 2 sequence. The latter intermediate was isolated as ferric tris-semiquinone in the case of 3,5-di-t-butyl-catechol (M.G.Weller and U.Weser, to be published).

Apparently, the protein environment induces significant changes in the iron-catechol moiety, leading to the "biologically usefull" dioxygenation rather than oxidation to quinone upon reaction with O_2. We here report a model study, introducing a simple ligand to replace the protein.

The ligand of choice would have to (i) exhibit a high stability constant with iron(III) to ensure ternary complex formation with catechol, (ii) be not more than 4- or 5-dentate to allow catechol to be coordinated additionally, (iii) be a "hard" base in order to minimize electron delocalization in the semiquinone chelate 2 (thus rendering

I. Bertini, R. S. Drago, and C. Luchinat (eds.), The Coordination Chemistry of Metalloenzymes, 273–278.

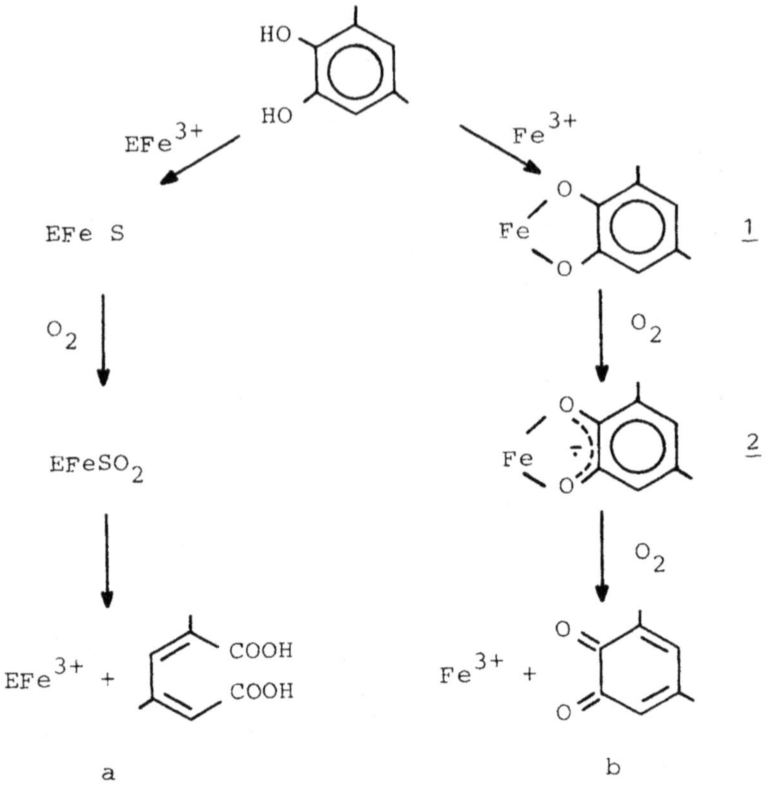

scheme 1

the formation of this "wrong" intermediate, which would subsequently yield quinone, less favourable).

One such ligand is nitrilotriacetate (NTA; 4-dentate, log K(Fe^{3+})= 15.9). And indeed, with ferric perchlorate, NTA, and 3,5-di-t-butylcatechol in aqueous borate buffer/DMF(5), a mixed complex $[Fe(NTA)(DBcat)]^{2-}$ 3 is formed, which upon incubation with dioxygen gave the lactone 5 (scheme 2) in high yields. Lactone 5 represents the Markovnikov-product of γ-lactonization of muconic acid 4, clearly indicating that ring cleavage has taken place.

In table 1 some ring cleavage reactions are listed (reactions 1-3). Apart from a few % of catechol and quinone, more than 80% lactone was found (6), even with 100-fold excess catechol (reaction 3): In the absence of NTA, only the quinone was detected as oxidation product (reaction 4). With Fe^{3+} also being absent, no reaction at all was observed, showing that borate, besides serving as a buffer, is also

scheme 2

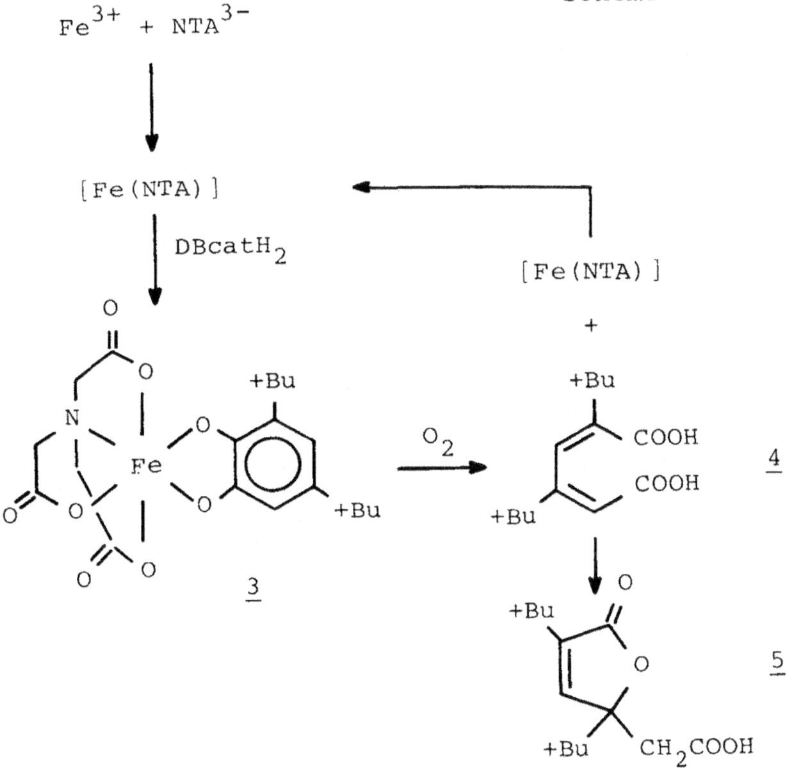

Table 1

Reaction[a]	$DBcatH_2$	NTA^{3-}	Fe^{3+}	time (days)	product %[b] DBcat	DBqu	lactone 5	turnover[c]
1	5	5	5	4	0	10	81	0.8
2	5	0.5	0.5	5	0	3	84	8.4
3	5	0.05	0.05	7	5	2	80	80
4	3	-	1	5	8	75	-	-
5	3	-	-	5	84	5	-	-

a) in 1 part aqueous 0.6M borate buffer pH 8.5; 2 parts DMF
b) based on original $DBcatH_2$; identification by ir, ^{1}H-nmr and analysis
c) mol ring cleavage product per mol Fe^{3+}

effectively protecting catechol from oxidation by forming the borate complex (reaction 5).

Thus, the system mimics the enzyme reaction, and it does so catalytically.

As to other inorganic dioxygenase models, several copper complexes have been reported to effect catechol cleavage (7). At least some of these, however, do not require dioxygen for the cleavage reaction, but for regeneration of the cupric ion, the latter being the oxidizing agent. Further, a mixture of 3,5-di-t-butylcatechol, Fe^{2+}, bipy and pyridine in THF gave some muconic acid besides quinone as the main product (8). Consequently, Fe(NTA) is the first inorganic system catalyzing the pyrocatechase reaction.

In addition to this functional analogy, there are spectroscopic similarities to the ferric centre of pyrocatechase: in the enzyme, upon binding of catechol to iron, a large incrase of long wavelength absorbance (around 650 nm, vs. λ max 440 nm for the native enzyme) is observed (3). This is paralleled by a change from orange Fe(NTA) to the blue-green $[Fe(NTA)(DBcat)]^{2-}$, λ max 660 nm in H_2O. As for the epr spectrum, the strong signal of pyrocatechase at $g = 4.3$ disappears with bound substrate; at 77 K the ES complex is epr-silent (3). The $g = 4.3$ signal of Fe(NTA) greatly diminishes upon coordination of DBcat and becomes superimposed on a new broad peak centred around $g = 4.2$; signals at $g = 9.o$, 7.55 and 5.5 are seen additionally (9). All signals are relatively week and would probably not be detectable in biological material.

Under anearobic conditions, the ternary complex was crystallized as a piperidinium salt and characterized as $[Fe(NTA)(DBcat)](pipH)_2$. Analysis, ir, vis and epr data are consistent with a ferric complex <u>3</u>, i.e. with chelated catechol rather than, e.g., monodentate catechol or semiquinone (6,10).

Due to the slow ring cleavage in the model system (see table), no information can yet be provided regarding the reaction mechanism. Que et al. (3) (see preceding communication) have proposed a mechanism for the pyrocatechase catalytic cycle, starting with an attack of dioxygen on iron(III)-coordinated monodentate catechol. Such a monodentate intermediate is of course also possible for the model system. This complex species would then accomodate coordination of superoxide or peroxide, respectively, to the iron, as proposed by Que. An example for this potential second intermediate could be formulated as $[Fe(NTA)(DBsq)(O_2^-\cdot)]^{2-}$ <u>6</u>.

In this context it seems of interest to note that neither catecholate nor semiquinone as monodentate ligands in ternary ferric salen complexes yielded ring cleavage upon oxidation; simple electron exchange, however, occurred quite readily (11). In these compounds, the tetragonal salen ligand apparently does not allow catechol and

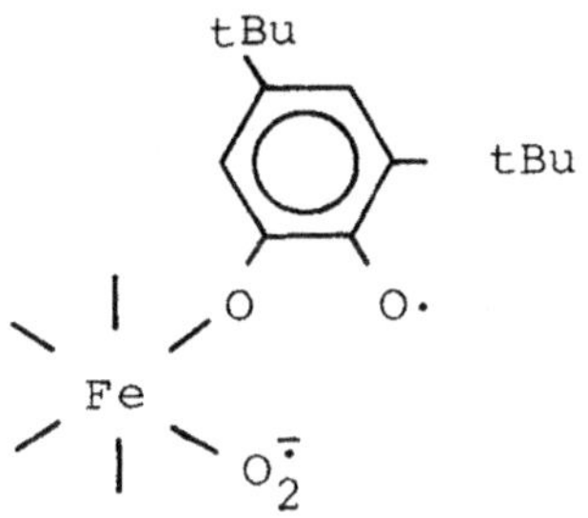

6

oxygen to be bound to the iron in adjacent positions, thus preventing reactions between these two components other than electron exchange. Vice versa, this may well be the reason for the non heme nature of ferric dioxygenases: both with $[Fe(NTA)(DBcat)]^{2-}$ and in, presumably, pyrocatechase, an intermediate of type 6 can reasonably be assumed; but not in complexes with tetragonal ligands like heme or salen coordinated to a ferric centre.

References

(1) Hayaishi,O., Katagiri,M., Rothberg,S. J.Am.Chem.Soc. 1955, 77, 5450-5451.

(2) Nozaki,M. Topics current Chem. 1979, 78, 145-186.

(3) Que,Jr.,L. Struct. Bonding (Berlin) 1980, 40, 39-72.

(4) Grinstead,R.R. Biochem. 1964, 3, 1308-1314.

(5) The solutions showed glass electrode readings of "pH" 9.1.

(6) Weller,M.G., Weser,U. J.Am.Chem.Soc., in press.

(7) Brown,D.G., Beckmann,L., Ashby,C.H., Vogel,G.C., Reinprecht,J.T., Tetrah. Lett. 1977, 1363-1364.
Tsuji,J., Takayanagi,H. J.Am.Chem.Soc. 1974, 96, 7349-7350.
Rogic,M.M., Demmin,T.R., Hammond,W.B. J.Am.Chem.Soc. 1976, 98, 7441-7443.
Rogic,M.M., Demmin,T.R. J.Am.Chem.Soc. 1978, 100, 5472-5487.
Demmin,T.R., Swerdloff,M.D., Rogic,M.M. J.Am.Chem.Soc. 1981, 103, 5795-5804.

(8) Funabiki,T., Sakamoto,H., Yoshida,S., Tarama,K. Chem.Commun. 1979, 754-755.

(9) Varian E 109 spectrometer, temperature 77K, frequency 9.24 GHz, modulation amplitude 10 G, microwave power 50 mW.

(10) Pierpont,C.G., Buchanan,R.M. Coord.Chem.Reviews, 1981, 38, 45-87.
Kessel,S.L., Emberson,R.M., Debrunner,P.G., Hendrickson,D.N. Inorg. Chem. 1980, 19, 1170-1178.
Harris,W.R., Carrano,C.J., Cooper,S.R., Sofen,S.R., Avdeef,A.E., McArdle,J.V., Raymond,K.N. J.Am.Chem.Soc. 1979, 101, 6097-6104 and ref. therein.
Anderson,B.F., Buckingham,D.A., Robertson,G.B., Webb,J., Murray, K.S., Clark,P.E. Nature 1976, 262, 722-724.
Salama,S., Stong,J.D., Neilands,J.B., Spiro,T.G. Biochem. 1978, 17, 3781-3785.
Hider,R.C., Mohd-Nor,A.R., Silver,J., Morrison,I.E.G., Rees,L.V.C. J.Chem.Soc. Dalton, 1981, 609-622.
Que Jr.,L., Heistand II,R.H. J.Am.Chem.Soc. 1979, 101, 2219-2221.
Brown,D.G., Johnson III,W.L. Z.Naturforsch. 1979, 34b, 712-715.

(11) Lauffer,R.B., Heistand II,R.H., Que Jr.,L. J.Am.Chem.Soc. 1981, 103, 3947-3949.

STEREOSELECTIVE PEROXIDATIC ACTIVITY OF IRON(III) COMPLEX IONS SUPPORTED ON POLYPEPTIDES

B. Pispisa

Istituto di Chimica Fisica, Università di Roma and Istituto Chimico, Università di Napoli

$[Fe(tetpy)(OH)_2]^+$ complex ions (tetpy = 2,2': 6',2":6",2"'-tetrapyridyl) anchored to $(L\text{-}Glu)_n$ (FeL) or $(D\text{-}Glu)_n$ (FeD) catalyze the H_2O_2-oxidation of L-ascorbate anion, L-dopa (3,4-dihydroxyphenylalanine) and L-adrenaline (epinephrine) at pH 7. It appears that the catalytic sequence involves 1) formation of a substrate-catalyst precursor complex and 2) electron transfer within this system. Oxidation of the lower valence metal chelate and substrate radical by H_2O_2 occurs in subsequent fast steps. "Renaturation" phenomena on the charged polypeptide matrices are observed upon progressive binding of iron(III) chelate ions. They are coupled with stereospecific effects in the catalysis in the sense that the larger is the amount of α-helix in the polymeric supports, the greater is the stereoselectivity. Implications of the stereochemistry of the precursor complex in the mechanism of the electron transfer reaction are discussed.

The understanding of the mode of action of metalloenzymes chiefly depends on the knowledge of the structural features of the protein surrounding the metal ions. Synthetic poly(α-aminoacids) can reproduce a great variety of structures and properties which characterize the proteins (1). It seems reasonable to assume that their combination with metal ions or metal derivatives exhibits some resemblance to metalloproteins. This possibility was investigated by examining the catalytic activity of iron(III) complex ions bound to polypeptides (2).

I. Bertini, R. S. Drago, and C. Luchinat (eds.), The Coordination Chemistry of Metalloenzymes, 279–290.

Iron-ion-catalyzed or complex-ion-catalyzed disproportionation of hydrogen peroxide (3) and oxidation of ascorbic acid by molecular oxygen or H_2O_2 (4) have been previously suggested as models for hemoproteins, such as catalases and peroxidases.

We have recently found a formal similarity between the decomposition reaction of H_2O_2 catalyzed by trans-$[Fe(tetpy)(OH)_2]^+$ ions (hereafter called FeTETPY, where tetpy = 2,2':6',2": 6",2"'-tetrapyridyl) (5) anchored to sodium poly(glutamate) $[(Glu)_n]$ and catalase. At variance with the results obtained using polymer-free FeTETPY ions (6), both mechanism and features of the intermediate were reminiscent of those observed with the enzyme.

More recently, the peroxidatic activity of FeTETPY-$(Glu)_n$ was investigated (7). The catalysis shows a high efficiency as compared to that observed with simple iron complexes (4). In addition, it proceeds stereoselectively when optically active substrates, such as L(+)-ascorbate anion, L-dopa and L-adrenaline, are oxidized in the presence of FeTETPY-L(Glu) (FeL) or FeTETPY-$(D\text{-}Glu)_n$ (FeD) system (Fig. 1 and 2), according to the reactions (7,8):

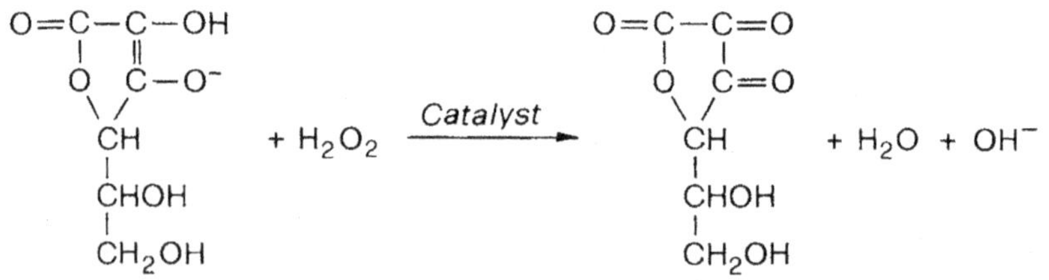

Ascorbate anion **Dehydroascorbic acid**

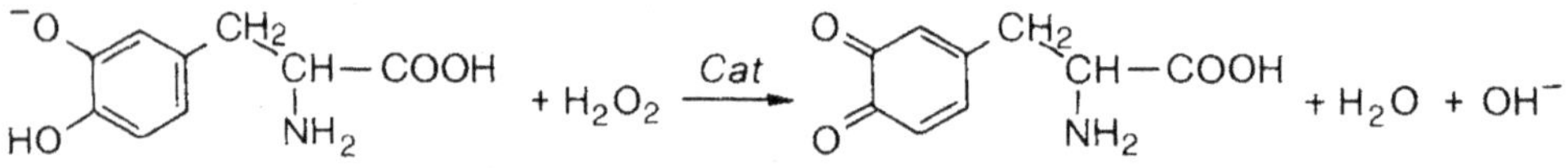

Dopa **Dopa-quinone**

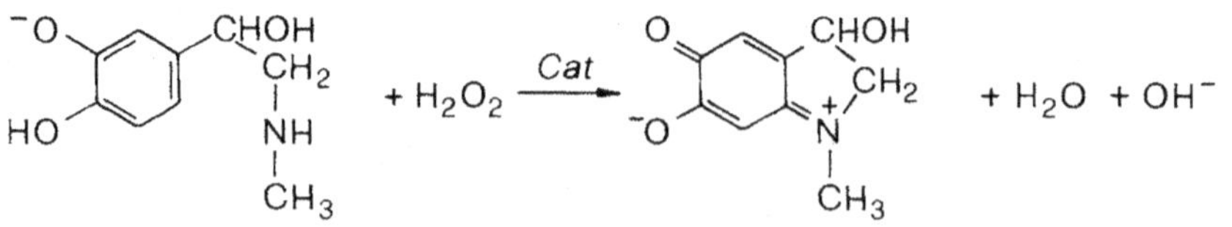

Adrenaline **Adrenochrome**

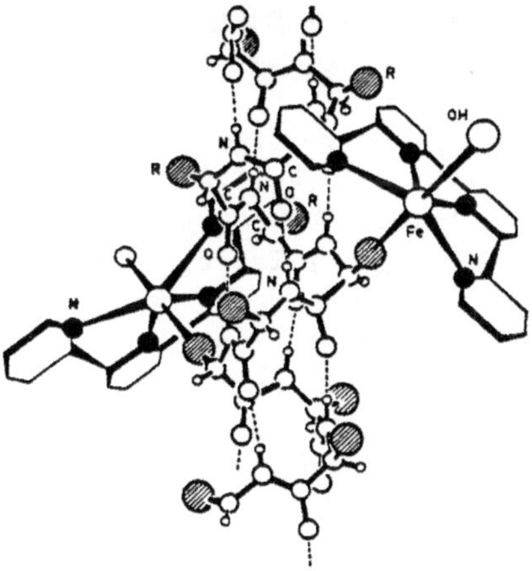

Fig. 1 - Molecular model of FeL catalytic system, at high [C] / [P] ratio ($R = -CH_2-CH_2-COO^-$). Actually, under these conditions, aggregates form. They are probably of the side to side type with a freezing of FeTETPY ions inbetween superimposed segments of $(L\text{-}Glu)_n$ helical chains.

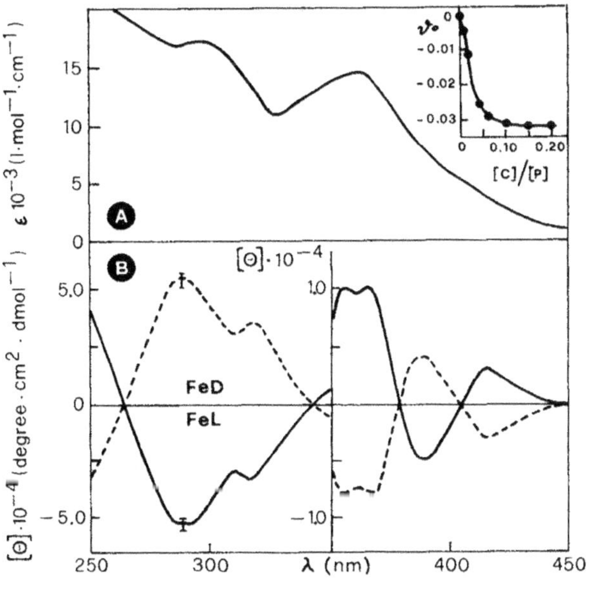

Fig. 2 - Absorption (A) and circular dichroism (B) spectra of FeTETPY-$(L\text{-}Glu)_n$ (FeL) and FeTETPY-$(D\text{-}Glu)_n$ (FeD) catalytic systems. [C] / [P] = 0.10, [P] = $5 \cdot 10^{-4}$ M, pH = 7.0 (tris buffer 0.05 M). The dichroic bands originate solely from the electronic transitions of the bound achiral complex molecules. Insert: variation of ellipticity at 287 nm as a function of complex-to-polymer-residue ratio of FeL; [C] = $3 \cdot 10^{-5}$ M, optical path length of 2 cm. In all cases the degree of association is higher than 93%.

Stereoselectivity corresponding to a 60% enantiomeric excess was observed in all cases when the α-helical fraction in the polypeptide matrices was as high as 0.7.

Structural characteristics of the catalysts - Chiroptical, dialysis equilibrium, phase-separation and viscosity data (9) on FeTETPY-$(Glu)_n$ solutions (pH 7) indicate that: i) progressive binding of complex ions determines a coil-to-α-helix transition in the charged polypeptide matrices, ii) FeTETPY ions form an inner-sphere complex with the polyelectrolyte, whose γ-carboxylate groups act as unidentate ligand, and iii) interchain interactions take place as the amount of bound complex ions increases, the iron molecules probably acting as bridging groups between $(Glu)_n$ chains (9b,10). Indeed, coacervation occurs well before a complete "neutralization" of the fixed charges on the polymer by the bound molecules is reached (9b).

The spectroscopic features of substrate-catalyst mixtures were also found to vary with the complex-to-polymer-residue ratio [C]/[P]. For instance, addition of dopa into FeL or FeD solutions at low [C]/[P] ratio (< 0.05) and in the disareated state gives rise to a new absorption band at around 340 nm (together with a shoulder at 325 nm), whose intensity increases with increasing the substrate concentration (Fig. 3 and 4). Since iron(III) ions chelate preferentially with the catechol end of dopa (11), we are inclined to think that this band originates from a substrate coordinated metal chelate (4,12), in which the O^- group of the substrate is bound to the iron via the "external" apical site.

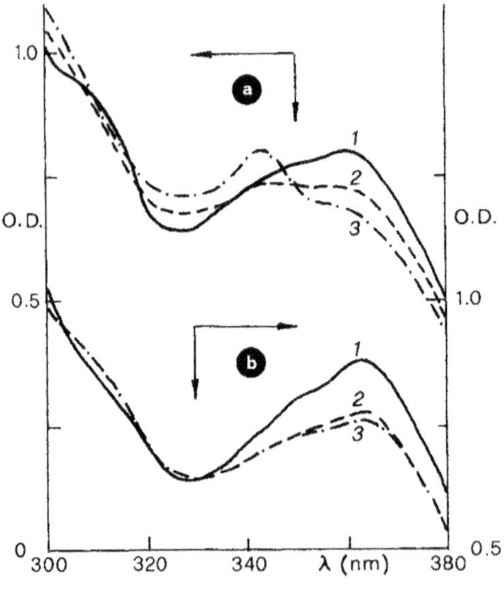

Fig. 3 - Absorption of FeL catalyst (curve 1), at [C]/[P] = 0.01 (a) and 0.20 (b), and of catalyst-dopa mixture (broken lines). Reading taken after 1 min (curve 2) and 15 min (3). [C] = $6\cdot10^{-5}$, [dopa] = $6\cdot10^{-5}$ M. Optical path length of 1 cm, pH = 7.0, tris buffer 0.013 M. (Dopa-quinone, the primary oxidation product of dopa, has absorption maxima at 304 and 475 nm).

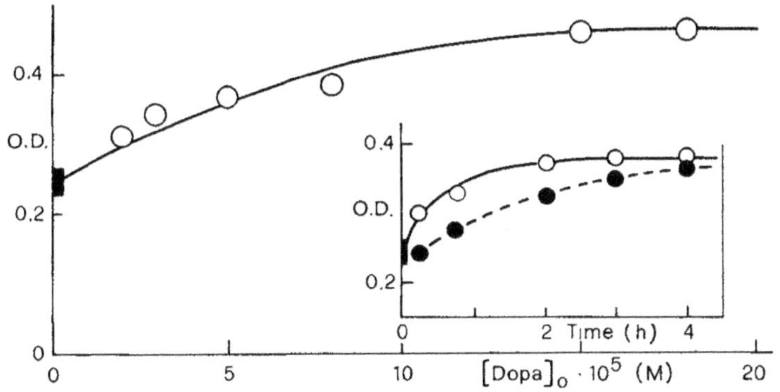

Fig. 4 - Absorption at 340 nm of FeL-dopa mixtures at different substrate concentrations. [C]/[P] = 0.02, [C] = $2 \cdot 10^{-5}$ M, optical path length of 1 cm. Reading taken after about 90 min. Insert: variation of absorption of catalyst-dopa mixture at 340 nm as a function of time; FeTETPY-(L-Glu)$_n$ at two [C]/[P] ratios: 0.02 (O) and 0.20 (●); [dopa] / [FeL] = 4; pH = 7.0 (tris buffer 0.013 M).

The formation of this adduct is, on the contrary, largely hindered when aggregation of polymer chains occurs, i.e. at high [C]/[P] ratio. Indeed, under these conditions, the 340 nm band reaches the maximum intensity only after a few hours (insert of Fig. 4), a finding which suggests that the formation of the intermediate is a diffusion-controlled process. Although we have no direct evidence to support this hypothesis, it is interesting to note that the intensity of the band increases linearly with the square root of time, suggesting a dependence of the phenomenon on a Fickian diffusion mechanism.

On the other hand, a plot of the initial rate of the catalytic oxidation of L-dopa as a function of the initial concentration of substrate, at [C]/[P] = 0.10 and 0.20, exhibits a Michaelis-Menten behavior (Fig. 5). This indicates the presence of a catalyst-substrate adduct under the experimental conditions in which the 340 nm intermediate does not form immediately. Indeed, the remarkable stereospecific effects observed at high [C]/[P] ratios (see later) imply the presence of a Michaelis complex in which dopa is bound to the asymmetric catalyst. Similar results were obtained with the other substrates.

It may be concluded that the complex-to-polymer-residue ratio is a critical parameter for the reactions under investigation in that it controls the structural features of the catalyst and then those of the precursor complex, too.

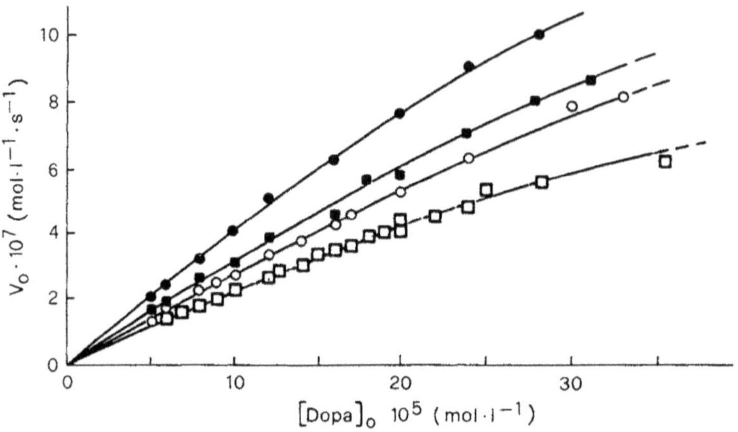

Fig. 5 - Initial rate of the catalytic oxidation of L-dopa as a function of substrate concentration. Catalyst: FeL (empty symbols) or FeD (filled symbols) at a complex-to-polymer-residue ratio of 0.10 (squares) and 0.20 (circles). T = 25.9°C, pH = 7.0 (tris buffer 0.05 M), [C] = $2 \cdot 10^{-5}$ M.

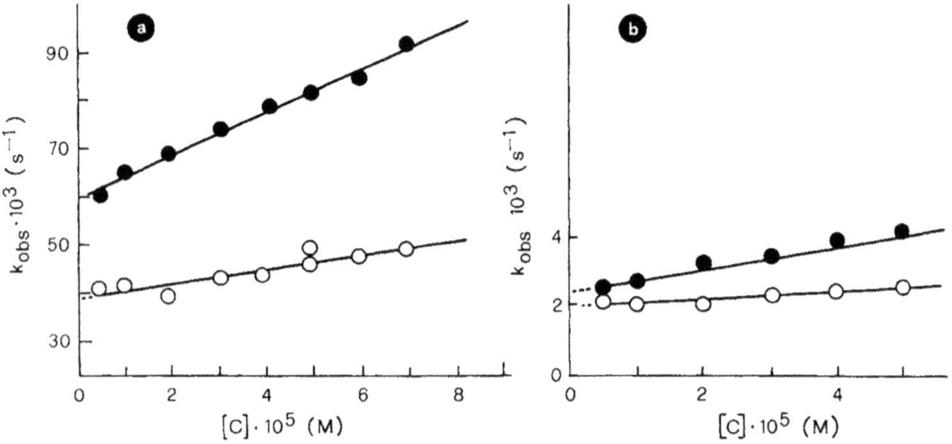

Fig. 6 - Catalytic effect for the oxidation of L-ascorbate anion (a) and L-dopa (b) in the presence of FeTETPY complex ions bound to $(L\text{-}Glu)_n$ (empty symbols) or $(D\text{-}Glu)_n$ (filled symbols) at [C]/[P] = 0.10 and pH = 7.0 (tris buffer 0.05 M). T = 25.9°C, $[H_2O_2]_o/[AH]_o$ = 100; $[AH^-]_o$ = ascorbate anion $1 \cdot 10^{-4}$ or dopa $2 \cdot 10^{-4}$ M. The slopes of the straight lines give the second-order rate constants k_{cat} ($M^{-1} \cdot s^{-1}$).

Kinetic data - Typical kinetic results of the H_2O_2-oxidation of the aforementioned substrates (AH^-) at pH 7 and $[H_2O_2]_o/[AH^-]_o = 100$ are reported in Fig. 6, where the pseudo-first order rate constant k_{obs} (s^{-1}) are plotted against complex concentration [C], at fixed [C]/[P] ratio of 0.10. The linear variation of rate with the concentration of polymer-supported chelate ions at 25.9°C indicates true catalytic behavior for the iron(III) compound. Furthermore, at all [C]/[P] ratios investigated but 0.01 the straight lines have intercepts which differ significantly from zero, i.e. $k_{obs} = k_o + k_{cat}$ [C]. The most plausible explanation of these results is that k_{obs} actually reflects contributions from parallel pathways. One (k_{cat}, $M^{-1}.s^{-1}$) refers to the electron transfer process from the substrate to the central metal ion, within a Michaelis adduct, and the other (k_o, s^{-1}) corresponds to a complex ion-uncatalyzed route to products, which becomes negligible as ([C]/[P]) → 0.

The observation (7) that the slope of the straight lines are $[H_2O_2]$-independent, within experimental errors, whilst the intercepts increase as $[H_2O_2]_o$ increases implies that, under the conditions used, only the rate of the "uncatalyzed" reaction is a function of hydrogen peroxide concentration.

From the results, the following empirical rate expression may be formulated (7):

$$-\frac{d[AH^-]}{dt} = k_{o_{app}}[AH^-][H_2O_2] + k_{cat}[AH^-][C]$$

where $k_{o_{app}}$ and k_{cat} are the second-order rate constants of the parallel reactions, which are complicated function of [C]/[P] ratio (see, for example, Fig. 7). For sake of brevity, we neglect here the uncatalyzed process, whose origin has been discussed elsewhere (7).

While precise details of the overall mechanism of the polymer-supported FeTETPY-catalyzed oxidation of the substrates remain to be determined, the main features of the reaction may be summarized as follows: a) formation of a substrate-catalyst precursor complex, b) electron transfer within this system, c) release of substrate radical from the reduced catalytic center, and d) oxidation of both the lower valence metal chelate and AH· (or $A^{\cdot}$) species by H_2O_2 in subsequent fast steps (4,7). $A^{\cdot}$ may also disproportionate very rapidly (13).

According to the data of Fig. 7, it is worth noting that at very low complex-to-polymer ratio ([C]/[P] ≈0.01) no stereoselectivity is observed. Under these conditions, the α-helical fraction of polypeptide matrices is lower than 0.05 (9b) and the configurational dissymmetry of the active sites is not able to impart any stereospecific effect in the catalysis. When the [C]/[P] ratio is increased, the

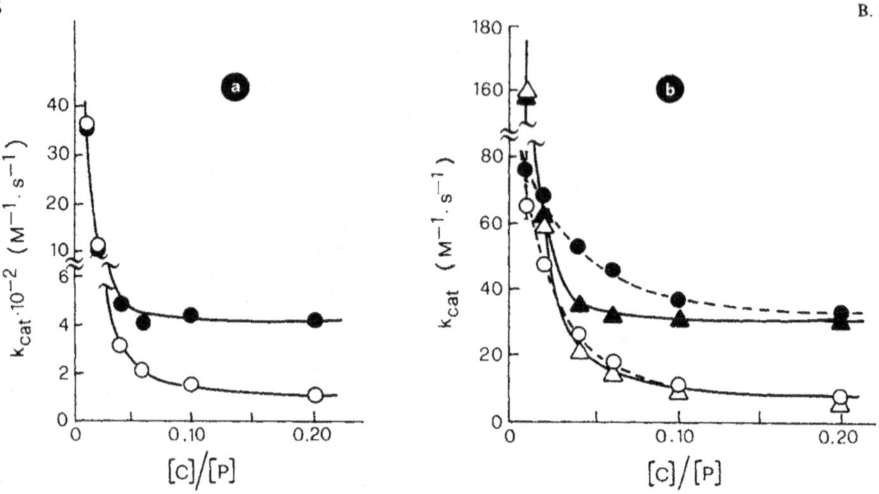

Fig. 7 - Variation of the second-order rate constants of the catalytic oxidation of L-ascorbate anion (a), L-dopa (b, broken lines) and L-adrenaline (b, full lines) as a function of [C]/[P] ratio. Catalyst: FeL (empty symbols) and FeD (filled symbols). T = 25.9°C, pH = 7.0 (tris buffer 0.05 M).

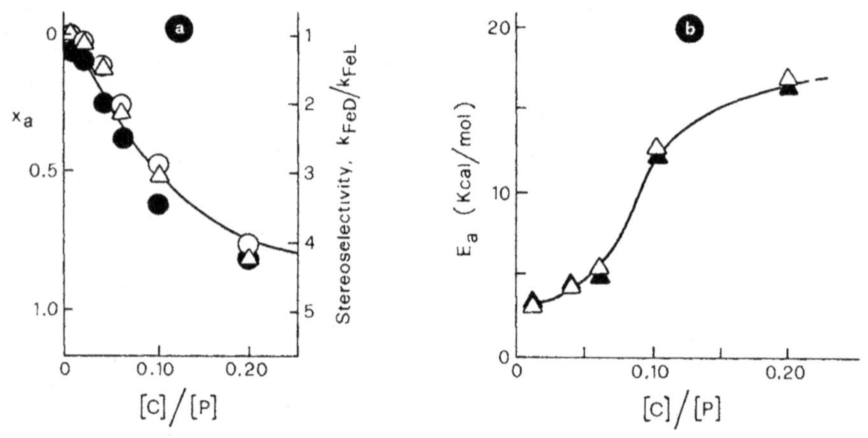

Fig. 8 - a) Variation of the stereoselectivity factor (k_{FeD}/k_{FeL}) of the catalytic oxidation of L-ascorbate anion (O), L-dopa (●) and L-adrenaline (Δ) at 25.9°C and of the α-helical fraction of polypeptide matrices (x_a, solid line), induced by the binding of FeTETPY ions, as a function of [C]/[P] ratio (see text); b) variation of the activation energy of the catalytic oxidation of L-ascorbate anion as a function of [C]/[P] ratio. The different symbols refer to the enantiomeric catalists used.

catalysis becomes stereoselective, the second-order rate constants k_{FeD} being definitely higher than k_{FeL}.

The coupling between binding-induced conformational phenomena in the charged polymeric supports by FeTETPY ions and stereospecific effects in the catalysis is shown in Fig. 8a. The stereoselectivity factors k_{FeD}/k_{FeL} are plotted against [C]/[P] together with the α-helical fraction of $(Glu)_n$ on binding of complex ions (x_a). The relationship between the parameter x_a and the amount of bound complex ions, under experimental conditions of binding equilibrium similar to those of this investigation, has been already reported by us (9b). It was essentially based on a statistical treatment by a two-state model for the polypeptide (14).

Finally, the change in activation energy of the catalytic oxidation of ascorbic acid with [C]/[P] ratio is illustrated in Fig. 8b. Similar S-shaped curves were observed with the other substrates. These findings indicate that the mechanism of the electron transfer catalysis changes with the conformational features of the polymeric supports.

Stereoselective and non-stereoselective electron transfer reaction: possible mechanisms - Spectroscopic data suggest that a mixed ligand metal chelate (4,13) forms when the substrate molecules are oxidized in the presence of FeTETPY-$(Glu)_n$ catalyst having a low [C]/[P] ratio. The formation of this type of adduct is also consistent with the fact that when a few complex ions are bound to $(Glu)_n$ chains at pH 7 the negatively-charged, random coil form of the polypeptide is largely populated (9). Under these conditions, the anionic substrate molecules can interact with the catalyst only through the bound, positively-charged FeTETPY ions, which then represent the primary site of binding for the reacting molecules. As a result, electron transfer may be regarded as taking place between the two species directly (see Scheme). The direct attack mechanism of the substrate on the central metal ion can account for the low activation energy of the reaction (4,15) (Fig. 8b). It is also consistent with the idea that a high conformational mobility of the precursor complex, owing to the size of the substituent in the substrates, should be reflected in a low stereoselectivity, as experimentally observed (Fig. 7 and 8).

On the other hand, both the conformational transition in $(Glu)_n$ and aggregation of the helical chains occur when the amount of bound complex ions is increased (9). Most of FeTETPY ions may be thought to be involved in bridging the polypeptide chains. Therefore, interactions between substrate molecules and the catalyst occur this time through the chiral residues of the polymer. For instance, hydrogen bonding between the γ-carboxylate groups of $(Glu)_n$ and, say, the amino group

PEROXIDATIC MECHANISMS:

I) Non-stereoselective (low [C]/[P] ratios)

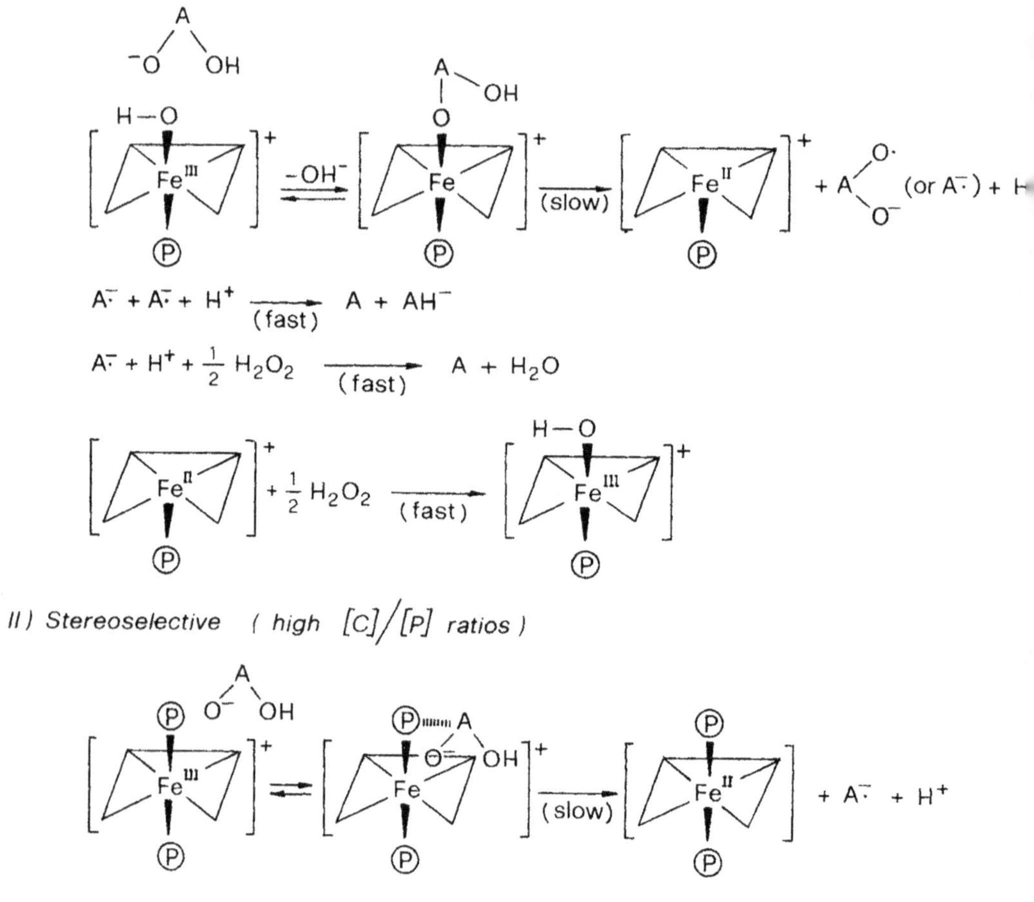

of dopa may take place, probably rigidifying the whole set of bonds in the precursor complex (16). This should enhance the difference in stability between the diastereomeric adducts or the difference in steric hindrances between the diastereomerically related transition states (17), both factors being able to cause the observed stereo-specific effects (Fig. 7 and 8). Indeed, we have not yet a definite answer to the question as to whether stereoselectivity is thermodynamically or kinetically controlled, or both.

According to molecular models, electron transfer between the substrate molecules bound to the helical polypeptide in the surrounding of the catalytic centers and the central metal ions can proceed by a remote attack mechanism, through the peripheral quaterpyridine ligand of the active sites (see Scheme). This is a viable mechanism for redox reactions, as already recognized in a number of cases involving ferriporphyrins or metalloproteins and different reductans (4c,18). We may therefore conclude with some confidence that the stereoselective pathway does not involve substitution on the iron center but that the reaction proceeds making use of an electron transfer site remote from the Fe(III) ion, possibly the π system of tetrapyridyl moiety. This is probably a route with an activation energy higher than that of the direct attack mechanism (see above), as experimentally observed (Fig. 8b).

In conclusion, there is a close relationship between structural features of the precursor complex and both mechanism and stereoselectivity of the electron transfer reactions investigated. It is attractive to consider the FeTETPY-$(Glu)_n$ system as a simple model for the allosteric effects attributed to enzymes. Accordingly, the progressive binding of complex ions determines a conformational transition in the charged polymeric matrix which, in turn, brings about a change in the mechanism of the oxidation reaction, leading to marked stereospecific effects in the catalysis.

I whish to thank Drs. M. Barteri, M. Farinella, M.V. Primiceri and Miss S. Nardelli for skilful assistance in the experimental work.

References

(1) Katchalski, E., Sela, M., Silman, H.I., Berger, A., in "The Proteins" Neurath, H.Ed., Academic Press, New York (1964), vol. 2, p. 405.
(2) Barteri, M., Pispisa, B.: 1976, Gazz. Chim. Ital., 106, p. 499; Barteri, M., Farinella, M., Pispisa, B.: 1977, Biopolymers, 16, p. 256.
(3) Jones, P., Wilson, I. "Catalases and Iron Complexes with Catalase-like Properties" in "Metal Ions in Biological Systems, vol. 7: Iron in Model and Natural Compounds" Siegel, H. Ed., Dekker Inc., New York (1978), chap. 5, and references therein.
(4) a) Grinstead, R.R.: 1960, J. Am. Chem. Soc., 82, p. 3464; b) Khan, M.M.T., Martell, A.E.: 1967, J. Am. Chem. Soc., 89, p. 4176 and 7104; c) Harris, F.L., Toppen, D.L.: 1978, Inorg. Chem., 17, p. 74.
(5) Branca, M., Pispisa, B., Aurisicchio, C.: 1976, J. Chem. Soc., Dalton, p. 1543; Cerdonio, M., Mogno, F., Pispisa, B., Vitale, S.: 1977,

Inorg. Chem., 9, p. 400.
(6) Barteri, M., Farinella, M., Pispisa, B., Splendorini, L.: 1978, J. Chem. Soc., Faraday Trans. I, 74, p. 288; Barteri, M., Farinella, M., Pispisa, B.: 1978, J. Inorg. Nucl. Chem., 40, p. 1277.
(7) Barteri, M., Pispisa, B., Primiceri, M.V.: 1980, J. Inorg. Biochem., 12, p. 167; Barteri, M., Pispisa, B.: 1982, Biopolymers, 21, p. 1093.
(8) Heacock, R.A.: 1959, Chem. Rev., 59, p. 181.
(9) a) Branca, M., Marini, M.E., Pispisa, B.: 1976, Biopolymers, 15,p. 2219; b) Branca, M., Pispisa, B.: 1977, J. Chem. Soc., Faraday Trans. I, 73, p. 213.
(10) Blauer, G.: 1964, Biochim. Biophys. Acta, 79, p. 547; Blauer, G., Alfassi, Z.B.: 1967, ibid., 133, p. 206.
(11) Gorton, J.E., Jameson, R.F.: 1968, J. Chem. Soc., A, p. 2615.
(12) Laurence, G.S., Ellis, K.J.: 1972, J. Chem. Soc., Dalton, p. 1667.
(13) Bielski, B.H.J., Richter, H.W.: 1975, Ann. N.Y. Acad. Sci., 258, p. 231.
(14) Monod , J., Wyman, J., Changeux, I.P.: 1965, J. Mol. Biol., 12, p. 88.
(15) Sutin, N. "Oxidation-Reduction in Coordination Compounds" in "Inorg. Biochem.", Eichhorn, G.L. Ed., Elsevier, Amsterdam (1973), vol. 2, p. 611.
(16) Beuder, H.L., Kézdy, F.J., Gunter, C.R.: 1964, J. Am. Chem. Soc., 86, p. 3714.
(17) Prelog, V.: 1953, Helv. Chim. Acta, 36, p. 308.
(18) Castro, C.E., Davis, H.F.: 1969, J. Am. Chem. Soc., 91, p. 5405; Yandell, J.K., Fay, D.P., Sutin, N.: 1973, ibid., 95, p. 1131; Toppen, D.L.: 1976, ibid., 98, p. 4023.

NMR STUDIES OF CYTOCHROMES

António V. Xavier

Centro de Quimica Estrutural, Lisboa, Portugal, and Gray Freshwater Biological Institute, University of Minnesota, U.S.A.

1. INTRODUCTION

Haem proteins cover a wide range of function in the biological systems. These functions include the transfer of oxygen, haemoglobins and myoglobins, the transfer of electrons, cytochromes, and the catalysis of oxidation-reduction reactions of bound substrates. From these proteins, cytochromes constitute the group which is most thoroughly characterized both from the point of view of their structure and of their physico-chemical properties (1-6). These studies have used practically all the spectroscopic techniques available to the biophysicist. The nuclear magnetic resonance (NMR) studies of cytochromes goes back to beginnings of the application of this technique to the characterization of biological systems.

This chapter reviews the most important information which was obtained by NMR in order to increase our knowledge about cytochromes. Keeping in line with the general theme of this book, particular attention is given to the characterization of the iron coordination and of how this coordination is adapted by the different cytochromes in order to best fit their biological function. In this respect, it is worth keeping in mind that the studies carried out in cytochromes, further than increasing our knowledge of the mechanisms of action of this protein, can also be used as models which will help to understand the role of haems in important enzymes (see other chapters in this book). Furthermore, the comparative NMR studies of cytochromes constitute a useful tool to probe the role of the polypeptide chain in modulating the chemical reactivity of the haem.

I. Bertini, R. S. Drago, and C. Luchinat (eds.), The Coordination Chemistry of Metalloenzymes, 291–311.

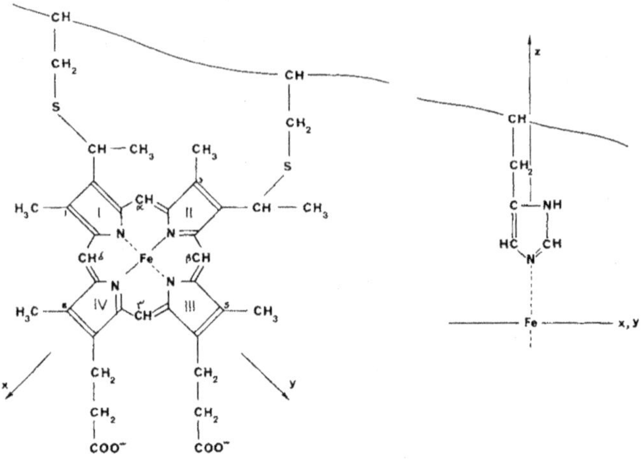

Figure 1. Structure of haem c. The sixth ligand is not shown in the figure. The four pyrrole rings are numbered I-IV, the four meso-positions α-γ and the peripheral positions with a methyl group are also indicated.

From the point of view of metal coordination, cytochromes can be differentiated by using different types of haems and different axial ligands to the iron. The porphyrins of cytochromes described in this chapter can either be a mesoporphyrin (Figure 1) as in the c type cytochromes, or a protoporphyrin, as in the b type ones. In the c type cytochromes the mesoporphyrin is covalently bound to the polypeptide chain by thioether bonds to two cysteinyl residues in a Cys-$(x)_n$-Cys-His where n is usually equal to two but can be three or four (3). Two cases are known of cytochromes c which use only one cysteinyl residue to form a thioether bond (3). In the b type cytochromes the protoporphyrin is not covalently bound to the polypeptide chain of the protein.

One of the axial ligands (the fifth) is in all cases known to be a hystidinyl residue. However, independently of the haem type, the sixth position can be used to bind either a methionyl or a hystidinyl residue, or it can even be unoccupied, as is the case in cytochromes c' (7,8). The number of combinations of different porphyrins and ligands found in cytochromes complicates their classification which has often been revised (3,9). This differentiation can strongly influence the properties of the cytochromes.

The ligand field strength observed for the haem iron in biological systems is such that modifications brought about by the sixth ligand can stabilize this ion in either high- or low-spin states. Moreover, the stability of the intermediate states is very close to that of the high- and low-spin states at the ligand field strength at which they crossover (10). This fact makes the haem properties very sensitive to subtle changes imposed in their coordination.

The NMR spectra of cytochromes is greatly influenced by the spin-state of the iron (11). If the crystal field strength is such that the high-spin states are stabilized, both the ferric (S = 5/2) and the ferrous (S = 2) forms are paramagnetic. Due to the high value of the magnetic moment and the fairly long electronic spin relaxation time of the high-spin iron ions (13), the NMR spectrum resonances of both the reduced and the oxidized forms of these cytochromes are quite broad, making it impossible to obtain detailed information from the main envelope of the protein spectrum. On the contrary both the NMR spectrum of low-spin ferricytochromes and ferrocytochromes are very well resolved, allowing the obtainment of detailed information about their structure and their kinetic properties. Furthermore, although the oxidized form is paramagnetic (S = 1/2), the reduced one is diamagnetic (S = 0). The presence of this intrinsic paramagnetic probe and its comparison with the diamagnetic blank, makes the low-spin cytochromes particularly suitable to be studied by high-resolution NMR spectroscopy. The shifts induced by the paramagnetic ferric ion are due to two different mechanisms (11). One is the contact, through bond, which is only felt by the nuclei of the paramagnetic ion ligands, the porphyrin and the axial ligands, and can give information about the unpaired electron density distribution. Another is the so-called pseudo-contact, which is a through space (dipolar) mechanism inducing shifts of the nuclei in the neighborhood of the ferric ion and which contains important structural information.

Another interesting differentiation relates to the wide range of redox potentials covered by the haems in different cytochromes as shown in Figure 2. Analysis of this figure shows that further than the influence of the chosen prosthetic group and axial ligands, there is an important role of the polypeptide in creating the proper environment so that the necessary mid-point redox potential of the iron is achieved.

This chapter will give particular attention on how NMR can follow modifications of the sixth ligand to the iron as well as follow how subtle constraints imposed to the mode of binding of this ligand can be used in order to control the midpoint redox potential of the iron, its spin state, and the electron density distribution in the haem.

2. CLASS II CYTOCHROMES *C*

Class II cytochromes *c* are a group of *c* type cytochromes which have their covalent haem attachment site near the *C*-terminus of the protein, by contrast to the Class I (e.g., mitochondrial cytochrome *c*) where the

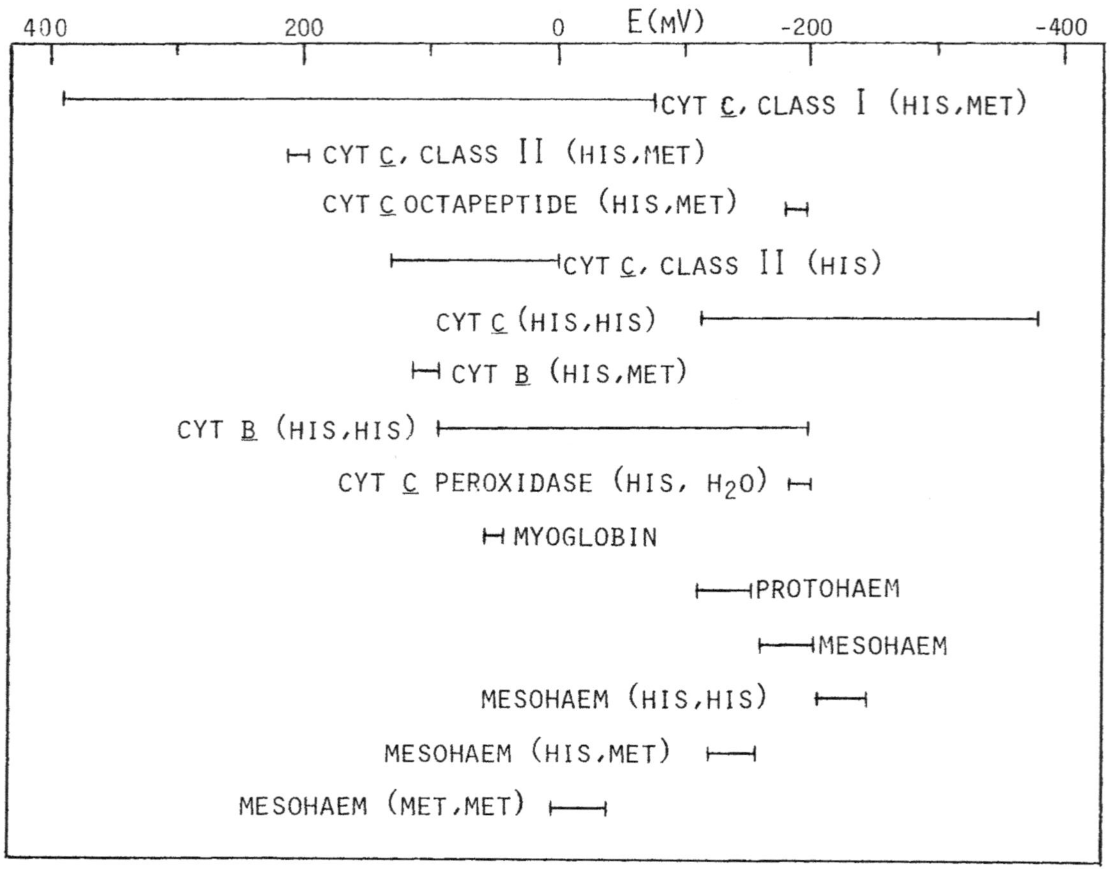

Figure 2. Approximate redox potentials of cytochromes, some haem proteins and model compounds (see (10) and other references given in the text). The axial ligands used are shown within brackets.

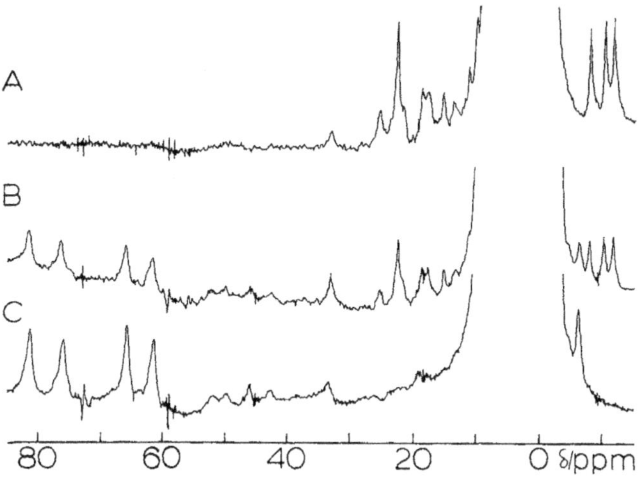

Figure 3. 270 MHz ^{1}H NMR spectra of R. rubrum cytochrome c' at pH* = 7.0 in D_2O and 303 K. The protein concentration is approximately 3 mM. Only the high- and low-field regions of the spectra are shown. Spectrum (A) is that of the fully reduced sample and spectrum (C) is that of the fully oxidized one. The middle spectrum (B) is that of a 50% reduced sample. (Adapted from reference 9).

covalent attachment is near the N-terminus (9).

The NMR studies of Class II cytochromes c clearly illustrates how the polypeptide chain can control the spin state of the iron.

The high- and low-field regions of the NMR spectra of cytochrome c' from Rhodospirillum rubrum (Figure 3) are similar to those of other high-spin iron porphyrin species (13-16). Although there is a severe broadening of the resonances induced by the high-spin iron, it is still possible to extract some useful information from these spectra (7). As it is shown in the middle spectrum, a sample of partially reduced cytochrome c' gives rise to separate resonances due to both oxidized and reduced molecules. This indicates that the electron exchange rate between haems is slow in the NMR time scale ($< 10^4$ s^{-1}). Very large paramagnetic shifts are observed in the spectrum of the ferricytochrome (lower spectrum). These are due to both contact and pseudocontact interactions between the protons nuclear spin and the electronic spin of the iron (13). The four larger peaks in the low-field region (between 50 and 85 ppm) with three-proton intensity each, arise from the four haem methyl groups. In the spectrum of ferrocytochrome c' (upper spectrum) the haem methyl group resonances are the three-proton intensity peaks, one at ≈ 22 ppm and the other three in the high-field region of the spectrum. This spectrum has several features in common with that of Aplysia deoxymyoglobin (16). These similarities and the presence of a broad resonance at 33 ppm in the region expected for the protons of a bound histidinyl residue (not easily observed in the spectrum shown in Figure 3) suggest that the fifth ligand is a histidine.

A pH titration of ferricytochrome c' over the range of 4.5-11 was followed by NMR (7). The chemical shifts of the haem methyl resonances were fitted to theoretical titration curves by assuming two pH-dependent equilibria with pK_a^* values of 5.8 and 8.7. Throughout the pH range studied, the rate of exchange is fast in the NMR time scale ($> 5 \times 10^4 s^{-1}$).

The 5.8 pH^* transition produces only minor shifts. The average paramagnetic shift changes for the four haem methyl resonances (all smaller than 2 ppm) is maintained throughout this ionization, showing that the perturbation does not cause an alteration of the total electron spin density in the porphyrin ring (11). This transition, not observed by other techniques, is similar to that observed in the NMR spectra of horseradish peroxidase (17), cytochrome c peroxidase (18), and deoxymyoglobin (19). Furthermore, it is not observed for Rhodopseudomonas palustris cytochrome c' which contains only one histidinyl residue. Thus, it probably arises from the ionization of the second histidine present in R. rubrum cytochrome c'.

The pH^* transition at 8.7 gives considerable changes throughout the NMR spectra, causing changes of up to 30 ppm in the resonances of the haem methyl groups. A similar transition is observed in NMR pH titrations of Rps. palustris and R. molischianum cytochromes c' (7,20), both of them containing only one histidine. These NMR observations, together with the above mentioned similarities between the NMR spectrum of ferrocytochrome c' and that of Aplysia deoxymyoglobin, which does not have a distal histidine, show that the second histidine residue of R. rubrum is not distal to the iron (7).

In order to investigate whether the high pK_a of cytochromes c' is caused by a water molecule coordinated to the iron, two other types of NMR studies were performed. The pH dependence of the water proton relaxation enhancement of the spin-lattice relaxation time, T_1, of cytochromes c' (7,20) was compared to that of metmyoglobin. Both these proteins exhibit the high pH transition. However, both the sign and the magnitude of the T_1 variation are different for the two proteins. Metmyoglobin shows a large decrease in the relaxation rate with increasing pH, caused by the ionization of the water ligand (21,22). Conversely a small increase is observed for cytochromes c' (7,20). Although these data must be taken with due caution (cf. the chapter by S. Koenig in this book), they indicate that the high pK_a is not associated with the ionization of a bound water molecule, contrary to the case observed in horse metmyoglobin and are consistent with the maintenance of the high-spin state at high pH.

This was confirmed in R. rubrum cytochrome c' by measurements of the molar paramagnetic susceptibility χ_M^P, at pH* 6.5 and 10.1 (7) using an NMR method (23). Table 1 shows the magnetic susceptibility data obtained (7). The determined values for the effective magnetic moment, μ_{eff}, agree with the calculated ones using the equations of Kurland and McGarvey (24). The fact that the magnetic susceptibility of R. rubrum cytochrome c' is not significantly affected by the high pK_a transition, as would be expected by the ionization of a bound water molecule, again indicates that water is not a sixth ligand.

Table 1
Magnetic Susceptibility data for R. rubrum cytochrome c' (7).

	pH*	χ_M^P (x 10^3)	μ_{eff}
ferricytochrome c'	6.5	12.3 ± 0.5	5.4 ± 0.1
ferricytochrome c'	10.1	14.4 ± 0.5	5.9 ± 0.1
ferrocytochrome c'	6.5	11.2 ± 0.5	5.2 ± 0.1

The determination of the X-ray structure of R. molischianum cytochrome c' (8), as well as resonance Raman data (25) and $H_2{}^{17}O$ EPR studies, confirmed that the sixth ligand site is unoccupied at neutral pH.

The involvement of a charged residue in the high pK_a has been further investigated by NMR studies, and confirmed by the observation of two resonances at very low-field (110 and 130 ppm) which only appear at high pH (26).

There remains however, an interesting controversy regarding the spin state of the iron of Class II cytochromes c. Maltempo has interpreted his data on Chromatium cytochrome c' in terms of a quantum mechanical admixture of S = 5/2 and S = 3/2 states (27-29). Furthermore, two of the Class II cytochromes c, cytochromes c_{556} isolated from Rps. palustris and Agrobacterium tumefaciens, which have an homologous amino acid sequence to the cytochromes c' (30-33), have been shown to be low-spin (30,31,33). The NMR spectra of their reduced form (20) show a three-proton intensity resonance above -3 ppm, typical of low-spin cytochromes with a methionyl residue as a sixth ligand (see below). However, the temperature dependence of the chemical shift of the haem methyl resonances of ferricytochrome c_{556} suggests that these cytochromes are not purely low-spin and that the amount of high-spin character increases with the temperature (20).

Based on an analysis of the X-ray structure of R. molischianum cytochrome c' (8), it has been pointed out that the S atom of Met 16 is only 0.4 nm away from the iron (20). Since all Class II cytochromes c have a methionyl residue between positions 10 and 25 (30-33), this residue could play an important role in controlling the haem spin state (19). This possibility was further used to speculate about the still unknown function of cytochromes c' and the fact that there is no clear evidence for the identification of cytochromes c' *in situ* (34). Thus, difficulties pointed out could be due to damages induced by purification and *in situ*, all Class II cytochromes c would be low-spin histidine-methionine cytochromes with electron transport function (19). However, the striking similarities between the NMR spectral features of cytochrome c', the globins and the peroxidases, should still be kept in mind. Again, the fact that subtle changes in the environment of the iron may induce modifications of its spin state and/or its redox potential (see Figure 2) also deserves further attention.

In this respect, it is interesting to refer to other observations. The first one refers to the studies of *Pseudomonas* cytochrome c peroxidase, an enzyme that contains two haem c moieties attached to a single polypeptide chain by covalent bonds (35). Optical absorption (36,37) and resonance Raman studies (38), have demonstrated that at room temperature one of the c hames is in the low-spin state and the other is in the high-spin state. NMR studies have also confirmed these observations (unpublished work from our laboratory). However, recent EPR studies at low temperature, 15 K, show that upon freezing the enzyme, most of the high-spin state haem goes into a low-spin state (39). The second one is the observation that the high-spin haem of *R. rubrum* cytochrome c' can be converted into a low-spin state using organic solvents (40).

The synthesis of model porphyrin complexes that mimic the active centre of these proteins (41-43) might help the further understanding of this problem.

3. LOW SPIN CYTOCHROMES

3.1 NMR Spectrum

As was pointed out earlier, the NMR spectra of low-spin cytochromes are faily well resolved both for the oxidized and for the reduced form of the protein. This is exemplified by the spectra of ferro- and ferricytochrome c_{551} isolated from *Pseudomonas perfectomarinus*, shown in Figure 4 (44). The influence of the presence of the paramagnetic ion is evident from the analysis of the spectrum of ferricytochrome c_{551} (lower spectrum). The resonances of the haem and axial ligands cover a range of chemical shifts which goes from approximately 35 to -45 ppm. The four resonances in the low-field end of the spectrum belong to the haem methyl groups. The resonances above -10 ppm are assigned to the bound methionyl residue. The spectrum of ferrocytochrome (upper spectrum) has a much narrower range of chemical shifts. However, due to the presence of the porphyrin some resonances are shifted out of the spectral envelope usually covered by diamagnetic proteins. The ring-current shifts induced by the porphyrin are particularly obvious for the haem meso protons, which appear between 9-10 ppm, and the protons of the sixth ligand methionyl residue, which appear at the high-field end of the spectrum.

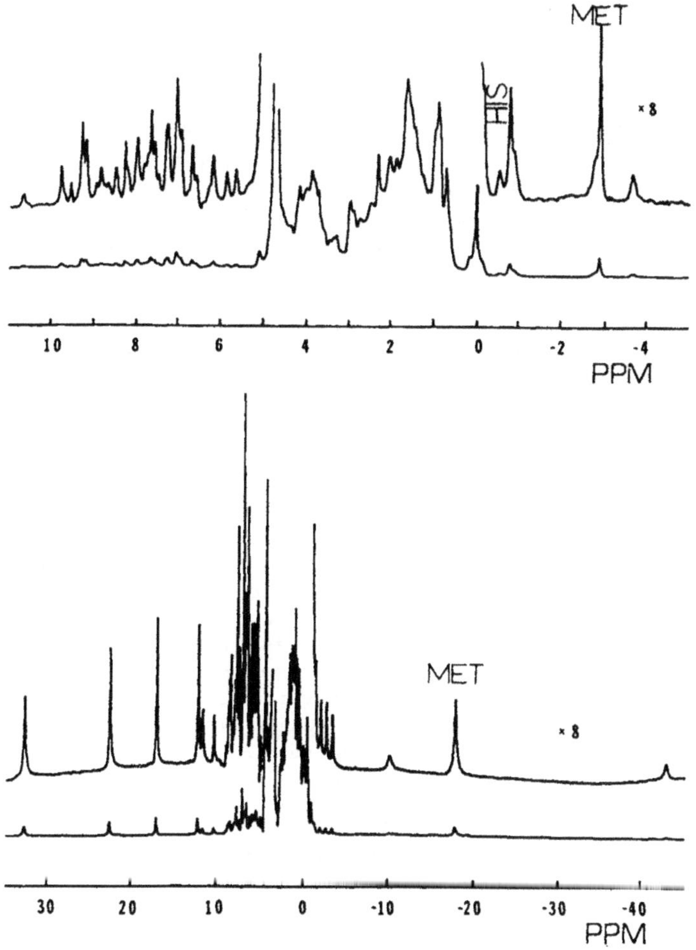

Figure 4. 300 MHz ^{1}H NMR spectra of cytochrome $\underline{c}_{551}$ from <u>Pseudomonas perfectomarinus</u> in the reduced (upper spectrum) and oxidized (lower spectrum) forms. The concentration of the sample is 1 mM. Other experimental conditions: temperature 300 K, pH* = 6.9 (oxidized) and 6.5 (reduced), chemical shifts are given in parts per million (ppm) from TSS. The spectra were obtained in a Bruker 300 CXP spectrometer equipped with an Aspect 2000 computer. 2048 scans were accumulated. (Adapted from reference 44).

The high resolution obtained has allowed a very thorough analysis of the spectrum of low-spin cytochromes c. The NMR spectra of horse heart ferro- and ferricytochrome c are those for which a larger amount of resonance assignments has been achieved. These assignments include the resonances of the haem group and of the axial ligands (45-49), of several exchangeable NH protons (50,51), and those of about 70-80% of the aromatic and aliphatic protons (52-59).

The detailed knowledge of these NMR spectra together with the geometrical information given by the magnitude of the chemical shifts induced in the spectrum of oxidized cytochrome (by pseudo-contact and ring-current) and those induced in the spectrum of reduced cytochrome (by ring-current only) make it possible to obtain the parameters necessary to study the structure of the protein in solution.

3.2 Structural Studies

The structural studies carried out by NMR have complemented quite considerably the structural data obtained by X-ray crystallography (5, 59,60). They have shown for the first time that in horse heart cytochrome c there is an oxidation linked conformation change affecting the region about ILE 57 and extending to the PHE 10 region (61,62).

The NMR studies are particularly useful to depict structural modifications by spectral comparisons. A comparative NMR study of tuna and horse heart cytochrome c has shown that the two proteins have identical polypeptide folding and that in general terms, there is a good agreement between the solution and the crystal structures of cytochromes c (63). This type of study has also been used to probe local conformational differences between several eukaryotic cytochromes c. By choosing cytochromes with different degrees of primary sequence homology, it was possible to probe the structural effects of amino-acid substitutions. By this method, Moore and Williams (64) have shown that there is no conformational change induced by the substitution of one amino-acid with similar size and chemistry, as is the case of cytochromes c from horse and donkey (THR 47 for SER 47). On the other hand, a comparison between the spectra of donkey and cow cytochromes c, which differ at two positions (LYS 60 and THR 89 for GLY 60 and GLY 89), shows that the substitution of a polar for a non-polar residue, produces a more widespread structural modification (64). Furthermore, the structural modifications are cooperative.

Comparisons of the NMR spectral parameters, together with sequence data, have also been used to study the structural homology between eukaryotic and prokaryotic c type cytochromes (65). Although mitochondrial cytochromes c are very similar to the Rhodospirillaceae ones, many features of the residues in the haem environment as well as in the polypeptide chain folding are found when comparing these proteins with Pseudomonas aeruginosa cytochrome c_{551} and Euglena gracilis cytochrome c_{552}. Again, a comparison of the low-potential cytochromes c_{553} from Desulfovibrio vulgaris with the other cytochromes c, shows important differences in the haem environments (65). This last point will be further discussed in a later section.

Other important information regarding the structural aspects of these cytochromes, which can be easily probed by NMR spectroscopy is the study of the mobility of individual amino acid residues, such as the rotation of the aromatic residues about their C_β-C_γ bond (66-68). These studies, when carried out at different temperatures and pH's, together with the information obtained from the induced chemical shifts, can also be readily used to study the stability of cytochrome c and to investigate the dynamic characteristics of different regions of the protein (58), its mode of folding and its stability (69,70).

3.3 Characterization of the Axial Ligands

Cytochromes having a methionyl residue as a sixth ligand to the haem give rise to a typical pattern in the NMR spectrum of their reduced form. The observation of a three-proton intensity singlet at approximately -3 ppm, together with four one-proton intensity resonance in this high-field region of the spectrum is an unequivocal demonstration that a methionyl residue is coordinated to the haem (11). These large up-field shifts are due to the ring-current effect induced by the porphyrin ring in the nearby protons sitting near the Z axis of the ring (see Figure 1).

The observation of this detailed spectral feature, provided the first evidence that cytochromes b can have a methionyl residue as a sixth ligand (71). Figure 5 shows the high-field spectral region of ferrocytochrome b_{562}. The three-proton intensity singlet at -2.98 ppm is assigned to the methyl group, and the four one-proton intensity resonances appearing from -1.46 to -3.22 ppm, were assigned to the β- and γ-methylene protons of the haem bound methionyl residue. The one-proton intensity singlet at 0.11 ppm is in the region expected for one of the protons of a histidinyl residue bound to the haem. These protons are also strongly shifted upfield by the ring-current effect of the porphyrin.

Again, in the case of cytochrome c_3 it was from the comparison of NMR spectra of cytochromes purified from different Desulfovibrio species, that it was first shown that both axial ligands are histidinyl residues (72).

NMR can further be used to distinguish between the two chiralities, R and S, of the axial methionyl residue. Using nuclear Overhauser enhancement (NOE) effects (73), Wuthrich and coworkers have shown that the axially bound methionine has different chirality in horse heart cytochrome c and in P. aeruginosa cytochrome c_{551}. This difference in the mode of attachment of the axial methionine was correlated with earlier observations that the electron spin density distribution in the four pyrrole rings of the porphyrin is different for these two proteins.

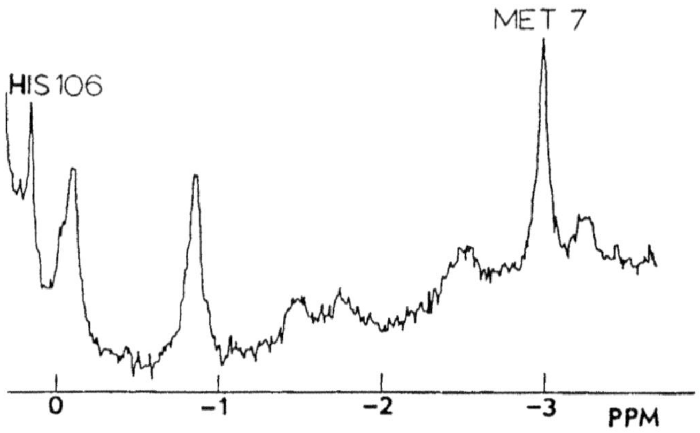

Figure 5. High-field region of the 270 MHz ^1H NMR spectrum of reduced cytochrome b562 (pH* = 7.0), in D_2O. The protein concentration is 1.5 mM. The sample was reduced with a slight excess of crystalline sodium dithionite, under argon. Dioxan was used as internal reference but the chemical shifts are given in parts per million (ppm) from TSS. (Adapted from reference 71).

The individual assignment of each of the four haem methyl groups in the ferricytochromes spectra (49), together with size of the paramagnetic shift induced in these nuclei (Table 2), shows that the

Table 2

Paramagnetic shifts of the haem methyl groups a)

Methyl group	Ring	Paramagnetic shift (74) Cytochrome c_{551}	cytochrome c
1	I	21.1	3.9
3	II	9.6	27.3
5	III	28.8	6.9
8	IV	14.4	31.7

a) see Figure 1

electronic haem structure is different for the two cytochromes (74). Indeed, the induced paramagnetic shift in the haem methyl group is strongly dominated by the contact interaction with the unpaired electron (11). A possible explanation for the difference observed was put forward on the basis that the lone pair orbital of the sulfur atom is directed towards the nitrogen atom of different pyrrole rings, depending on the chirality of the bound methionine (74).

A combination of these NMR studies and X-ray crystallography data (75,76) was used to support a mechanism for electron transfer of cytochromes c (77,78) through the haem edge exposed to the solvent (73). As shown in Table 2 the high spin density is mainly localized in pyrrole rings II and IV for horse heart cytochrome c and in rings III and I for cytochrome c_{551}. This electron spin density distribution fits quite well the electron transfer mechanism proposed, since the X-ray structures of these two proteins (75,76) show that in cytochrome c_{551}, due to a deletion of the peptide fragment 39 to 59, ring III should be accessible to solvent. Conversely, in mammalian cytochrome c ring II is the most exposed one.

3.4 Control of the Redox Potential

Some of the features that can control the haem redox potential have been successfully studied by NMR. As shown in Figure 2, the cytochromes cover a wide range of mid-point redox potentials. The redox potential of cytochromes with two histidinyl residues as axial ligands is usually lower than that of cytochromes with one histidinyl and one methionyl axial ligand. The redox potential of cytochrome b_5 is lower than that of cytochrome b_{562} and the redox potential of the haems of cytochromes c_3 and c_7 (with two histidinyl residues as axial ligands) is much lower than those found for the other c type cytochromes. The absence of a sixth ligand also leads to a lower redox potential than the thioether ligation. In accord with this fact, the redox potential of cytochrome c' is lower than that of the low-spin Class II cytochromes c.

A more subtle control of the redox potential may also be achieved by the constrains imposed by the protein into the magnitude of the iron-sulfur bond length of methionine bound haems. Moore and Williams (79) have found an interesting correlation between the redox potential of cytochromes c and the chemical shift induced in the bound methionine protons by the ring-current of the porphyrin. These shifts, being dependent on the distance between the nuclei and the haem increase with the decrease of that bond length. The correlation was interpreted as being due to a modification of the ability of sulfur to donate electrons to the iron.

The possibility that the protein folding might control not only the bond length of the iron-sulfur but even the binding and release of methionine, as is exemplified by the NMR study of Class II cytochromes c (20), suggests that the role of a proximal/distal methionine might be an important one and justifies a further look into the possible dynamic features of this axial ligand. Another interesting example that justifies this observation, is the replacement of the sixth methionine ligand by histidine or lysine on reduction of the haem iron both in the low-spin haem of cytochrome c peroxidase (38) and in Thermus thermophilus cytochrome c_{552} (80).

The involvement of cytochromes in the coupled transport of electrons and protons has also been probed by NMR. A pH dependence of the midpoint redox potential has been reported for several prokaryotic c cytochromes (81) as well as b cytochromes (82-84) and cytochrome c oxidase (85-87). This redox-Bohr effect, whose implication has recently been reviewed (88), has been followed by NMR spectroscopy in R. rubrum cytochrome c_2 (89), P. aeruginosa cytochrome c_{551} (90) and Desulfovibrio cytochromes c_3 (91). In cytochrome c_{551} the pH dependence of the chemical shifts of several protons were fitted to ionization constants which are different for the two oxidation states, $pK_a^{red} \simeq 6.2$ and $pK_a^{ox} \simeq 7.3$ (90). In particular, these two constants are observed for the same residue in the different oxidation states, MET 61, showing unequivocally that there is a pH dependence of the midpoint redox potential.

As a word of warning it should be pointed out that with increasing pH (i.e., with decreasing the redox potential) there is small downfield shift of the methyl group resonance of MET 61 (90) contrary to the shift expected from the correlation described above (40). This observation demonstrates that the magnitude of the Fe-S bond is not the only factor controlling the redox poential (92-94).

The pH dependence of D. vulgaris cytochrome c_3 is discussed in the next Section, together with the possible physiological relevance of this effect.

4. ELECTRON TRANSFER MECHANISMS

High resolution NMR has been thoroughly used to study the mechanisms involving the electron exchange of cytochromes, either as isolated proteins (91, 95-97), or bound to other electron transfer proteins (98), and between cytochromes and inorganic complexes (99-101). The NMR studies take advantage of the paramagnetic perturbation of the spectral parameters and are based on measurements of line broadening, spin-lattice relaxation times, and saturation transfer experiments performed in solutions containing a mixture of oxidation states. The methods and relevant equations have been reviewed by Gupta and Mildvan (102) and their application is quite straightforward for cytochromes with only one haem and consequently only two oxidation states.

Cytochromes c_3, having four haems per molecule, can exist in sixteen different oxidation states: one with the four haems reduced, four with only one of the haems oxidized, six with two of the haems oxidized, four with three of the haems oxidized, and finally one with all the haems oxidized. Thus, the complex redox equilibria of cytochrome c_3 can be described by 32 Nernst equations between these 16 oxidation states. A redox titration of these cytochromes can be described by grouping the different oxidation states in five steps which, starting with the fully reduced molecules (Step 0), are generated by a sequential loss of one electron. Thus, the 4 oxidation states with only one haem oxidized (Step I) are first generated, and so on until the fully oxidized state (Step IV) is reached (44,91).

Furthermore, the equilibria between the oxidation states can involve two types of one-electron exchange mechanisms: an intermolecular, between haems of different molecules and involving the equilibria between oxidation states belonging to different redox steps; and an intramolecular, between haems of the same molecule and involving the equilibria between states belonging to the same redox step (91).

Figure 6 shows the low-field region of the NMR spectra obtained during a redox titration of *D. vulgaris* cytochrome c_3, to where the methyl group resonances of the four haems are shifted by the paramagnetic low-spin ferric ions. Since there are four haems, the total paramagnetic shift induced for each haem methyl group is a contribution of four components. These are the large paramagnetic shifts induced by the ferric ion of the haem to which the methyl group belongs (of contact and pseudo-contact origin) and the shifts induced by any of the three other ferric haems (of pseudo-contact origin only). Since the intramolecular electron exchange rate is fast on the NMR time scale ($> 5x10^4$ s^{-1}), the spectra of each of the three intermediate redox steps (Steps I, II and III) are a weighted average of the spectra of the oxidation states belonging to each redox step (44). Furthermore, the fact that as the redox titration proceeds, the resonances due to these intermediate steps first grow and then decrease (see Figure 6), indicates that the intramolecular electron exchange rate is slow on the NMR time scale ($< 5x10^5$ M^{-1} s^{-1}), for the temperature and concentration used (91).

In order to identify all the resonances which belong to one and the same haem methyl group in the different redox steps, a network of saturation transfer experiments (103) was performed at different stages of oxidation (91,104). An example of these assignments is presented in Figure 6 for the haem methyl group M_2^i (the second one counting from the low-field end of the spectrum of the fully oxidized protein-top spectrum). The resonance due to this methyl group in the spectrum of the fully reduced protein, M_2^0 (the upperscript referring to the redox step), appears in the usual position for a diamagnetic haem methyl group at ≃ 3 ppm, and is not shown in the figure. As the reoxidation proceeds, Step I spectrum starts to be generated and the resonance M_2^I appears at 8 ppm. This resonance is not easily depicted since it is still in the main envelope of the protein spectra. Here is a situation where saturation transfer difference spectroscopy is of double usefulness. In fact, the difference spectra only shows the resonances that are coupled together by exchange processes (91). Following up with the reoxidation procedure, this resonance moves subsequently to resonances M_2^{II} at 14.8 ppm, M_2^{III} at 28.2 ppm, and grows into its final position M_2^{IV} at 29.6 ppm, in the spectrum of the fully oxidized protein (Step IV).

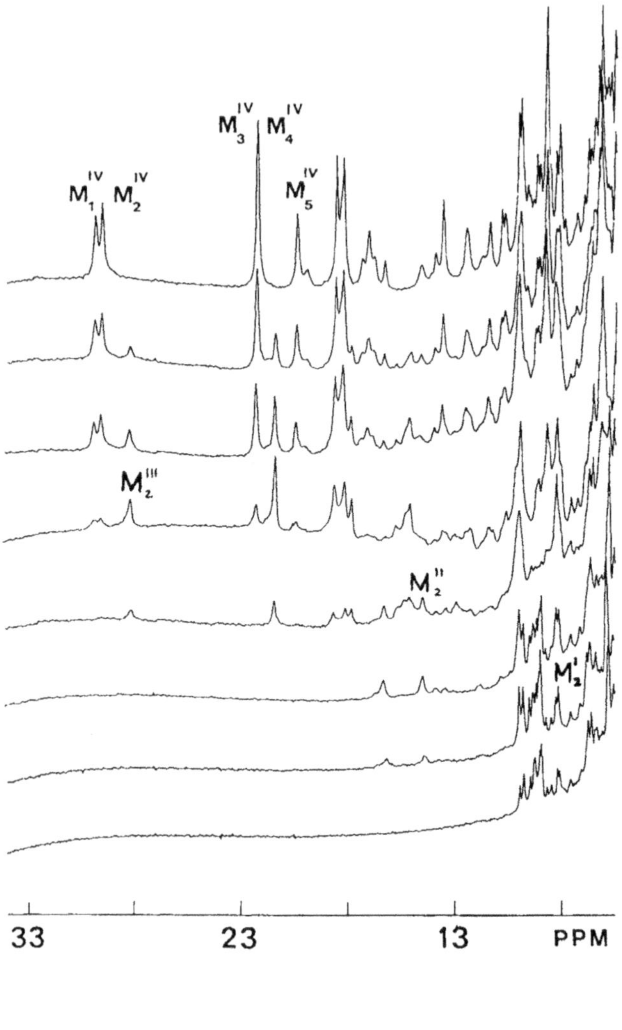

Figure 6. NMR redox titration of Desulfovibrio vulgaris cytochrome c_3, starting with the fully reduced spectrum (bottom) and going to the fully oxidized one (top). The figure shows only the low-field region of the 300 MHz spectra, taken at 298 K and pH* = 6.8, in D_2O. Some relevant resonances are assigned in the figure. (Adapted from reference 91).

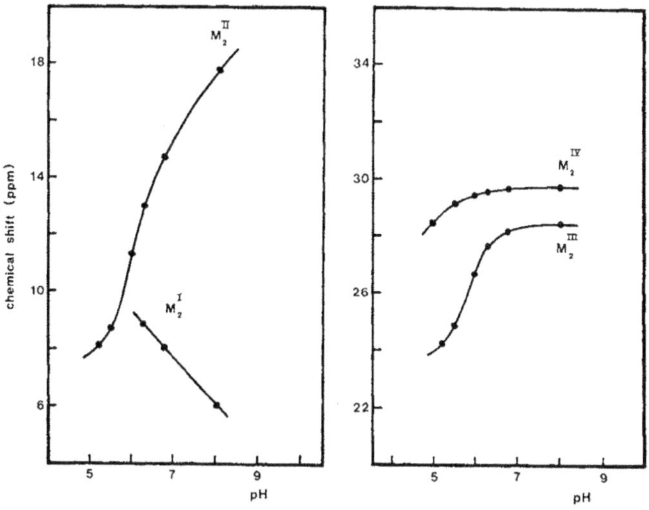

Figure 7. pH* dependence of the chemical shifts of haem methyl group M_2^i of D. vulgaris cytochrome c_3. (Adapted from reference 91).

As was stated earlier, the chemical shift of the resonance due to any of the intermediate redox steps, M_2^i (i = I, II, or III), is a weighted average of the paramagnetic shifts induced by the paramagnetic (oxidized) haems of all the oxidation states belonging to that step. Any change in the difference between the mid-point redox potentials of the four haems results in a modification of the relative populations of these oxidation states and consequently, in the change of the chemical shift of the resonances due to the intermediate redox steps. Conducting the redox titrations at different pH's, it was possible to demonstrate that the mid-point redox potential of at least one of the haems is pH dependent (91). Indeed, Figure 7 shows that there is a much larger shift with the pH for the resonances of a particular haem methyl group, M_2^i, in the intermediate redox steps, than that observed for the resonance of the same group in the spectrum of the fully oxidized protein, M_2^{IV}.

There is, however, a controversy concerning the physiological relevance of the pH dependence of the redox potential of cytochromes c. This is due to the fact that the redox potential of cytochrome c_2 in situ is reported to be pH independent (105). However, since cytochrome c_3 is an indispensable coupling factor for the Desulfovibrio species hydrogenase (106), the pH dependence of its redox properties should have physiological relevance (91).

Another important question concerning the mechanism of electron transfer is that of the specific recognition between donor and acceptor proteins. Here again NMR spectroscopy can be used to probe specific interactions (98,107,108) as well as alterations of electron exchange rates (98).

ACKNOWLEDGEMENTS

I am indebted to my colleagues who have contributed to the work described here, Drs. M.H. Emptage, G.R. Moore, I. Moura, J.J.G Moura, J. LeGall, H. Santos, J. Villalain, R.J.P. Williams and J.M. Wood. The studies in my laboratories were supported by NIH Grant GM25879, and the JNICT, the INIC and the Gulbenkian Foundation (Portugal).

REFERENCES

(1) Lemberg, R. and Barrett, J.: 1973, "Cytochromes", Academic Press, N.Y.
(2) Ferguson-Miller, S. et al.: 1979, "The Porphyrins" (ed. Dolphin, D.) Academic Press, N.Y., p. 149.
(3) Timkovich, R.: 1979, "The Porphyrins" (ed. Dolphin, D.) Academic Press, N.Y., p. 241.
(4) Mathews, F.S. et al.: 1979, "The Porphyrins" (ed. Dolphin, D.) Academic Press, N.Y., p. 108.
(5) Dickerson, R.E.: 1980, Sci. Am. 242, p. 136.
(6) Salemme, F.R.: 1977, Ann. Rev. Biochem. 46, p. 299.
(7) Emptage, M.H. et al.: 1981, Biochem. 20, p. 58.
(8) Webber, P.C. et al.: 1980, Nature, 286, p. 302.
(9) Ambler, R.P. et al.: 1979, Nature, 278, p. 661.
(10) Williams, R.J.P.: 1971, Cold Spring Harbour Symp. Quant. Biol. 36, p. 53.
(11) Wüthrich, K.: 1976, "NMR in Biological Research: Peptides and Proteins", North-Holland Publ. Co., Amsterdam.
(12) Moore, G.R. and Williams, R.J.P.: 1976, Coord. Chem. Rev. 18, p. 125.
(13) Walker, F.A. and LaMar, G.N.: 1973, Ann. N.Y. Acad. Sci., 206, 328.
(14) Goff, H. and LaMar, G.N.: 1977, J. Am. Chem. Soc. 99, p. 6599.
(15) Iizuka, T. et al.: 1976, FEBS Lett., 64, 156.
(16) Wüthrich, K. et al.: 1975, J. Magn.Res. 19, p. 111.
(17) Morishima, I. et al.: 1977, Biochem. 16, p. 5109.
(18) Satterlee, J.D. and Erman, J.E.: 1980, Arch. Biochem. Biophys. 202, p. 608.
(19) LaMar, G.N. et al.: 1978, Biochim. Biophys. Acta, 357, p. 270.
(20) Moore, G.R. et al.: 1982, Eur. J. Biochem. 123, p. 73.
(21) Fabry, T.L. et al.: 1971, "Probes of Structure and Function of Macromolecules" (ed. Chance, B. et al.) Academic Press, N.Y.
(22) Mildvan, A.S. et al.: 1971, "Probes of Structure and Function of Macromolecules" (ed. Chance, B.) Academic Press, N.Y., p. 205.
(23) Phillips, W.D. and Poe, M.: 1972, Methods Enzymol. 24, p. 304.
(24) Kurland, R.J. and McGarvey, B.R.: 1970, J. Magn. Res. 2, p. 286.
(25) Spiro, T.G. et al.: 1979, J. Am. Chem. Soc. 101, p. 2648.

(26) LaMar, G.N. *et al.*: 1981, J. Am. Chem. Soc. 103, p. 4405.
(27) Maltempo, M.M.: 1974, J. Chem. Phys. 61, p. 2540.
(28) Maltempo, M.M. *et al.*: 1974, Biochim. Biophys. Acta, 342, p. 290.
(29) Maltempo, M.M. and Moss, T.H.: 1976, Q. Rev. Biophys. 9, p. 181.
(30) Meyer, T.E. *et al.*: 1975, J. Biol. Chem. 250, p. 8416.
(31) Van Beeumen, J. *et al.*: 1980, Eur. J. Biochem. 107, p. 475.
(32) Ambler, R.P.: 1973, Biochem. J. 135, p. 751.
(33) Ambler, R.P. *et al.*: 1979, Nature, 278, p. 661.
(34) Bartsch, R.G.: 1978, "The Photosynthetic Bacteria" (ed. Clayton, R.K. and Sistron) Plenum, N.Y., p. 249.
(35) Ellfolk, N. and Soininen, R.: 1970, Acta Chem. Scand. 24, p. 2126.
(36) Soininen, R. and Ellfolk, N.: 1973, Acta Chem. Scand. 27, p. 35.
(37) Rönberg, M. and Ellfolk, N.: 1979, Biochim. Biophys. Acta, 581, p. 325.
(38) Rönberg, M. *et al.*: 1980, Biochim. Biophys. Acta, 626, p. 23.
(39) Aasa, R. *et al.*: 1981, Biochim. Biophys. Acta, 670, p. 170.
(40) Emptage, M.H.: 1978, Ph.D. Thesis, University of Illinois, Urbana, U.S.A.
(41) Dolphin, D.H. *et al.*: 1977, Inorg. Chem. 16, p. 711.
(42) Reed, C.A., *et al.*: 1979, J. Am. Chem. Soc. 101, p. 2948.
(43) Reed, C.A. *et al.*: 1980, J. Am. Chem. Soc. 102, p. 2302.
(44) Xavier, A.V. *et al.*: 1980, "Coordination Chemistry Environment in Iron Containing Proteins and Enzymes" (ed. Dolphin, D., *et al.*), D. Reidel Publishing Company, Dordrecht, *in press*.
(45) Wüthrich, K.: 1969, Proc. Natl. Acad. Sci. USA 63, p. 1071.
(46) McDonald, C.C. *et al.*: 1969, Biochem. Biophys. Res. Commun. 36, p. 442.
(47) Redfield, A.G. and Gupta, R.K.: 1971, Cold Spring Harbor Symp. Quant. Biol. 36, p. 405.
(48) McDonald, C.C. and Phillips, W.D.: 1973, Biochem. 12, p. 3170.
(49) Keller, R.M. and Wüthrich, K.: 1978, Biochim. Biophys. Acta 533, p. 195.
(50) Stellwagen, E. and Shulman, R.G.: 1973, J. Mol. Biol. 75, p. 683.
(51) Patel, D.J. and Canuel, L.L.: 1976, Proc. Natl. Acad. Sci. USA 73, p. 1398.
(52) Moore, G.R. and Williams, R.J.P.: 1980, Eur. J. Biochem. 103, p. 493.
(53) Moore, G.R. and Williams, R.J.P.: 1980, Eur. J. Biochem. 103, p. 503.
(54) Keller, R.M. *et al.*: 1973, FEBS Lett. 36, p. 151.
(55) Cohen, J.S. *et al.*: 1974, J. Biol. Chem. 249, p. 1113.
(56) Dobson, C.M. *et al.*: 1975, FEBS Lett. 51, p. 60.
(57) Moore, G.R. and Williams, R.J.P.: 1975, FEBS Lett. 53, p. 334.
(58) Campbell, I.D. *et al.*: 1976, FEBS Lett. 70, p. 96.
(59) Dickerson, R.E. et al.: 1976, J. Mol. Biol. 100, p. 473.
(60) Dickerson, R.E. and Timkovich, R.: 1975, "The Enzymes" (ed. Boyer, P.) 11, Academic Press, N.Y., p. 397.
(61) Moore, G.R. *et al.*: 1980, J. Inorg. Biochem. 12, p. 1.
(62) Moore, G.R. *et al.*: 1980, J. Inorg. Biochem. 13, p. 347.
(63) Moore, G.R. and Williams, R.J.P.: 1980, Eur. J. Biochem. 103, p. 533.

(64) Moore, G.R. and Williams, R.J.P.: 1980, Eur. J. Biochem, 103, p. 543.
(65) Cookson, D.J. et al.: 1978, Eur. J. Biochem. 83, p. 261.
(66) Campbell, I.D. et al.: 1975, Proc. R. Soc. London, Ser. B189, p. 503.
(67) Wüthrich, K. and Wagner, G.: 1975, FEBS Lett. 50, p. 265.
(68) Cave, A. et al.: 1975, FEBS Lett. 65, p. 190.
(69) Moore, G.R. and Williams, R.J.P.: 1980, Eur. J. Biochem. 103, p. 513.
(70) Moore, G.R. and Williams, R.J.P.: 1980, Eur. J. Biochem. 103, p. 523.
(71) Xavier, A.V. et al.: 1978, Nature, 275, p. 245.
(72) McDonald, C.C. et al: 1974, Biochem. 13, p. 1952.
(73) Senn, H. et al.: 1980, Biochem. Biophys. Res. Commun. 92, p. 1362.
(74) Keller, R.M. and Wüthrich, K.: 1978, Biochem. Biophys. Res. Commun. 83, p. 1132.
(75) Takano, T. et al.: 1978, Proc. Natl. Acad. Sci. USA 75, p. 2674.
(76) Almassy, R.J. et al.: 1978, Proc. Natl. Acad. Sci. USA 75, p. 2674.
(77) Salemme, F.R. et al.: 1973, J. Biol. Chem. 248, p. 7701.
(78) Pettigrew, G.W.: 1982, FEBS Lett. 86, p. 1416.
(79) Moore, G.R. and Williams, R.J.P.: 1977, FEBS Lett. 79, p. 229.
(80) Kihara, H. et al.: 1978, Biochim. Biophys. Acta, 532, p. 337.
(81) Pettigrew, G.W. et al.: 1975, Biochim. Biophys. Acta, 430, p. 197.
(82) Straub, J.P. and Colpa Boonstra, J.P.: 1962, Biochim. Biophys. Acta, 60, p. 650.
(83) Urban, P.F. and Klingenberg, M.: 1969, Eur. J. Biochem. 9, 519.
(84) Wilson, D.F. et al.: 1972, Arch. Biochem. Biophys. 151, p. 112.
(85) Wilson, D.F. et al.: 1972, Biochim. Biophys. Acta, 256, p. 277.
(86) Van Gelder, B.F. et al.: 1977, "Structure and Function of Energy Transducing Membranes" (ed. van Dam, K. et al.) Elsevier/North-Holland, Amsterdam, N.Y., p. 61.
(87) Artzabanov, V. Yu. et al.: 1978, FEBS Lett. 87, p. 180.
(88) Papa, S.: 1982, J. Bioenerg. Biomembr., 15, p. 69.
(89) Smith, G.M.: 1979, Biochem. 18, p. 1628.
(90) Moore, G.R. et al.: 1980, Biochim. Biophys. Acta, 590, p. 261.
(91) Moura, J.J.G. et al.: 1982, Eur. J. Biochem., in press.
(92) Kassner, R.J.: 1973, J. Am. Chem. Soc. 95, p. 2674.
(93) Stellwagen, E.: 1978, Nature, 275, p. 74.
(94) Kraut, J.: 1981, Biochem. Soc. Trans. 9, p. 197.
(95) Kowalsky, A.: 1965, Biochem. 4, p. 2382.
(96) Gupta, R.K. and Redfield, A.G.: 1970, Biochem. Biophys. Res. Commun. 41, p. 273.
(97) Gupta, R.K. and Koenig, S.H.: 1971, Biochem. Biophys. Res. Commun. 45, p. 1134.
(98) Moura, J.J.G. et al.: 1977, FEBS Lett. 81, p. 275.
(99) Stellwagen, E. and Shulman, R.G.: 1973, J. Mol. Biol. 80, p. 559.
(100) Boswell, A.P. et al.: 1982, Eur. J. Biochem. 124, p. 289.
(101) Eley, C.G.S. et al.: 1982, Eur. J. Biochem. 124, p. 295.
(102) Gupta, R.K. and Mildvan, A.S.: 1978, Methods Enzymol. 54, p. 154.
(103) Gupta, R.K. and Redfield, A.G.: 1970, Science 169, p. 1204.
(104) Moura, I. et al.: 1980, Cienc. Biol. (Portugal) 5, p. 189.

(105) Prince, R.C. and Dutton, P.L.: 1977, Biochim. Biophys. Acta, 459, p. 573.
(106) LeGall, J. et al.: 1982 "Iron Sulfur Proteins" (ed. Spiro, J.G.) John Wiley and Sons, N.Y., p. 177.
(107) Moura, I. et al.: 1980, Cienc. Biol. (Portugal) 5, p. 195.
(108) Boswell, A.P. et al.: 1980, Biochem. Soc. Trans. 8, p. 637.

THE STRUCTURE OF THE METAL CENTERS IN CYTOCHROME c OXIDASE

Sunney I. Chan, Craig T. Martin, Hsin Wang, Gary W. Brudvig, and Tom H. Stevens

A. A. Noyes Laboratory of Chemical Physics, 127-72, California Institute of Technology, Pasadena, California 91125

ABSTRACT

Progress toward elucidation of the structure of the metal centers in cytochrome c oxidase will be reviewed. Our studies are based on low-temperature electron paramagnetic resonance (EPR) spectroscopy. We have used nitric oxide (NO) extensively to probe the O_2 reduction site of the enzyme. In addition, we have isolated auxotrophs of Saccharomyces cerevisiae in order to incorporate isotopically substituted amino acids into the yeast protein. This latter approach, in conjunction with low-temperature EPR and electron nuclear double resonance (ENDOR) spectroscopy, has allowed unambiguous information on the structure of two of the four metal centers.

INTRODUCTION

Cytochrome c oxidase mediates the transfer of electrons between reduced cytochrome c and molecular oxygen, reducing dioxygen to two molecules of water. The overall reaction is given by:

$$4 \text{ ferrocytochrome } c + O_2 + 4H^+ \rightarrow 4 \text{ ferricytochrome } c + 2H_2O.$$

The free energy for this reaction at physiological pH is -50 kcal per mole of O_2 reduced (1). The reduction of O_2 is extremely efficient, with the turnover rate approaching 400 sec^{-1} (2).

Cytochrome c oxidase is a transmembrane protein embedded in the inner membrane of the mitochrondrion. All indications are that the oxidase receives its reducing equivalents from cytochrome c on the

I. Bertini, R. S. Drago, and C. Luchinat (eds.), The Coordination Chemistry of Metalloenzymes, 313–328.

intermembrane space, i.e., from the cytosol side, and the four H^+ that are used up per turnover originate from the matrix. In this manner, four negative charges are transferred across the inner membrane and a transmembrane pH and/or potential gradient is established. Evidence is mounting that the enzyme is also a proton pump (3) with protons pumped across the inner membrane from the matrix to the cytosol side during the electron transport and/or O_2 reduction process. The number of vectorial protons (four to eight) pumped per turnover, however, remains elusive (3,4).

Because of the importance of cytochrome c oxidase in cellular respiration and energy conservation, the structure of the protein has naturally received considerable attention. Impressive progress has been made toward deciphering the primary sequence (5-7) and the subunit structure (8) of the enzyme. A glimpse of the three-dimensional folding and assembly of the protein (Figure 1) has also emerged from electron microscopy imaging studies (9). The structures of the four metal centers that are intimately associated with the electron transfer, O_2 reduction, and possibly proton pumping, are beginning to unfold (10-13). Information on the location of the metal centers in the enzyme and their relative proximity to one another is expected to be forthcoming.

In this report, we will review our efforts toward elucidating the structure of the four metal centers in cytochrome c oxidase. Our studies are based largely on low-temperature electron paramagnetic resonance (EPR) spectroscopy. We have used nitric oxide (NO) extensively to probe the structure of the O_2 reduction site. In addition,

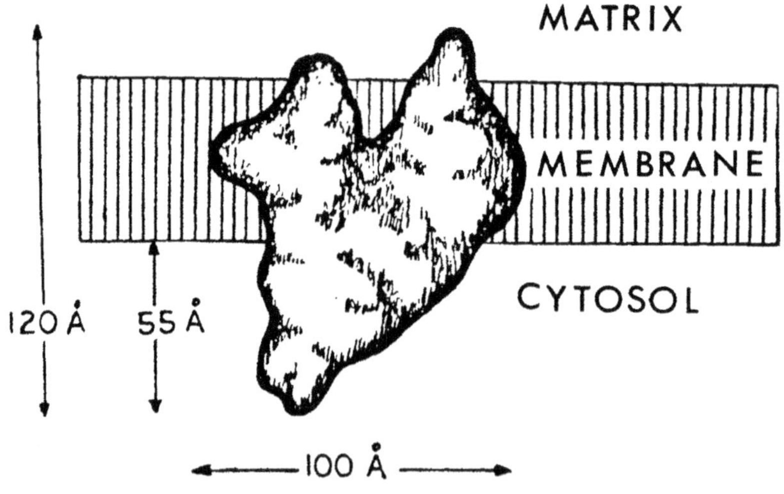

Figure 1. Three-dimensional folding and assembly of oxidized cytochrome c oxidase as revealed by electron microscopy imaging.

we have manipulated the yeast system Saccharomyces cerevisiae in order to incorporate isotopically substituted amino acids into the yeast protein. This approach, in conjunction with low-temperature EPR and electron nuclear double resonance (ENDOR) spectroscopy, has provided unambiguous information on the ligands of two of the four metal centers. Our work is still incomplete. Nevertheless, we feel that an important beginning has been made and we are optimistic that a molecular picture of the structure and function of cytochrome c oxidase is on the horizon.

THE METAL CENTERS

Cytochrome c oxidase contains two heme a's and two copper ions, all of which are inequivalent. Cytochrome a_3 and Cu_B form the oxygen reduction site, while cytochrome a and Cu_A participate in electron transfer. We now describe in turn results relating to the structure of each of these metal centers.

Cytochrome a

There seems general agreement that cytochrome a accepts electrons from ferrocytochrome c and transfers them to the Cu_A center. The cytochrome c binding site(s) is known to be located on the cytosol-side of the inner mitochondrial membrane. Since the electron transfer between ferrocytochrome c and cytochrome a is known to be rapid, the location of cytochrome a must be near the cytosol side. The reduction potential of cytochrome a is similar to that of ferricytochrome c (14), which is expected if cytochrome a is the point of entry of the electrons into the protein and degradation of the redox free energy is to be minimized in this electron transfer step.

All indications are that cytochrome a is six-coordinate and low spin, with nitrogens from two neutral imidazoles as axial ligands. Such a structure would be consistent with the proposed electron transfer function of cytochrome a. The magnetic circular dichroism (MCD) and EPR spectra of cytochrome a demonstrate clearly that cytochrome a is low spin (15,16). There have been two lines of evidence supporting two histidine imidazoles as axial ligands. Babcock et al.(17) have compared the optical and resonance Raman spectra of cytochrome a with those of related low-spin heme a models. Similarly, Peisach (18) has compared the EPR g-values of cytochrome a with those of low-spin heme a models with known axial ligands. The Peisach compilation of g-values is reproduced in Table I. The g-values (3.0, 2.21, 1.45) of cytochrome a (Figure 2) are reproduced only by the bis neutral-imidazole complex of heme a.

Copper A

Copper A is the low potential copper in cytochrome c oxidase and it has been suggested that the functional role of this metal center is to transfer electrons from cytochrome a to the dioxygen reduction site.

TABLE I. The Relation between EPR Parameters and Structure of Low Spin Ferric Heme Compounds (Taken from Ref. 18)

Compound	Axial Ligands		g Values		
Glycera Hb $MeNH_2$	imid	RNH_2	3.30	1.98	1.20
Leg Hb pyridine	imid	pyr	3.26	2.10	0.82
Cytochrome *c*	$imid^o$	met	3.07	2.26	1.25
Bis imid heme	$imid^o$	$imid^o$	3.02	2.24	1.51
Bis $imid^-$ heme	$imid^-$	$imid^-$	2.80	2.26	1.72
MbOH	$imid^-$	OH^-	2.55	2.17	1.85
Cyt. $P\text{-}450_{cam}$	$imid^o$	RS^-	2.45	2.26	1.92
Cyt. *c* oxidase	$imid^o$	$imid^o$	3.0	2.2	1.5

Its reduction potential is similar to that of cytochrome *a* (14). Since the first reduction potential drop occurs between Cu_A and the O_2 reduction site, Chan *et al*. (19) have proposed that this redox energy may be conserved and Cu_A might serve as the proton pump of the enzyme.

The structure of the Cu_A center has been of considerable interest because this copper is highly unusual (20). The EPR signal of the Cu_A center (Figure 2) is atypical of Cu(II) in that no copper hyperfine splittings are clearly resolved and the g-values (2.18,2.03,1.99) are quite small; in fact, one g-value is below that of the free electron. Two mechanisms have been proposed to explain the unusual EPR properties. One possibility is that the unpaired electron spin resides primarily on an

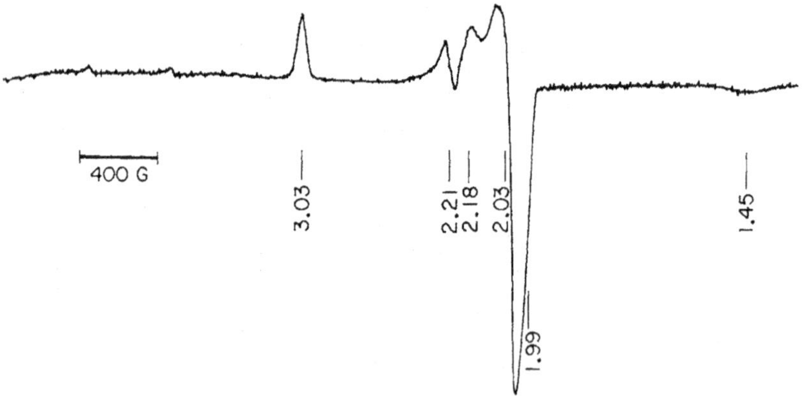

Figure 2. X-band EPR spectrum of oxidized beef heart cytochrome *c* oxidase at 10 K.

associated ligand (21). This mechanism calls for extensive electron charge delocalization from the involved associated ligand onto the copper ion. The g-values for the Cu_A center EPR signal are, in fact, typical of those of thiyl radicals (22), and it was originally suggested that this EPR signal might be due to a disulfide interacting with a copper ion (23) or due to a sulfur radical (24). A second possibility is that the orbital containing the unpaired electron is a copper hybrid 3d orbital with strong admixtures of 4s and 4p character (25). In this case, the unpaired electron would reside primarily on the copper and the unusual EPR properties would result from distorting the copper into a near tetrahedral geometry which allows mixing of copper 4s and 4p orbitals with the 3d ground state.

Recent ENDOR studies (26) have revealed a small copper hyperfine interaction associated with the Cu_A center, but this copper hyperfine interaction is nearly isotropic. This result indicates that both the isotropic hyperfine interaction and the anisotropic distributed dipole interaction between the unpaired electron and the copper nucleus are small. The implication is that <u>either</u> the unpaired electron spin density resides primarily on an associated ligand sufficiently removed from the copper ion, <u>or</u> the unpaired electron is localized on a copper ion with cubic (or higher symmetry) coordination. However, the Cu ENDOR results, i.e., the small anisotropy in the copper hyperfine interaction, place limits on the extent of any d-p mixing. It is possible to ascertain the relative amounts of 3d and 4p character in a hybrid orbital necessary to eliminate the anisotropic copper hyperfine interaction by calculating the distributed dipole interaction of an electron in a 3d or 4p orbital with the copper nucleus. The calculations show that it is necessary to include about three times as much 4p(z) character as $3d(x^2-y^2)$ or 3d(xy) character to obtain an isotropic copper hyperfine interaction (27). Such a large mixing is unreasonable in view of the fact that the $3d^84p$ configuration lies about 125,000 cm^{-1} above the $3d^9$ configuration for divalent copper (28). Thus, although mixing of copper 3d and 4p orbitals could, in principle, account for the observed EPR properties of the Cu_A center, this possibility is unreasonable from the standpoint of energetics.

We subscribe to the view that the EPR properties of the Cu_A center are best accounted for by delocalization of the unpaired electron spin onto an associated ligand. In fact, X-ray absorption edge data indicate that Cu_A in the oxidized protein is in a highly covalent environment (29) and might even be reduced (30) to Cu(I). Extensive delocalization of spin onto an associated ligand in the oxidized enzyme would predict that Cu_A remains mainly in the Cu(I) state both in the oxidized and reduced forms of the Cu_A center, since upon reduction an associated ligand would be the actual electron acceptor rather than the Cu ion itself. This prediction seems to be born out by the observation that the "1s-4s" transition in the X-ray edge spectrum of Cu_A does not change substantially when the enzyme is reduced (30).

If our picture of the structure of the Cu_A center is correct, the question remains as to what ligands among the various amino acid residues can render to Cu_A the unusual EPR properties manifested by this metal center. Cysteines come to mind; however, at least two cysteines must be involved since ligation of one cysteine to a divalent copper does not lead to sufficient electron transfer from the cysteine sulfur to account for the observed EPR properties (cf. "blue coppers" (31)). However, it is well known that ligation of one cysteine to a $3d^9$ copper renders the metal center more susceptible to reduction by a second cysteine sulfur (23).

These considerations led us to propose some years ago (21) that the Cu_A center in cytochrome c oxidase consisted of a Cu(I) ion ligated by two cysteine sulfurs, one a cysteinate and the other a sulfur radical. We have now undertaken experiments to test this model by incorporating cysteine substituted with ^2H at the β-methylene carbon ((^2H)Cys) into yeast cytochrome c oxidase using a cysteine auxotroph of the yeast system Saccharomyces cerevisiae. As part of a larger overall effort to define the ligands of the various metal centers, a histidine auxotroph was also isolated and used to incorporate histidine substituted with ^{15}N at both imidazole ring positions ((^{15}N)His) into the yeast enzyme. The strategy here is to exploit the differences in nuclear spin or nuclear magnetic moment of isotopes to identify magnetic interactions between the unpaired electron spin and the atom in question. Thus, perturbation of the EPR spectrum for the metal center by the isotopic substitution, or in the case where the hyperfine interaction is too small to be discernible in the EPR, modification of the ENDOR spectrum, may be used to obtain unambiguous information about the involvement of the particular amino acid at the site.

In the specific case of Cu_A, we have confirmed the involvement of at least one histidine and one cysteine as ligands. We have compared the EPR spectra of native, (^2H)Cys, and (^{15}N)His yeast cytochrome c oxidase. Although subtle differences were noted between the EPR spectra of native and the isotopically substituted yeast proteins, the recent ENDOR studies (12), carried out in collaboration with Dr. C. P. Scholes of SUNY at Albany, were unequivocal in this conclusion. The ENDOR spectra of the (^2H)Cys and (^{15}N)His yeast oxidase are compared with that of the native yeast enzyme in Figure 3. We note that the spectrum observed for the native yeast oxidase is very similar to that of the beef heart protein (32). The two signals seen at 21.7 and 19.7 MHz are due to strongly coupled protons and correspond to proton hyperfine couplings of 16.2 and 12.2 MHz. That these protons have origin in the β-CH_2 of cysteine may be ascertained by noting the intensity change in the ENDOR signal in this region upon the deuterium substitution. The two signals seen at 7.1 and 9.2 MHz in the native enzyme are split by twice the characteristic ^{14}N Zeeman energy ($2\nu(^{14}N) = 2.00$ MHz) and can be assigned to a ^{14}N with hyperfine coupling of 16 MHz. This assignment was confirmed by the ENDOR spectrum of the (^{15}N)His yeast enzyme. Upon comparison of the ENDOR spectra for the (^{15}N)His and native yeast protein, we see that only the intensities of the ^{14}N ENDOR signals at 7 and 9 MHz are substantially reduced by the isotopic substitution of the imidazole ring nitrogens. On the basis of the difference in nuclear

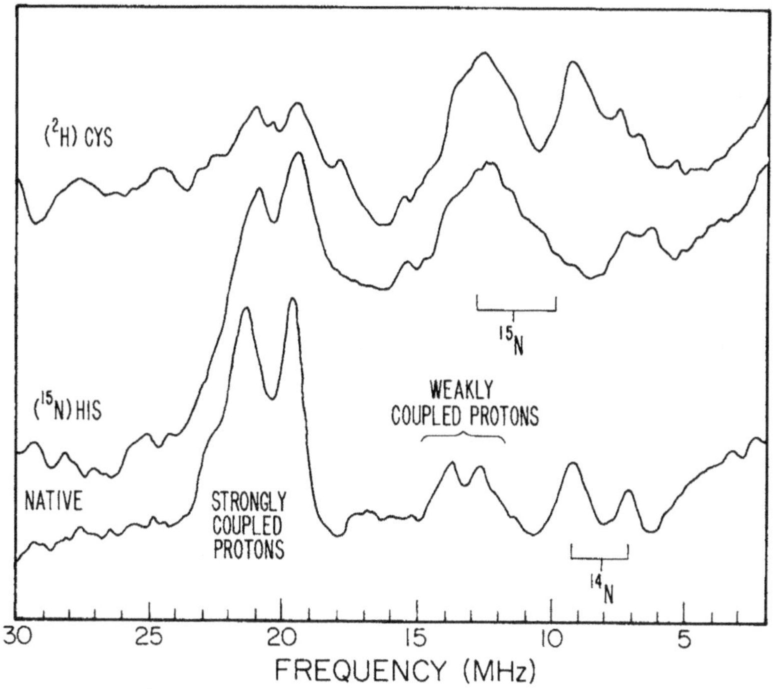

Figure 3. ENDOR spectra of native, (^{15}N)His and (^{2}H)Cys yeast cytochrome c oxidase observed at g = 2.04, microwave frequency 9.12 GHz and temperature 2.1 K.

magnetic moments between the two nitrogen isotopes, substitution of the ^{14}N nucleus by ^{15}N would replace the ^{14}N ENDOR signals at 7 and 9 MHz by ^{15}N ENDOR signals at 10 and 13 MHz, respectively. The increased intensity observed in the 10-15 MHz region in the ENDOR spectrum of the (^{15}N)His protein is consistent with the appearance of these ^{15}N ENDOR signals. From these observations, it is clear that there is at least one histidine and one cysteine ligand to Cu_A in cytochrome c oxidase.

The issue of whether there are one or two cysteines associated with Cu_A remains unsettled. In principle, the number of cysteines could be inferred from the number of proton ENDOR signals and/or the magnitudes of the proton hyperfine interactions affected by the deuterium substitution. In our experiments, only the two strongly coupled proton ENDOR signals clearly disappear upon deuterium substitution, although the signal-to-noise in the 10-15 MHz region of the ENDOR spectra does not allow us to rule out the existence of weakly coupled β-CH_2 protons should there exist a second cysteine ligand. In any case, from the well-known equations describing hyperconjugation in sulfur radicals (33), and by comparison with model compounds (22), we can estimate the unpaired spin density (ρ_S^π) in the sulfur 3p(z) orbital(s). If we assume that the strongly coupled proton hyperfine interactions arise from two protons on two different cysteine ligands to Cu_A, one cysteine sulfur would have ρ_S^π ranging from 0.14 to 0.37, while for the other cysteine sulfur, ρ_S^π would range from 0.23 to 0.52 depending on the choice of dihedral angle between the strongly coupled β-CH_2 proton and the 3p(z) orbital of the sulfur in each case. We make the assumption here that the other proton on each methylene carbon exhibits a negligible (less than 3 MHz) hyperfine coupling. A simpler situation is that these hyperfine couplings arise from protons on the same cysteine ligand. In this case, only two distinct orientations of the β-CH_2 protons relative to the sulfur 3p(z) orbital are possible. In the one case, the resultant unpaired spin density on sulfur would be 0.23, and in the other, 0.83. The latter value is more in agreement with X-ray absorption edge and EXAFS data on the Cu_A site. It is important to note that the assignment of the strong proton hyperfine couplings to methylene protons on the same cysteine ligand does not preclude the existence of a second cysteine ligand to Cu_A. In fact, if ρ_S^π is as high as 0.83, a second cysteine ligand is necessary to facilitate this large delocalization of spin from the copper to the first cysteine (i.e., a delocalization of charge from cysteine to copper). If this were so, the magnitude of the hyperfine coupling to these more distant protons would be expected to be very small.

Experiments are currently in progress to incorporate cysteine substituted with ^{13}C at the methylene carbon ((^{13}C)Cys into yeast cytochrome c oxidase using the cysteine auxotroph which we have isolated. In contrast to the hyperfine interaction of the β-methylene protons, the ^{13}C hyperfine interaction is relatively insensitive to the dihedral angle between the C-H bond and the sulfur 3p(z) orbital bearing the unpaired electron and hence should provide a more direct indication of ρ_S^π.

Cytochrome a_3-Copper B

Cytochrome a_3-Cu_B constitutes the oxygen reduction site of the enzyme. The structure of this site has remained elusive because antiferromagnetic coupling between the high spin heme and the d^9 copper yields a S = 2 ground state and renders the site EPR silent in the oxidized enzyme (34). Even in the reduced enzyme, where the copper is d^{10}, the high spin ferrous heme has a S = 2 ground state. To obtain EPR signals from the cytochrome a_3-Cu_B site, it is necessary to convert the site into a half-integral spin system which has a ground state Kramer's doublet (35). We recently showed that the binding of NO to the site under various conditions allowed EPR signals to be observed from both cytochrome a_3 and Cu_B (10).

While it is well known that a variety of ligands including CO, NO, CN^-, F^-, formate, etc., bind to cytochrome a_3 (36), we have obtained evidence that NO can bind to Cu_B as well (10,37). Three complexes of NO with cytochrome c oxidase can be prepared: (i) the reduced enzyme plus NO, which exhibits a nitrosylferrocytochrome a_3 EPR signal (38,39), (ii) the oxidized enzyme plus NO, which exhibits a high-spin ferricytochrome a_3 EPR signal (10), and (iii) the one-quarter reduced NO-bound enzyme (10), which exhibits a triplet EPR signal due to the interaction of the electron spins on nitrosylferrocytochrome a_3 (S = ½) and Cu_B(II) (S = ½).

The EPR signal observed when NO is added to reduced cytochrome c oxidase is typical of that observed for other nitrosylferroheme proteins such as cytochrome c, hemoglobin, or cytochrome c peroxidase (40). In particular, nitrogen superhyperfine splittings are observed for both the NO nitrogen and an endogenous nitrogen bound axially to cytochrome a_3 opposite the NO. The g-values for this complex as well as magnitudes of the ^{14}N hyperfine splittings indicate that the endogenous nitrogen ligand of cytochrome a_3 is most likely from an imidazole nitrogen. Unambiguous identification of histidine as the endogeneous axial ligand to cytochrome a_3 (11) was obtained from the nitrosylferrocytochrome a_3 complex prepared from the ((^{15}N)His) yeast oxidase described earlier.

When ^{14}NO is bound to ferrocytochrome a_3 of the yeast enzyme, the EPR signal of the complex exhibits a nine-line hyperfine pattern which can be interpreted in terms of the superposition of three sets of three lines arising from two nonequivalent nitrogens (I = 1) interacting with the unpaired electron. The larger of the two hyperfine coupling constants is 20.8 G and the smaller 6.9 G. When ^{15}NO is used in this experiment, the ^{15}NO-bound protein exhibits an EPR spectrum with g-values identical with those of the ^{14}NO-bound species, but with a hyperfine pattern consisting of two sets of three lines. This pattern is consistent with the presence of one ^{14}N and one ^{15}N nitrogen bound axially to cytochrome a_3 with a 28.2 G splitting for the ^{15}N and a 7.0 G splitting for the ^{14}N ligand. These spectra of the NO-bound native yeast protein are compared with those of the ^{14}NO- and ^{15}NO-bound (^{15}N)His yeast cytochrome c oxidase in Figure 4. It is apparent from this comparison that the hyperfine patterns have been altered upon (^{15}N)His substitution. The (^{15}N)His ^{15}NO-bound protein hyperfine pattern consists of two sets of doublets, with a ^{15}NO nitrogen splitting of 27.5 G and a splitting of about 12 G for the ^{15}N nitrogen of the histidine. The (^{15}N)His ^{14}NO-bound protein hyperfine pattern consists of three sets of doublets, with splitting of 21 G and 10.2 G for the ^{14}NO and histidine ^{15}N nitrogen, respectively. Thus, the substitution of (^{15}N)His for (^{14}N)His in cytochrome c oxidase results in the involvement of a (I = ½) ^{15}N nucleus rather than a (I = 1) ^{14}N nucleus in the nitrosylferrocytochrome a_3 EPR signal. These studies provide unequivocal identification of histidine as the endogenous fifth ligand to cytochrome a_3.

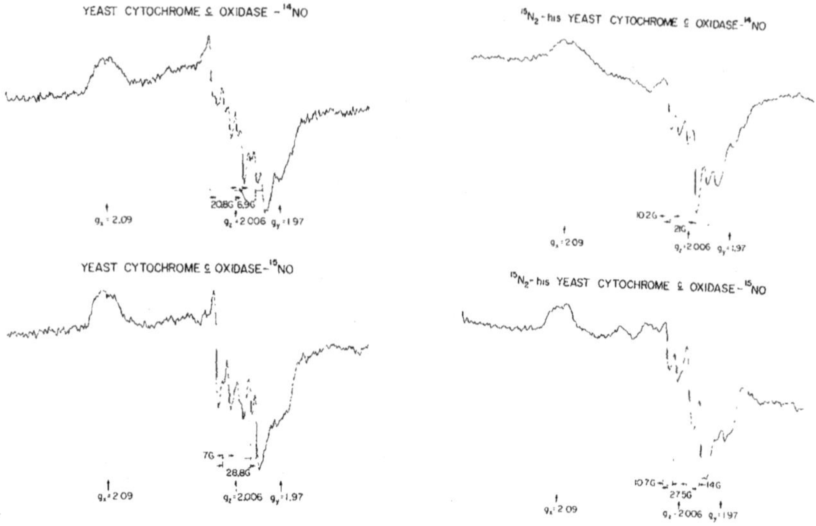

Figure 4. X-band EPR spectra of NO-bound reduced native and (^{15}N)His yeast cytochrome c oxidase at 30 K.

The above results pertain to the reduced enzyme, and one might ask whether a histidine is also the fifth ligand to cytochrome a_3 in the oxidized enzyme. We do not possess evidence to the contrary. In the course of our work, we have shown that NO also binds to the oxidized enzyme. The site of binding appears to be Cu_B as the NO binding breaks the coupling between the cytochrome a_3 and Cu_B to unveil a high spin heme signal in the EPR (Figure 5) without a corresponding Cu(II) EPR signal from Cu_B. The signals at $g \simeq 6$ arise from the $M_s = \pm \frac{1}{2}$ component (Kramer's doublet) of the $S = 5/2$ spin manifold when the applied field is in the plane of the porphyrin ring. The intensity of these signals is temperature dependent and could be used to measure the zero field splitting D for the high spin heme, which could, when compared with those of known hemes, in turn be used to infer the nature of the fifth ligand. In the absence of exogenous ligands, we measured D to be 9 cm^{-1}, and in the presence of F^-, $D \simeq 6\ cm^{-1}$ (41). These zero-field parameters are indicative of a high spin heme with an axial histidine.

Neither the EPR signal from nitrosylferrocytochrome a_3 nor the high spin cytochrome a_3 signal provide any information regarding the position of the histidine vis-a-vis Cu_B. A bridging imidazole has been proposed to provide the exchange coupling between the two metal centers (42). However, the triplet EPR signal observed from the one-quarter-reduced NO-bound enzyme provides a measure of the interaction

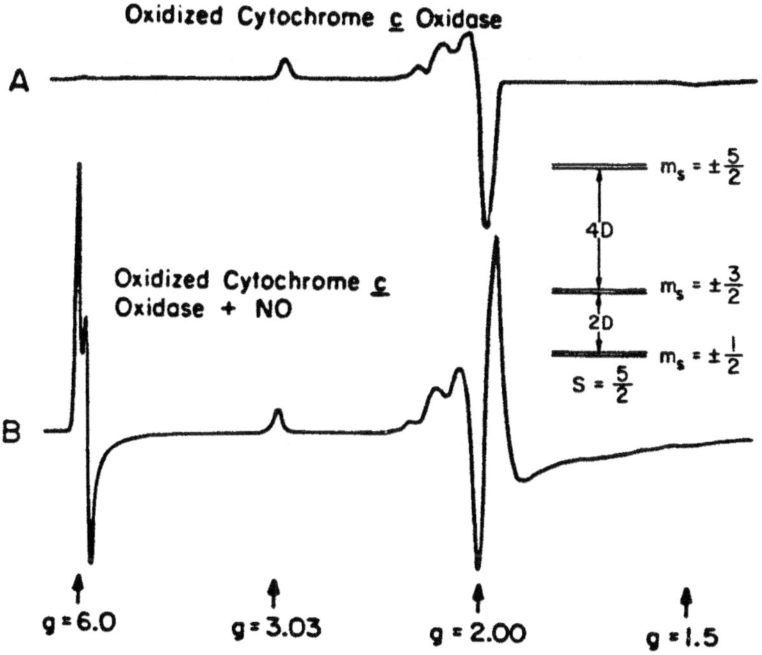

Figure 5. X-band EPR spectra of native oxidized beef heart cytochrome c oxidase in the absence and presence of NO. (Inset): Energy level diagram of the S = 5/2 spin manifold.

between cytochrome a_3 and Cu_B in the nitrosylferrocytochrome a_3, $Cu_B(II)$ complex and indicates that NO must bind between these two metal ions in this complex; in other words, the histidine imidazole of cytochrome a_3 is distal to Cu_B.

In order to obtain the one-quarter-reduced NO-bound enzyme, we exploited the high reduction potential of cytochrome a_3 to activate the disproportionation of N_3^- in the presence of NO according to the overall reaction (10)

$$\text{ferricytochrome } a_3 + N_3^- + 2NO \rightarrow \text{nitrosylferrocytochrome } a_3 + N_2 + N_2O \; .$$

The optical spectrum of the resultant nitrosylferrocytochrome a_3 suggests that the imidazole is still in place. The fact that a triplet EPR signal is observed for this complex indicates that NO binds to the ferrocytochrome a_3 and the unpaired e^- on the NO is sufficiently close to interact magnetically with $Cu_B(II)$. The zero-field splitting D of the triplet can be estimated from the breadth of the $\Delta M_s = \pm 1$ transition. We obtained a value for $|D|$ of 0.07 cm^{-1}. If we assume a purely dipolar

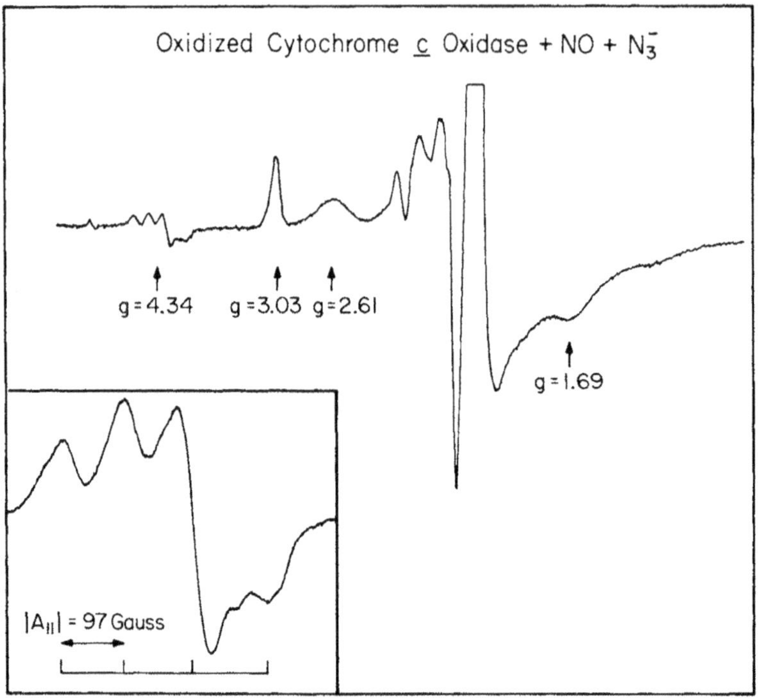

Figure 6. X-band EPR spectrum of the nitrosylferrocytochrome $\underline{a}_3$, Cu_B(II) species at 7 K. (Inset): Magnified view of the half-field transition region.

interaction between the two spins, we estimate the distance between the two "spin centers" to be ~ 3.4 Å.

The triplet state from the nitrosylferrocytochrome $\underline{a}_3$, Cu_B(II) species exhibits copper hyperfine splittings on the half-field transition at g_1 ~ 4. The magnitude of the copper hyperfine interaction, ~0.02 cm^{-1}, is characteristic of a tetragonal geometry for Cu_B. This conclusion seems also supported by the recent work of Reinhammar and Malmström (43). This result would argue against a cysteine ligand for Cu_B, at least in this state of the enzyme, in contrast to recent EXAFS studies (13). Preliminary ENDOR data (B. Hoffman and B. Reinhammer, private communication) indicate that there are two nitrogens associated with Cu_B. Comparative ENDOR studies of the native and the (^{15}N)His yeast oxidase are in progress to ascertain whether these nitrogens have origin in histidine imidazoles.

A model for oxidized cytochrome $\underline{c}$ oxidase is shown in Figure 7. In this model, Cu_B(II) is tetragonal and is linked to ferricytochrome

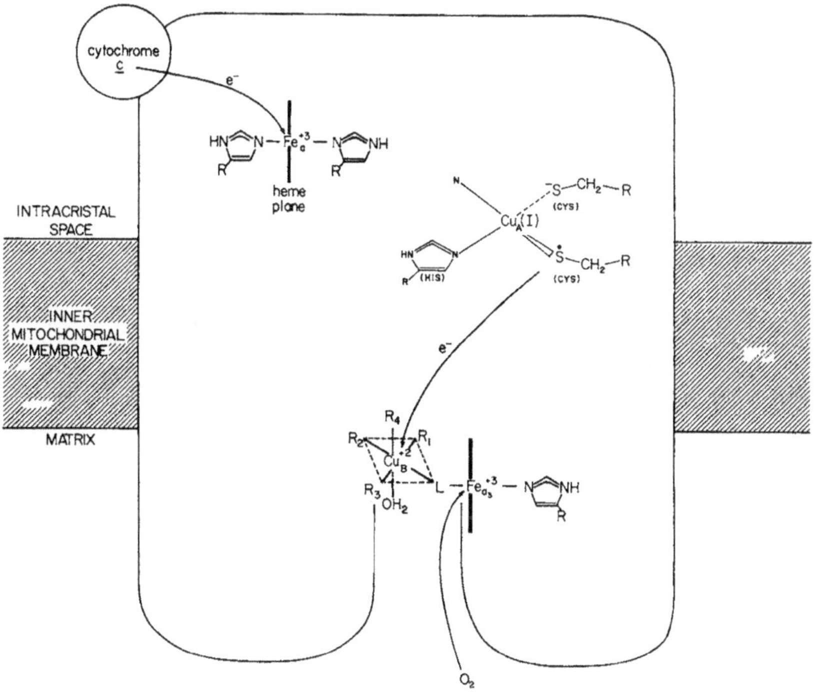

Figure 7. A model for oxidized cytochrome c oxidase.

$\underline{a}_3$ via an equatorial ligand, L, which mediates the strong antiferromagnetic exchange interaction between the two metal ions. The identity of L remains elusive, but we now have evidence suggesting that the nature of the bridging ligand or its involvement in the mediation of the electrostatic exchange varies with the various states of the oxidized enzyme (41). Presumably the bridging ligand falls away upon reduction of these metal centers to expose the oxygen binding site on cytochrome $\underline{a}_3$.

SUMMARY

We have reviewed the progress that has been made toward elucidating the structure of the metal centers in cytochrome c oxidase. In our work, we have employed NO to probe the structure of the O_2 reduction site and manipulated the yeast system Saccharomyces cerevisiae to provide direct incorporation of isotopically substituted amino acids into the enzyme. These approaches, in conjunction with low-temperature EPR and ENDOR spectroscopies, have proven to be extremely powerful for obtaining unambiguous information on the ligands of the metal centers. These structural results, when combined with primary sequence data from various species of cytochrome c oxidase, should permit one to locate the ligands on the various subunits of the protein and eventually to map the spatial distribution of the metal centers.

ACKNOWLEDGMENTS

This work was supported by Grant GM-22432 from the National Institute of General Medical Sciences, U.S. Public Health Service, and by BRSG Grant RR07003 awarded by the Biomedical Research Support Grant Program, Division of Research Resources, National Institutes of Health. C. T. Martin, G. W. Brudvig, and T. H. Stevens were recipients of National Research Service Awards (5T32GM-07616) from the National Institute of General Medical Sciences. This article is Contribution No. 6653 from the Division of Chemistry and Chemical Engineering, California Institute of Technology.

REFERENCES

(1) Lehninger, A. L.: 1975, in Biochemistry, Worth Publishers, New York.
(2) Wikström, M., Krab, K., and Saraste, M.: 1981, in Cytochrome Oxidase, Academic Press, London.
(3) Wikström, M., and Krab, K.: 1979, Biochim. Biophys. Acta 549, pp. 177-222.
(4) Lehninger, A. L., Reynafarje, B., Davies, P., Alexandre, A., Villalobo, A., and Beavis, A.: 1981, in Mitrochondria and Microsomes (Lee, C. P., Schatz, G., and Dallner, G., eds.), Addison-Wesley, Menlo Park, California, pp. 459-479.
(5) Anderson, S., Bankier, A. T., Barrell, B. G., de Bruijn, M. H. L., Coulson, A. R., Drovin, J., Eperon, I. C., Nierlich, D. P., Roe, B. A., Sanger, F., Schreier, P. H., Smith, A. J. H., Staden, R., and Young, I. G.: 1981, Nature 290, pp. 457-465.
(6) Yasunobu, K. T., Tanaka, M., Haniu, M., Sameshima, M., Reimer, N., Eto, T., King, T. E., Yu, C., Yu, L., and Wei, Y.: 1979, in Cytochrome Oxidase (King, T. E., Orii, Y., Chance, B., and Okunuki, K., eds.) Elsevier, Amsterdam, pp. 91-101.
(7) Steffens, G. J., and Buse, G.: 1979, in Cytochrome Oxidase (King, T. E., Orii, Y., Chance, B., and Okunuki, K., eds.) Elsevier, Amsterdam, pp. 153-159.
(8) Capaldi, R. A.: 1979, in Membrane Proteins in Energy Transduction (Capaldi, R. A., ed.) Marcel Dekker, New York, pp. 201-227.
(9) Henderson, R., Capaldi, R. A., and Leigh, J. S.: 1977, J. Mol. Biol. 112, pp. 631-648.
(10) Stevens, T. H., Brudvig, G. W., Bocian, D. F., and Chan, S. I.: 1979, Proc. Natl. Acad. Sci. U.S.A. 76, pp. 3320-3324.
(11) Stevens, T. H., and Chan, S. I.: 1981, J. Biol. Chem. 256, pp. 1069-1071.
(12) Stevens, T. H., Martin, C. T., Wang, H., Brudvig, G. W., Scholes, C. P., and Chan, S. I.: 1982, J. Biol. Chem., in press.

(13) Powers, L., Chance, B., Ching, Y., and Angiolillo, P.: 1981, Biophys. J. 34, pp. 465-498.
(14) Andréasson, L.-E.: 1975, Eur. J. Biochem. 53, pp. 591-597.
(15) Babcock, G. T., Vickery, L. E., and Palmer, G.: 1976, J. Biol. Chem. 251, pp. 7907-7919.
(16) van Gelder, B. F., and Beinert, H.: 1969, Biochim, Biophys. Acta 189, pp. 1-24.
(17) Babcock, G. T., Callahan, P. M., Ondrias, M. R., and Salmeen, I.: 1981, Biochemistry 20, pp. 959-966.
(18) Peisach, J.: 1978, in Frontiers of Biological Energetics (Dutton, P. L., Leigh, J. S., Jr., and Scarpa, A., eds.) Vol. 2, Academic Press, New York, pp. 873-881.
(19) Chan, S. I., Bocian, D. F., Brudvig, G. W., Morse, R. H., and Stevens, T. H.: 1979, in Cytochrome Oxidase (King, T. E., Orii, Y., Chance, B., and Okunuki, K., eds.) Elsevier, Amsterdam, pp. 177-188.
(20) Beinert, H.: 1966, in The Biochemistry of Copper (Peisach, J., Aisen, P., and Blumberg, W. E., eds.) Academic Press, New York, pp. 213-234.
(21) Chan, S. I., Bocian, D. F., Brudvig, G. W., Morse, R. H., and Stevens, T. H.: 1978, in Frontiers of Biological Energetics (Dutton, P. L., Leigh, J. S., Jr., and Scarpa, A., eds.) Vol. 2, Academic Press, New York, pp. 883-888.
(22) Hadley, J. H., Jr., and Gordy, W.: 1977, Proc. Natl. Acad. Sci. U.S.A. 74, pp. 216-220.
(23) Hemmerich, P.: 1966, in The Biochemistry of Copper (Peisach, J., Aisen, P., and Blumberg, W. E., eds.) Academic Press, New York, pp. 15-34.
(24) Peisach, J., and Blumberg, W. E.: 1974, Arch. Biochem. Biophys. 165, pp. 691-708.
(25) Greenaway, F. T., Chan, S. H. P., and Vincow, G.: 1977, Biochim. Biophys. Acta 490, pp. 62-78.
(26) Hoffman, B. M., Roberts, J. E., Swanson, M., Speck, S. H., and Margoliash, E.: 1980, Proc. Natl. Acad. Sci. U.S.A. 77, pp. 1452-1456.
(27) Chan, S. I., Brudvig, G. W., Martin, C. T., and Stevens, T. H.: 1982, in Electron Transport and Oxygen Utilization (Ho, C., ed.) Elsevier, Amsterdam, in press.
(28) Moore, C. E.: 1952, in Atomic Energy Levels, Vol. II, United States Department of Commerce, National Bureau of Standards.
(29) Powers, L., Blumberg, W. E., Chance, B., Barlow, C. H., Leigh, J. S., Jr., Smith, J., Yonetani, T., Vik, S., and Peisach, J.: 1979, Biochim. Biophys. Acta 546, pp. 520-538.
(30) Hu, V. W., Chan, S. I. and Brown, G. S.: 1977, Proc. Natl. Acad. Sci. U.S.A. 74, pp. 3821-3825.
(31) Fee, J. A.: 1975, Struct. Bonding (Berlin) 23, pp. 1-60.
(32) Van Camp, H. L., Wei, Y. H., Scholes, C. P., and King, T. E.: 1978, Biochim. Biophys. Acta 537, pp. 238-246.
(33) Wertz, J. E., and Bolton, J. R.: 1972, in Electron Spin Resonance, McGraw-Hill, New York.
(34) Tweedle, M. F., Wilson, L. J., García-Iñiguez, L., Babcock, G. T., and Palmer, G.: 1978, J. Biol. Chem. 253, pp. 8065-8071.

(35) Abragam, A., and Bleaney, B.: 1970, in Electronic Paramagnetic Resonance of Transition Ions, Oxford Press, London.
(36) Wilson, D. F., and Erecińska, M.: 1978, Meth. Enzymol. 53, pp. 191-201.
(37) Brudvig, G. W., Stevens, T. H., and Chan, S. I.: 1980, Biochemistry 19, pp. 5275-5285.
(38) Blokzijl-Homan, M. F. J., and Van Gelder, B. F.: 1971, Biochim. Biophys. Acta 234, pp. 493-498.
(39) Stevens, T. H., Bocian, D. F., and Chan, S. I.: 1979, FEBS Lett. 97, pp. 314-316.
(40) Yonetani, T., Yamamoto, H., Erman, J. E., Leigh, J. S., Jr., and Reed, G. H.: 1972, J. Biol. Chem. 247, pp. 2447-2455.
(41) Brudvig, G. W., Stevens, T. H., Morse, R. H., and Chan, S. I.: 1981, Biochemistry 20, pp. 3912-3921.
(42) Palmer, G., Babcock, G. T., and Vickery, L. E.: 1973, Proc. Natl. Acad. Sci. U.S.A. 73, pp. 2206-2210.
(43) Reinhammar, B., Malkin, R., Jensen, P., Karlsson, B., Andréasson, L.-E., Aasa, R., Vänngård, and Malmström, B. G.: 1980, J. Biol. Chem. 255, pp. 5000-5003.

MODELS OF METALLOENZYMES: PEROXIDASE AND CYTOCHROME P-450

John T. Groves* and Thomas E. Nemo

Department of Chemistry
The University of Michigan
Ann Arbor, Michigan 48109

ABSTRACT

Evidence has emerged over the last decade to support an intermediate heme-ironoxo complex as the active oxygen species in the catalytic cycle of cytochrome P-450. Studies with deuterated substrates have shown that a large isotope effect and significant rearrangement accompany the aliphatic hydroxylation mediated by this enzyme. Model studies with chromium, manganese and iron porphyrins have shown that the latter stages of oxygen transfer can be simulated outside of the protein. An oxochromium(V) porphyrin complex has been characterized by its e.s.r. spectrum. An X-ray crystal structure of our oxochromium(IV) complex has been determined. An oxomanganese(V) porphyrin complex has been shown to hydroxylate alkanes under mild conditions. An iron(IV) porphyrin cation radical species which has spectroscopic features similar to those reported for peroxidase compound I has been shown to transfer oxygen to hydrocarbon substrates.

INTRODUCTION

The peroxidases and cytochrome P-450, a monooxygenase, are both heme-containing enzymes which have received sustained attention over the past two or three decades. The most enigmatic chemical aspect of the peroxidases has been the unusual high-valent states of these enzymes. Whereas for cytochrome P-450, the wide variety of oxidative transformations mediated by this enzyme have lacked clear analogy in organic chemistry.

The purpose of this chapter is to discuss the current state of the mechanism of cytochrome P-450 (1). Recent advances have shown that the high-valent peroxidase intermediates may have relevance to the so-called "active oxygen species" of cytochrome P-450. Results with model porphyrin complexes which support this view will also be discussed.

I. Bertini, R. S. Drago, and C. Luchinat (eds.), The Coordination Chemistry of Metalloenzymes, 329–341.

In 1958, Klingenberg (2) and Garfinkel (3) independently reported the observation of a previously unrecognized absorption at 450 nm upon reducing crude extracts of liver microsomes with sodium dithionite under an atmosphere of carbon monoxide. In the mid 1960's, Omura and Sato (4) established that the anamolous 450 nm absorption was due to a reduced heme binding carbon monoxide, which upon denaturation irreversibly shifted to 420 nm. Omura and Sato referred to the species responsible for these absorptions as cytochrome P-450 and cytochrome P-420. Estabrook and coworkers (5), while investigating microsomes of the adrenal cortex, initially linked the hydroxylation of 17-hydroxyprogesterone with the source of the 450 nm absorption.

Approximately 12 years ago, Coon, et al. (6) solubilized and resolved the liver microsomal (lm) enzyme and three constituents were obtained from the purification procedure: (1) a soluble form of $P\text{-}450_{lm}$, (2) a soluble form of NADPH-cytochrome $P\text{-}450_{lm}$ reductase, and (3) a soluble phospholipid. Reconstitution of these three fractions in the presence of NADPH and molecular oxygen caused the oxidation of organic substrates according to the reaction below (7).

$$RH + O_2 + NADPH + H^+ \rightarrow R\text{-}OH + H_2O + NADP^+$$

Gunsalus and coworkers (8) discovered that strains of Pseudomonas putida grown on d-camphor provided a source of the enzyme, termed $P\text{-}450_{cam}$, that is soluble and easily purified. $P\text{-}450_{cam}$ requires an iron-sulfur protein designated as putidaredoxin and the flavo-protein, NADH-putidaredoxin reductase for its catalytic activity (9). $P\text{-}450_{cam}$ efficiently and specifically catalyzes the hydroxylation of camphor at the 5-exo position.

For the purposes of discussion, consider the mechanism for oxygen activation and transfer for cytochrome P-450 outlined in Scheme I. Beginning with the resting ferric state, two successive, one-electron reductions of the heme center can effect the binding and reduction of molecular oxygen to a peroxoiron(III) state. Heterolytic cleavage of the O-O bond of a peroxoiron(III) intermediate would lead to an iron(V)oxo complex which could also be called an "iron(III)oxene" complex or a ferrate. By analogy to the known chemistry of chromates and permanganates, such a complex would be expected to be reactive toward hydrocarbons (10).

Support for this description of the active oxidant has come from a variety of sources. For example, $P\text{-}450_{lm}$ has been demonstrated to oxidize organic compounds in the absence of O_2, NADPH, and the reductase enzyme if exogenous peroxidic compounds are present. This effect was first observed in the oxidation of drugs by Kadlubar, et al. (11) and by Rahimtula and O'Brien (12), using microsomal preparations. Employing purified enzymes, Coon, et al. (13) established that aliphatic hydroxylation occurred in the absence of molecular oxygen and the reductase system in the presence of cumene hydroperoxide.

Scheme I

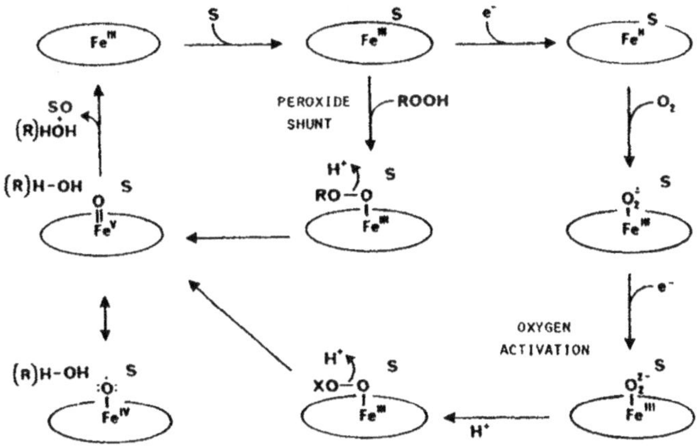

Similarly, peroxy acids (13), sodium periodate (14), sodium chlorite (13), iodosylbenzene (15), iodosylbenzene diacetate (16) and pyridine-N-oxide (17) are effective as oxygen sources. Iodosylbenzene was particularly interesting since it has only one oxygen and it suggested that the active oxidant might be formed directly according to the equation below (15).

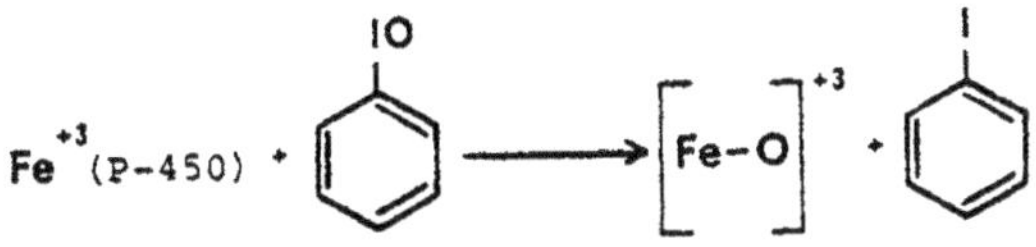

Sligar, et al. (18) have shown that for reduced P-450_{cam} in the absence of putidaredoxin and the reductase enzyme, exposure to $^{18}O_2$ and dihydrolipoic acid resulted in ^{18}O incorporation into lipoic acid and hydroxylated camphor. These observations suggested that acylation of an iron-oxygen complex may be involved in the activation of molecular oxygen (19).

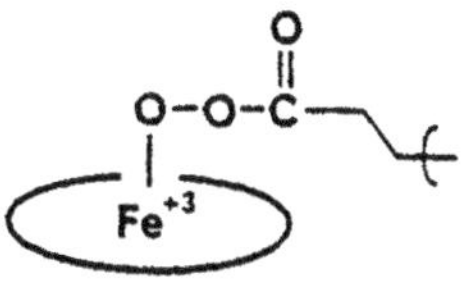

The subsequent details of oxygen transfer to the substrate are obscure and, in some aspects, controversial. In particular, two modes of iron(III)-peroxide decomposition may be considered: (a) a heterolytic path to generate a reactive iron-centered oxidant (20) and (b) a homolytic path to produce a reactive oxyradical species (1c).

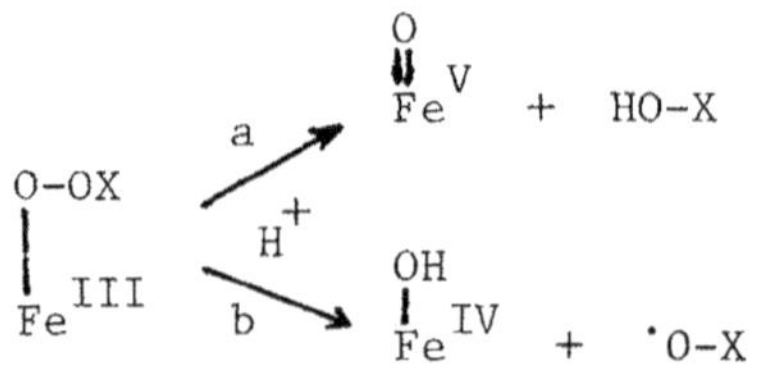

Much useful information about the nature of the active oxidant and the mechanism of the oxidation process has been obtained by studying the oxidation products of simple organic molecules. We have reported that the oxidation of norbornane with the reconstituted liver microsomal cytochrome P-450 system produced a 3.4:1 mixture of exo- and endo-norborneol, respectively. By contrast, oxidation of exo,exo,exo,exo-2,3,5,6-tetradeuteronorbornane gave a 0.76:1 ratio for the same products. Two important conclusions emerge from these results. First,

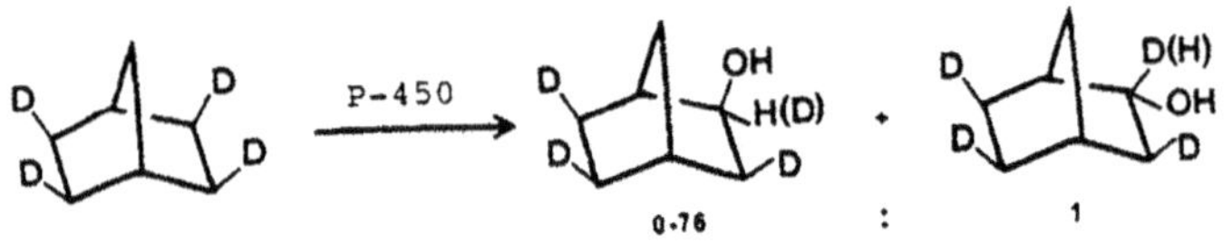

aliphatic hydroxylation has occurred with a large hydrogen isotope effect (11.5±1) and, furthermore, the hydroxylation was not totally stereospecific. Indeed, the results require partial epimerization concomitant with hydroxylation.

A large isotope effect has also been observed for the competitive intramolecular hydroxylation of protium and deuterium containing benzylic sites (22). And recently, cytochrome P-450_{cam}, which produces only 5-exo-hydroxycamphor has been shown to remove either the exo or endo hydrogens (23). The most reasonable mechanism to explain these results is homolytic hydrogen abstraction to give a transient carbon radical intermediate (Scheme II).

Scheme II

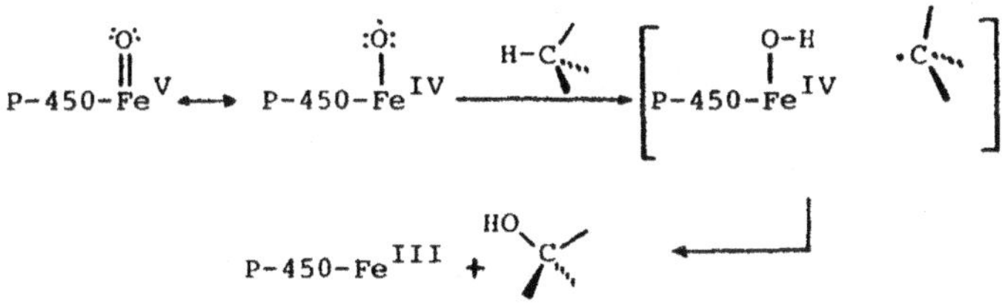

Similar conclusions were reached from an examination of the hydroxylation of 1,2-dideuterocyclohexane. The cume hydroperoxide mediated hydroxylation of this substrate resulted in significant allylic rearrangement.

These results can be interpreted in terms of an initial hydrogen abstraction to produce an allylic radical, partial rearrangement, and then capture of the radical by the iron(IV) hydroxo species, as just described for the norbornane oxidation. If the allylic radical was longlived, providing unlimited freedom of movement of the substrate at the active site can occur, equal amounts of the allylic isomers would have been obtained (24) (Scheme III).

Dolphin, et al (25) have discussed a similar mechanism for the oxidation of Dieldrin in which an intermediate carbon radical was captured by the proximate double bond of the substrate.

The ω-hydroxylation of octane with a chiral methyl (H,D,T) by rat liver microsomes containing cytochrome P-450 has been investigated by Caspi (26). The conclusion of these investigators was that hydroxylation of the chiral methyl occurred with a significant hydrogen isotope effect but with retention of configuration at carbon. Although the cause of this difference is uncertain, the norbornyl and allyl radicals, for which rearrangement was evident, are both more stable and more sterically hindered than would be the case for the hydroxylation of a methyl group.

Scheme III

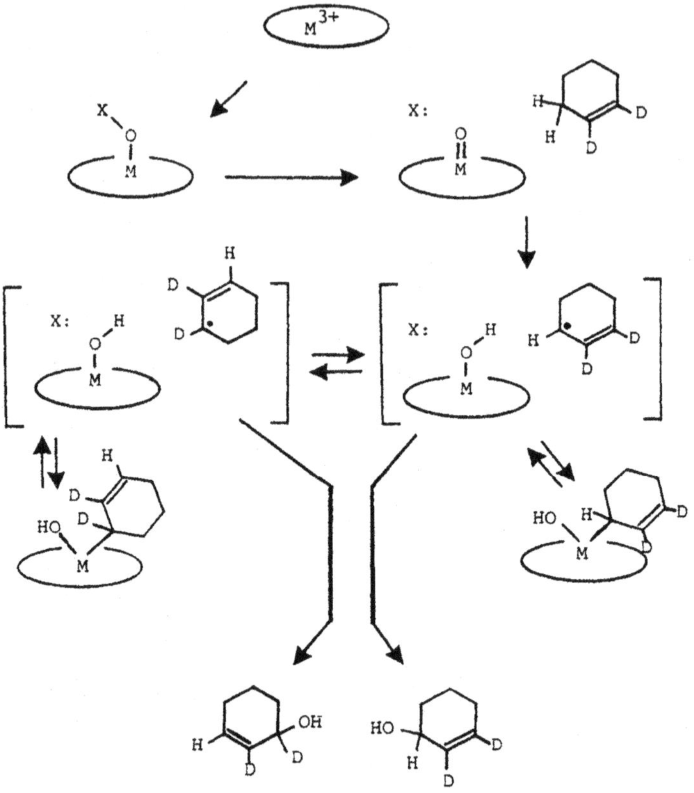

MODEL SYSTEMS

Chromium and Manganese Porphyrins

An important development in the search for simple chemical models of cytochrome P-450 action was the discovery that iron (27), chromium (28) and manganese (29) porphyrins would catalyze oxygen transfer from iodosylbenzene to simple hydrocarbons (30).

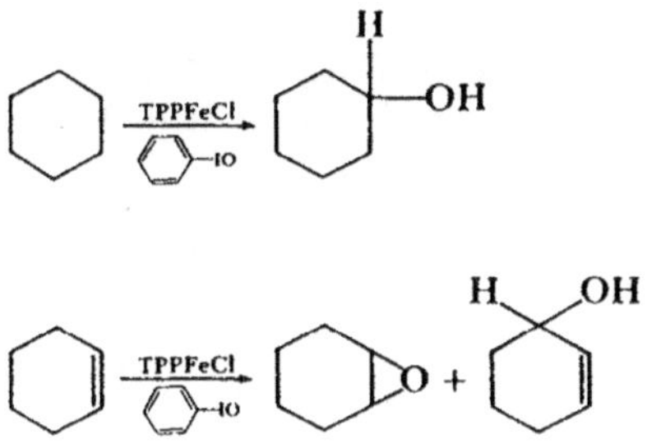

In the case of the chromium porphyrins, the higher valent oxides have been isolated and characterized (31,32). The oxidation of chlorotetraphenylporphyrinatochromium(IV) [Cr(TPP)Cl] with iodosylbenzene led to the formation of a reactive red intermediate which we have characterized as a chromium(V) complex (1). The visible spectrum (Figure 1) revealed a blue-shifted Soret band. More definitive was the e.s.r spectrum of 1 which gave a well-resolved signal at room temperature (Figure 2-a). The nine-line pattern centered at g = 1.982 is

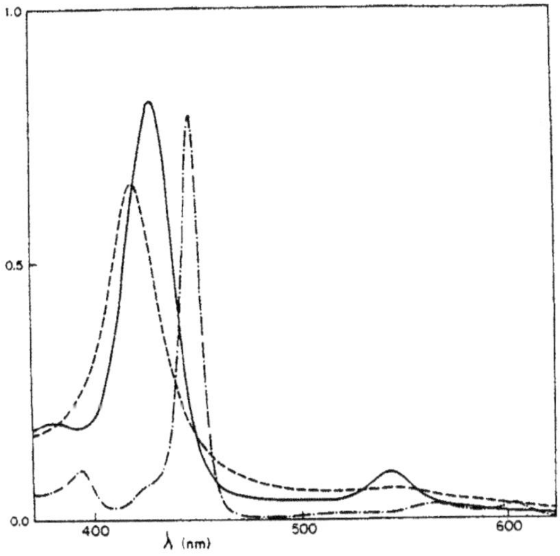

Figure 1. Visible spectra of Cr(TPP)Cl(—·—), CrO(TPP) (2) (—) and CrO(TPP)Cl (1) (---) in methylene chloride.

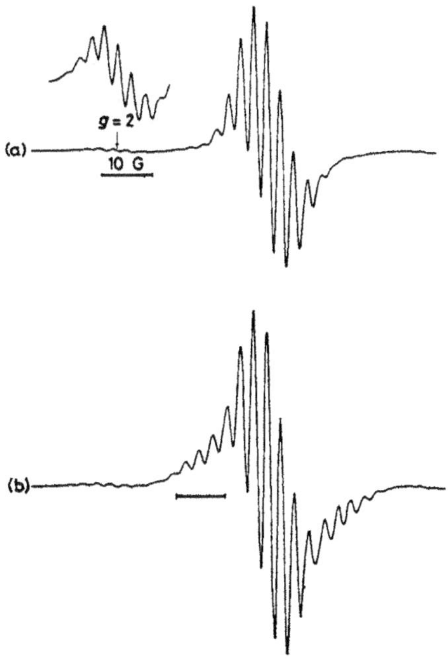

Figure 2. The e.s.r. spectra of 1 and [^{17}O]-1.

characteristic of a metal-centered radical with a pyrrole nitrogen hyperfine interaction of 0.285 mT. The conspicuous satellites are the result of the 9% abundant, I=3/2 ^{53}Cr nucleus (a_{Cr}=2.3 mT). A similar spectrum has been observed for Cr(TPP)NO which should also have a singly occupied d_{xy} orbital (33). Upon oxidation of Cr(TPP)Cl with ^{17}O-iodosylbenzene, additional lines in the e.s.r. spectrum due to ^{17}O splitting are evident (Figure 2-b). Computer simulation of this spectrum gave a good fit with one ^{17}O attached to chromium. The assignment of this oxygen as a chromyl species, Cr=O, was based on the ir spectrum of the corresponding oxochromium(V) perchlorate which showed a band at 972 cm^{-1} (34).

Upon standing solutions of **1** decomposed to an extraordinarily stable, bright red material which was shown to be the corresponding oxochromium(IV) complex **2** (35). The X-ray crystal structure of the tetra-p-tolyl derivative of **2** (31) (Figure 3) revealed a saddle-shaped porphyrin with the chromium atom 0.469 Å above the plane described by the four pyrrole nitrogens. The chromium-oxygen bond distance was very short (1.572 A) consistent with a formal triple bond.

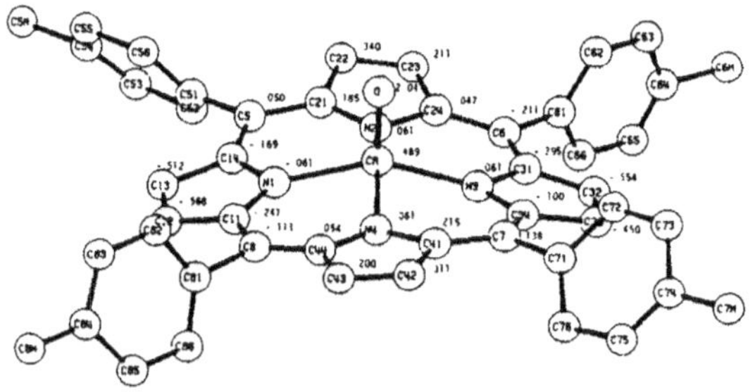

Figure 3

The oxidation of Mn^{III}(TPP)Cl with iodosylbenzene led to the isolation of a 1:1 adduct of the two starting materials (3). Faraday measurements indicated a triplet ground state for this material and on this basis we favor an oxomanganese(V)-iodobenzene complex formulation for **3**, although other electronic distributions are possible. This oxomanganese complex, **3**, was extraordinarily reactive toward hydrocarbons under mild conditions. Cyclohexane gave a 70% yield of cyclohexanol and cyclohexyl chloride (2.5:1) while for adamantane the combined yield of alcohols and chlorides was 80%.

The mechanism of this aliphatic functionalization became evident from an examination of norcarane. Stoichiometric or catalytic oxidation of norcarane gave a mixture of 2-norcaranyl and 3-cyclohexenyl methyl derivatives. A mechanism consistent with the observations is shown in Scheme IV. Hydrogen atom abstraction from norcarane by the chloro-oxomanganese(V) (**3**) complex would give a chloro-hydroxymanganese(IV)

Scheme IV

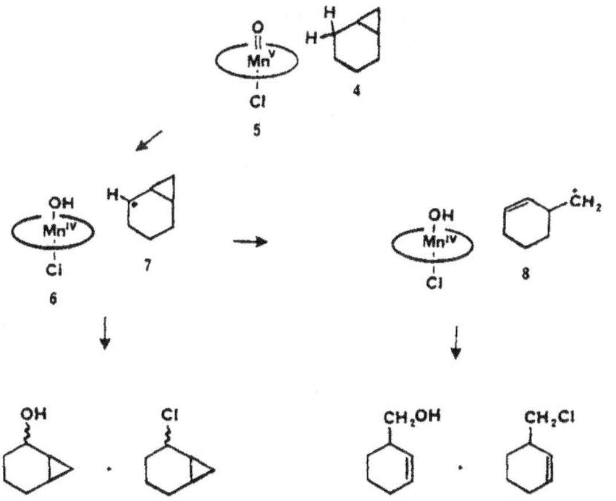

complex and a norcaranyl free radical. Rapid chlorine or hydroxyl transfer to this radical would give the observed norcaranyl products. Opening of the exocyclic cyclopropyl bond is a characteristic of the cyclopropylcarbinyl radical and atypical of the corresponding carbocation (36). Thus, free radical cyclopropyl ring opening and chlorine or hydroxyl transfer would give the observed rearrangement products.

Further evidence of the free radical nature of this oxomanganese species was obtained by examining the stereochemistry of epoxidation. Whereas trans-stilbene gave only trans-stilbene oxide (53% yield), cis-stilbene gave a mixture of cis- and trans-stilbene oxide (1:1.6) in 88% yield. Carbon-carbon bond rotation in an intermediate manganese(IV) carbon radical species would explain this result (Scheme V).

Scheme V

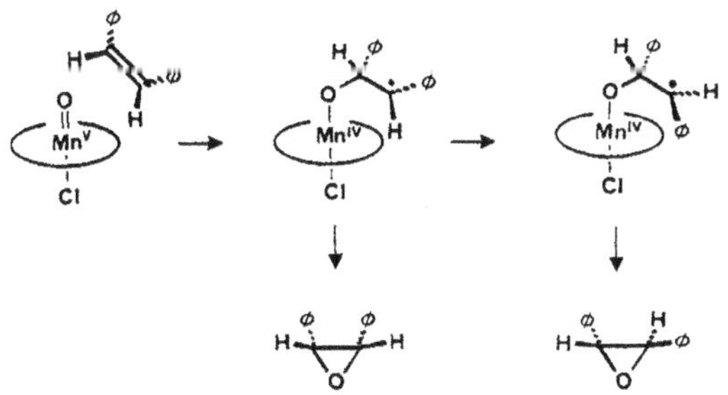

High-Valent Iron-Porphyrins

The oxidation of Fe(TPP)Cl or Fe(TTP)Cl with iodosylbenzene gave a transient, red intermediate which decomposed in less than a minute at 25°C (37). Oxidation of chloro-tetramesitylporphyrinatoiron(III) with m-chloroperoxybenzoic acid at -80° gave a reactive green intermediate (4) which we have characterized spectroscopically (38). The ^1H NMR spectrum of 4 showed very low-field absorbances for the methyls (δ 10, 24, 27) and the meta-protons (δ 68) of the pendant aryl group. Further, no ^{13}C resonance could be found for the meso-carbon of the porphyrin ring when that position was enriched with ^{13}C (39). These data indicate large spin and charge densities at the meso-carbons as would be expected for a porphyrin π-cation radical. The magnetic susceptibility (4.2 μ_B) was close to the value expected for three compared electrons. The Mössbauer spectrum of 4 showed a sharp quadrupole doublet centered at 0.05 mm/s (ΔEq=1.49 mm/s). Taken together the results favor a low spin iron(IV)-porphyrin cation radical for 4.

This green species was found to react even at -80° with olefins to give epoxides. Thus, quenching of 4 with norbornene gave 78% norbornene oxide. If $H_2{}^{18}O$ was added to the solution of 4 prior to the addition of norbornene, the epoxide was found to contain 99% ^{18}O.

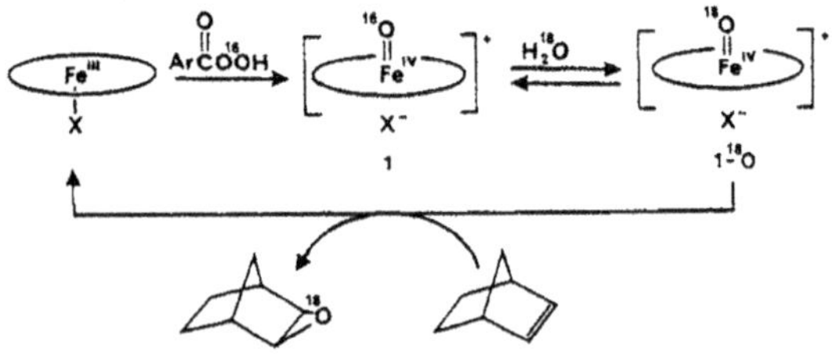

Thus, this new compound which has spectral characteristics of compound I of peroxidase (40) has shown chemical reactivity consistent with its intermediacy in the oxygen activation cycle of cytochrome P-450.

ACKNOWLEDGMENT

Support of this research by the National Institutes of Health and the National Science Foundation is gratefully acknowledged.

REFERENCES

(1) (a) Coon, M. J. and White, R. E.: "Dioxygen Binding and Activation by Metal Centers," Spiro, T. G., Ed., John Wiley and Sons, New York, 1980, p. 73; (b) Groves, J. T.: "Advances in Inorganic Biochemistry," Eichhorn, G. L. and Marzilli, L. G., Eds., Elsevier and North-Holland, New York, 1979, p. 369; (c) White, R. E. and Coon, M. J.: 1980, Ann. Rev. Biochem. 49, p. 315; (d) Ullrich, V.: 1979, Topics in Current Chemistry 83, p. 67; (e) De Pierre, J. W. and Ernster, L.: 1977, Ann. Rev. Biochem. 46, p. 201; (f) Gunsalus, I. C. and Sligar, S. G.: 1978, Adv. Enzymology 47, p. 44; (g) Griffin, B. W., Peterson, J. A. and Estabrook, R. W.: "The Porphyrins," Dolphin, D., Ed., Vol. VII, Academic Press, New York, 1979, p. 333.

(2) Klingenberg, M.: 1958, Arch. Biochem. Biophys. 75, p. 376.

(3) Garfinkel, D.: 1958, Arch. Biochem. Biophys. 77, p. 493.

(4) Omura, T. and Sato, R.: 1964, J. Biol. Chem. 239, p. 2370; 1965, ibid. 239, p. 2379.

(5) Estabrook, R. W., Cooper, D. Y. and Rosenthal, O.: 1963, Biochem. Z. 338, p. 741.

(6) (a) Lu, A. Y. H. and Coon, M. J.: 1960, J. Biol. Chem. 243, p. 1331; (b) Coon, M. J. and Lu, A. Y. H.: "Microsomes and Drug Oxidations," Gillette, J. R., Connly, A. H., Cosmides, G. J., Estabrook, R. W., Fouts, J. R. and Mannering, G. J., Eds., Academic Press, New York, 1969, p. 151.

(7) Lu, A. Y. H., Junk, K. W. and Coon, M. J.: 1969, J. Biol. Chem. 244, p. 3714.

(8) Katagari, M., Ganguli, B. N. and Gunsalus, I. C.: 1968, J. Biol. Chem. 243, p. 3543.

(9) Gunsalus, I. C.: 1968, Hoppe-Seyler's Z. Physiol. Chem. 349, p. 1610.

(10) Groves, J. T.: "Mechanisms of Metal-Catalyzed Oxygen Insertion" in "Metal Ion Activation of Dioxygen," Spiro, T. G., Ed., John Wiley and Sons, New York, 1980, pp. 125-161.

(11) Kadlubar, F. F., Morton, K. C. and Ziegler, D. M.: 1973, Biochem. Biophys. Res. Commun. 54, p. 1255.

(12) Rahimtula, A. D. and O'Brien, P. J.: 1974, Biochem. Biophys. Res. Commun. 60, p. 440; Hrycay, E. G. and O'Brien, P. J.: 1972, Arch. Biochem. Biophys. 153, p. 480.

(13) Nordblom, G. D., White, R. E. and Coon, M. J.: 1976, Arch. Biochem. Biophys. 175, p. 524.

(14) Hrycay, E. G., Gustafsson, J. A., Ingelman-Sundberg, M. and Ernster, L.: 1975, Biochem. Biophys. Res. Commun. 66, p. 209.

(15) Lichtenberger, F., Nastainczyk, W. and Ullrich, V.: 1976, Biochem. Biophys. Res. Commun. 70, p. 939.

(16) Gustafsson, J.-A., Rondahl, L. and Bergman, J.: 1979, Biochemistry 19, p. 865; Berg, A., Ingelman-Sundberg, M. and Gustafsson, J.-A.: 1979, J. Biol. Chem. 254, p. 5264.

(17) Sligar, S. G., Kennedy, K. A. and Pearson, D. C.: "Oxidases and Related Redox Systems," King, T. E., Mason, H. S. and Morrison, M., Eds., University Park Press, Baltimore, 1980.

(18) Sligar, S. G., Kennedy, K. A. and Pearson, D. C.: 1980, Proc. Natl. Acad. Sci. U.S.A. 77, p. 1240.

(19) Hamilton, G.: 1969, Adv. Enzymol. 32, p. 55.

(20) (a) Groves, J. T. and Van Der Puy, M.: 1974, J. Am. Chem. Soc. 96, p. 5274; (b) 1976, ibid. 98, p. 529; (c) Groves, J. T. and McClusky, G. A.: 1976, J. Am. Chem. Soc. 98, p. 859.

(21) Groves, J. T., McClusky, G. A., White, R. E. and Coon, M. J.: 1978, Biochem. Biophys. Res. Commun. 81, p. 154.

(22) Hjelmeland, L. M., Aronoro, L. and Trudell, J. R.: 1977, Biochem. Biophys. Res. Commun. 76, p. 541.

(23) Gelb, M. H., Heimbrook, D. C., Malkönen, P. and Sligar, S. G.: 1982, Biochemistry 21, p. 270.

(24) Groves, J. T., Akinbote, O. F. and Avaria, G. E.: "Microsomes, Drug Oxidations, and Chemical Carcinogenesis," Coon, M. J., Conney, A. H., Estabrook, R. W., Gelboin, H. V., Gillette, J. R. and O'Brien, P. J., Eds., Vol. I, Academic Press, New York, 1980, p. 523.

(25) Dolphin, D., Addison, A. W., Cairns, M., DiNello, R. K., Farrell, N. P., James, B. R., Paulson, D. R. and Welborn, C.: 1979, Int. J. Quantum Chem. 16, p. 311; based on a scheme proposed by Bedford, C. T. in "Foreign Compound Metabolism in Mammals," Vol. 3, The Chemical Society, London, 1975, p. 402.

(26) Shapiro, S., Piper, J. U. and Caspi, E.: 1982, J. Am. Chem. Soc. 104, pp. 2301-2305.

(27) (a) Presented by JTG at the Second Pingree Park Conference, "Biochemical and Clinical Aspects of Oxygen," Sept. 24-29, 1978; (b) Groves, J. T., Nemo, T. E. and Myers, R. S.: 1979, J. Am. Chem. Soc. 101, p. 1032.

(28) Groves, J. T. and Kruper, W. J., Jr.: 1979, J. Am. Chem. Soc. 101, p. 7613.

(29) Groves, J. T., Kruper, W. J., Jr. and Haushalter, R. C.: 1980, J. Am. Chem. Soc. 102, p. 6375.

(30) (a) Chang, C. K. and Kuo, M. S.: 1979, J. Am. Chem. Soc. 101, p. 3413; (b) Hill, C. L. and Schardt, B. C.: 1980, J. Am. Chem. Soc. 102, p. 6374.

(31) Groves, J. T., Kruper, W. J., Jr. and Haushalter, R. C.: 1982, Inorg. Chem. 21, pp. 1363-1368.

(32) Groves, J. T. and Haushalter, R. C.: 1981, J. Chem. Soc., Chem. Commun., pp. 1165-1166.

(33) Wayland, B. B.; Olson, L. W. and Siddiqui, Z. U.: 1976, J. Am. Chem. Soc. 98, p. 94.

(34) Takahashi, T., unpublished results.

(35) (a) Budge, J. R., Gatehouse, B. M. K., Nesbit, M. C. and West, B. O.: 1981, J. Chem. Soc., Chem. Commun., p. 370; (b) Buchler, J. W., Lay, K. L., Castle, L. and Ullrich, V.: 1982, Inorg. Chem. 21, pp. 842-844.

(36) (a) Friedrich, E. C. and Holmstead, R. L.: 1976, J. Org. Chem. 37, p. 2550; (b) Friedrich, E. C. and Jassawalla, J. D. C.: 1979, ibid. 44, p. 4224.

(37) Groves, J. T., Kruper, W. J., Jr., Nemo, T. E. and Myers, R. S.: 1980, J. Mol. Catal. 7, pp. 169-177.

(38) Groves, J. T., Haushalter, R. C., Nakamura, M., Nemo, T. E. and Evans, B. J.: 1981, J. Am. Chem. Soc. 103, p. 2884.

(39) Nakamura, M., unpublished results.

(40) Hewson, W. D. and Hager, L. P.: 1979, Porphyrins 7, p. 295 and references therein.

COORDINATION CHEMISTRY OF CYTOCHROME P-450 AND HEME MODELS

Mansuy D.

Laboratoire de Chimie de l'Ecole Normale Supérieure, 24, rue Lhomond, 75231 Paris Cedex 05, France

1. INTRODUCTION : OCCURRENCE AND FUNCTION OF CYTOCHROME P450-DEPENDENT MONOOXYGENASES.

The enzymes which catalyse the direct incorporation of dioxygen into organic compounds are widely distributed in nature. Depending on whether they catalyze the incorporation of both oxygen atoms of dioxygen or only one oxygen atom into the substrate, the enzymes are called respectively dioxygenases and monooxygenases, and involve the following stoichiometry :

$$RH + O_2 \longrightarrow [ROOH] \quad \text{dioxygenases}$$

$$RH + O_2 + 2e^- + 2H^+ \longrightarrow ROH + H_2O \quad \text{monooxygenases}$$

Whereas dioxygenases are only able to oxidize particular substrates which are chemically activated by themselves by the presence of functional groups in their molecule, monooxygenases are able to hydroxylate a wide range of organic compounds including inert alkanes (1). Most monooxygenases receive electrons from an external reducing agent (different from the substrate to be hydroxylated) and are multienzymatic systems.
Many of them contain the same type of biocatalyst, a hemoprotein called cytochrome P450, and it is noteworthy that almost all monooxygenases already known to catalyze alkanes hydroxylation contain this cytochrome.

Actually, cytochromes P450 are widely distributed in living organisms like mammals, fish, birds, yeasts and plants. In man, they exist in several organs and tissues with a maximum concentration in the liver and variable concentrations in kidneys, lungs, brain, skin... Concerning their specificities towards substrates and stereoselectivity of the hydroxylation reaction, one should distinguish two classes of cytochrome P450dependent monooxygenases. The first ones are involved in the metabolism of endogenous compounds such as steroids in mammals, or terpenes in bacteria and

I. Bertini, R. S. Drago, and C. Luchinat (eds.), The Coordination Chemistry of Metalloenzymes, 343–361.

plants, and are highly specific. They only tolerate small variations in the structure of the substrate and hydroxylate it with a very high stereospecificity. The second ones seem in charge of the degradation of foreign compounds ("xenobiotics") and have to hydroxylate a broad spectrum of lipophilic compounds, making them more hydrophilic and facilitating their eliminination from the body. Because of their function, the latter cytochromes P450 are considerably less specific than the former ones (1).

These cytochromes P450 play a key role in pharmacology and toxicology. Since they control the elimination rate and thus the plasma levels of drugs, they have a great influence on the variation of the therapeutic effects of these drugs as a function of time.

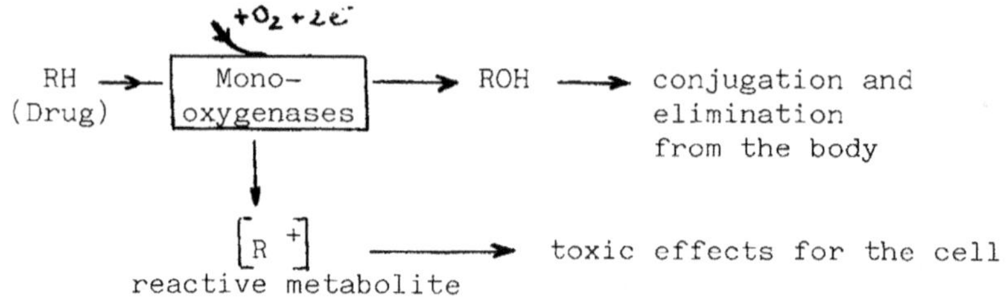

Moreover, in order to be able to activate C-H bonds, cytochromes P450 involve in their catalytic cycle a very reactive iron complex. When reacting with certain drugs or environmental compounds, this reactive iron complex forms reactive electrophilic metabolites which are able to alkylate cell macromolecules.
Such phenomenons are at the origin of the secondary toxic effects of some drugs and of the carcinogenic effects of several xenobiotics.

In order to determine at the molecular level, the nature of the interactions between organic compounds and the various iron-cytochrome P450 complexes involved in the catalytic cycle of this hemoprotein, and to understand their implications in pharmacology and toxicology, it is very useful to work both on the enzymatic system and on iron-porphyrin models (2). The coordination chemistry of those cytochromes P450 that are involved in xenobiotics detoxification, is particularly rich, for the two following main reasons : (i) they exhibit an active site very accessible for an extremely wide range of even very large organic compounds (ii) they involve, in their catalytic cycle, several intermediate complexes with different oxidation states of the iron, some of them being able to react with a great variety of substrates.

2. THE CATALYTIC CYCLE OF CYTOCHROME P450

2.1. Components of the monooxygenase system

The electrons necessary for monooxygenase reaction to take place, are transferred from NADPH or NADH, to the iron of cytochrome P450 via an electron transfer chain. For instance, in detoxifying monooxygenases present in several tissues in mammals and localized in the cellular subfraction called microsomes, cytochromes P450 receive electrons from NADPH via a flavoprotein, and, to a lesser extent, from NADH via another flavoprotein and cytochrome b_5 (1).

NADPH ⟶ NADPH-cytochrome P450 reductase ⟶ cyt. P450

NADH ⟶ NADH-cyt. b_5 reductase ⟶ cyt. b_5 ⟶ cyt. P450

In bacterial monooxygenases or in specific steroid hydroxylases of adrenal gland, another component, an iron-sulfur protein, transfers electrons from the flavoprotein to cytochrome P450.

2.2. Structure of cytochrome P450

It is now established that the various cytochromes P450, so far characterized, which have different specificities for substrates, exhibit differences in molecular weight, amino acid composition and terminal amino acids. However, all these cytochrome P450 forms exhibit the following common characteristics : (i) they all contain an heme, iron-protoporphyrin IX (Fig. 1), bound to a single polypeptide chain with a molecular weight between 40 000 and 60 000.

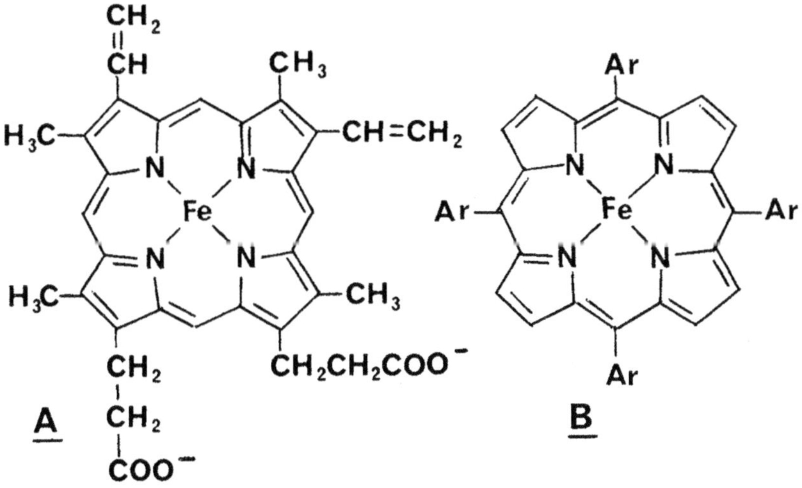

Figure 1. Formula of iron-protoporphyrin IX (A) and of the synthetic iron-tetraphenyl porphyrin (TPP), Ar = C_6H_5, (B), used very often in model studies.

(ii) their spectral properties are almost identical, and only very small, but significant, differences between the various forms have been reported, indicating an essentially identical coordination sphere of the heme (1).

A main characteristic feature of cytochrome P450, when compared to other hemoproteins, is the nature of its axial endogenous ligand, a cysteinate residue from the protein. The presence of this ligand has been infered from various spectroscopic studies on cytochrome P450 itself and from model studies (see following section) (2).

Thus, in order to understand the interactions of cytochrome P450-iron with various organic compounds, as described in the following chapters, one can use the schematic representation of cytochrome P450 active site indicated on Figure 2. In addition to its aforementioned characteristics, it is important to know that the active site involves, in close proximity to the iron, an especially hydrophobic part of the protein, where the substrates are bound, very often by hydrophobic interactions only. At least in the case of microsomal cytochromes P450, this active site is largely opened for organic compounds.

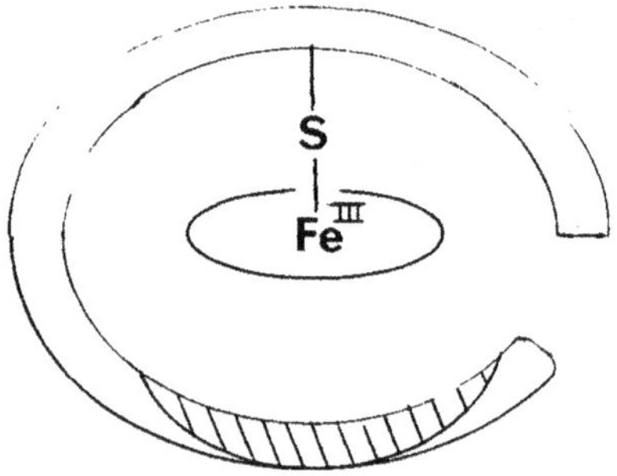

Figure 2. Schematic representation of the cytochrome P450 active site.

2.3. Catalytic cycle of dioxygen activation and substrates hydroxylation by cytochrome P450 (1,2)

In its resting state, two forms of cytochrome P450 are in equilibrium : a hexacoordinate low-spin iron (III) complex, the sixth ligand of which is not completely known, but seems to be an OH-containing molecule, and a pentacoordinate high-spin iron(III) complex with the cysteinate as only axial ligand (Fig. 3). Various factors such as temperature or ionic strength affect the position of this spin-state equilibrium. In particular, the binding of a substrate on the hydrophobic part of the protein shifts the equilibrium toward the high-spin state, favoring the next step of the catalytic cycle, the iron reduction, since the redox potential of the high-spin form is about 100mV higher than that of the low-spin form (3).

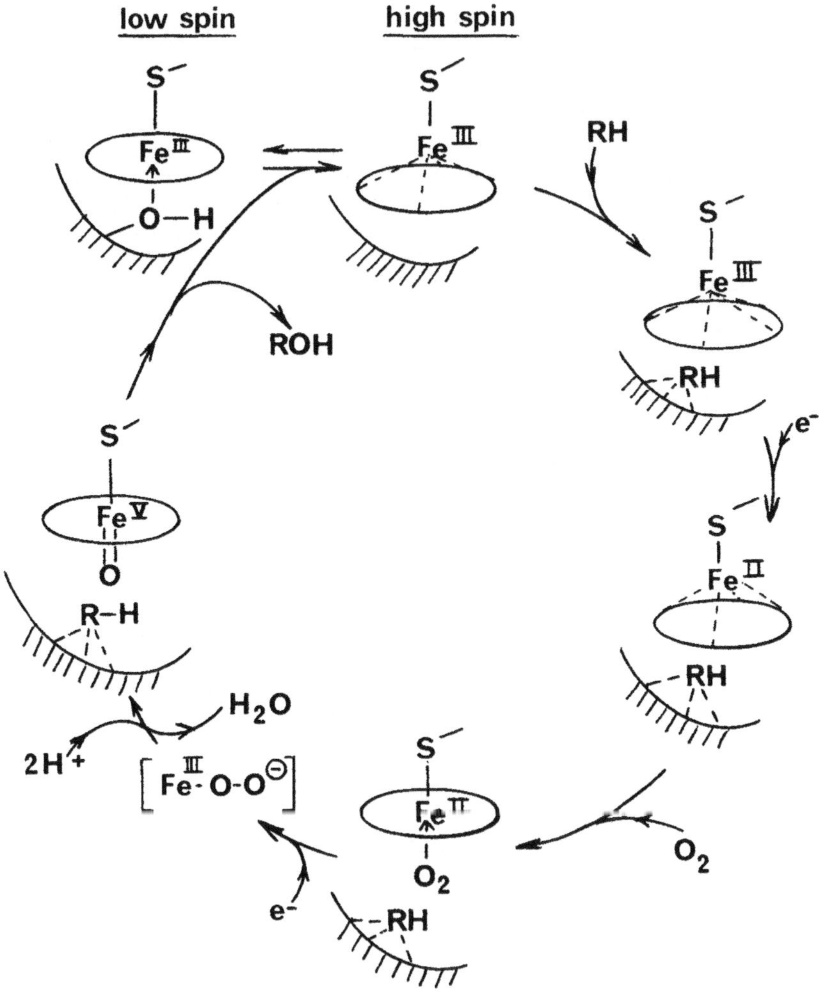

Figure 3. The catalytic cycle of substrates hydroxylation by cytochrome P450.

Cytochrome P450, in its reduced state, is a first reactive intermediate since its iron(II) is considerably electron-enriched not only by the porphyrin ring but also by its thiolate ligand. It is important to realize that cytochrome P450 is, first of all, an electron-transfering agent. Most often, it transfers electrons toward dioxygen. However, it can also transfer electrons toward several substrates when these substrates are able to compete efficiently with dioxygen. This occurs with reducible substrates, particularly when the local oxygen concentration is low. Cytochrome P450 is thus able to catalyse the reduction of halogenated compounds, epoxides, nitroaromatics and amine-oxides (2).

Cytochrome P450-iron(II) has a vacant coordination position and is able to bind several ligands including dioxygen. The oxy-complex is generally unstable, its dissociation leading easily to Fe(III) and $O_2^{\cdot -}$ (1). In the catalytic cycle, it receives an extra electron from NADPH leading to a very reactive oxygenating complex (the active oxygen cytochrome P450 complex). The structure of this complex is not precisely known presently, since its very short lifetime has made very difficult any direct spectral studies. The possible nature and the reactivity of this complex will not be discussed here since this will be done by other lecturers. One possible structure, the oxo form $Fe^{V}=O$, has been indicated in Fig. 3, and is only a very formal representation, other electronic configurations being possible and even more likely (1). Whatever its structure may be, the active oxygen cytochrome P450 complex is able to insert its oxygen atom into several substrates including alkanes, performing alkanes and aromatics hydroxylation, olefins and aromatics epoxidation, and N- or S-oxidations.

Whereas the structure of the cytochrome P450 active oxygen complex is still a matter of controversy, those of the other intermediate complexes of the catalytic cycle are comparatively well-known. Their structures, schematically represented in Fig. 3, are derived from several spectral studies which have been performed on cytochromes P450 of different origins. Moreover, most of their unique spectral characteristics due to the presence of a thiolate axial ligand, have been reproduced in thiolate-iron-porphyrin model complexes (2). Some characteristic spectral data of the cytochrome P450 complexes and of their synthetic models are compared in table 1.

Good synthetic models for the first three states of the catalytic cycle have been obtained in the crystalline state, and their structure established by X-ray studies (table 2), giving a first good idea of the iron coordination chemistry during the catalytic cycle.

Table 1. Spectral parameters of the complexes involved in the cytochrome P450 catalytic cycle and some of their synthetic models.

	Ref.	Electronic spectra λ (nm)				EPR (g values)		
$P450Fe^{III}$(low spin)	(4a)	417	535	571		2.45	2.26	1.91
$Fe^{III}(P)(pNO_2C_6H_4S)(DMF)$	(5)	420	533	567		2.46	2.28	1.90
$P450Fe^{III}$(high spin)	(4a)	391	520	540	646	8	4	1.8
$Fe^{III}(P)(pNO_2C_6H_4S)$	(5)	391	517	540	646	7.2 (with pCl C_6H_4S)	4.8	1.9
$P450Fe^{II}$	(4a)	408	540					
$[Fe^{II}(P)(nBuS)]^-$	(6)	408	540	560				
$P450Fe^{II}\ O_2$	(4b)	350	418	552				
$[Fe^{II}(TPpivP)(SC_6HF_4)(O_2)]^-$	(7)	364	427	562	610			

P = protoporphyrin IX-dimethylester. Cytochrome P450 from Pseudomonas Putida. TPpivP = Tetraphenylpivaloylporphyrin.

Table 2. Structural data of synthetic models for the first three states of cytochrome P450 catalytic cycle and for cytochrome P450 Fe^{II}CO.

	d_{Fe-S}(Å)	$d_{Fe-porphyrin\ core}$	ref
$Fe^{III}(TPP)(C_6H_5S)(C_6H_5SH)$	2.27	≃ 0	(8)
$Fe^{III}(PPDME)(pNO_2C_6H_4S)$	2.32	0.43	(5)
$Fe^{II}(TPP)(C_2H_5S)^-$	2.36	0.52	(9)
$Fe^{II}(TTP)(C_2H_5S)(CO)^-$	2.35	≃ 0	(9)

Moreover, the very recent IR study of a model of oxy-cytochrome P450 (table 1) revealed the appearance of the O_2 stretch at 1139 cm^{-1} (7) which is in favor of an end-on dioxygen binding mode.

The great similarity of the spectral characteristics of cytochrome P450 complexes and their thiolate-heme models, and the unique ability of thiolate ligands to produce the peculiar spectral characteristics of some cytochrome P450 complexes leave little doubt on the presence of an axial cysteinate ligand in these complexes.

3. CYTOCHROME P450-IRON COMPLEXES FORMED BY DIRECT INTERACTION WITH ORGANIC COMPOUNDS

In a very general manner, an organic compound containing an hydrophobic part in its molecule will enter the active site of cytochrome P450 and will bind to the hydrophobic part of the protein (Fig. 3). Compounds containing certain functional groups are also able to bind to the iron. Thus, compounds containing an accessible oxygen atom act as sixth ligands for high-spin ferric cytochrome P450, leading to visible spectra almost superimposable to that of native cytochrome P450 in its low-spin state (10,11). This suggests that the second endogenous axial ligand of low-spin ferric cytochrome P450 contains an OH or OR function, and could be water or the OH group of an amino-acid residue (1,2).
Moreover, various compounds containing other accessible heteroatoms such as N,S or P also bind to cytochrome P450-iron (III), but give low-spin hexacoordinate complexes exhibiting visible spectra significantly different from that of resting low-spin cytochrome P450 (table 3). Whereas ferric complexes of oxygen-containing ligands display a Soret peak around 417-420 nm, complexes of pyridine or imidazole derivatives, such as metyrapone (2-methyl-1,2-di-3-pyridy-1-propanone) or 1-phenyl-imidazole, which are well-known inhibitors of cytochrome P450, exhibit Soret peaks around 425 nm. Ferric cytochrome P450 also binds isocyanides and thioethers, leading to Soret peaks around 425-430 nm, and phosphines and thiolates, leading to characteristic hyperporphyrin spectra with a split Soret peak (maxima around 380 and 460 nm). When comparing spectral data of cytochrome P450 ferric complexes with various ligands, one notes great changes in optical absorption especially in the Soret region, but only small variations in EPR g-values (1). This makes optical spectroscopy a very good method for the study of substrates interaction with cytochrome P450, and even for binding constants determinations.

Table 3. Visible spectra parameters of cytochrome P450* - ligand complexes (after (1)).

	λ (nm)					
$P450Fe^{III}$ (high spin)	391		520		540	646
$P450Fe^{III}$ (low spin)native	417		535		571	
+ octylamine	423		538		574	
+ 1-Me-imidazole	424		540		578	
+ Metyrapone	421		536			
+ Ethylisocyanide	430		548			
+ octylmethylsulfide	426		537		572	
+ benzylthiolate	377	465		557		
+ diethylphenylphosphine	377	454		556		

* cytochrome P450 from Pseudomonas Putida

In its ferrous state, cytochrome P450 also binds the previously mentioned compounds leading to low spin ferrous complexes characterized by a Soret peak around 445 nm for nitrogen-containing molecules and 450-460 nm for isocyanides and phosphines. Contrary to cytochrome P450-iron(III), cytochrome P450-iron(II) binds CO, leading to an hyperporphyrin spectrum (1,2) involving the 450 nm peak, which was at the origin of the name of cytochrome P450, but fails to bind oxygen-containing ligands and thiolates.

In order to illustrate the implications of cytochrome P450-iron-ligand complexes formation in pharmacology or toxicology, figure 4 gives a schematic view of the interaction between cytochrome P450 and an alkaloid, ellipticine, which is used as an antitumor agent. This compound binds by its pyridinic nitrogen atom to cytochrome P450-iron, and, presumably because of favorable interactions with the protein active site, it exhibits a great affinity for the various forms of microsomal cytochrome P450 (12). By preventing dioxygen fixation, it strongly inhibits various microsomal monooxygenase activities (12, 13). Since it displays an especially high affinity for the cytochromes P450 involved in the formation of reactive metabolites from polyaromatic hydrocarbons, it was shown to act as a good inhibitor of the mutagenic and carcinogenic effects of these hydrocarbons (14, 15).

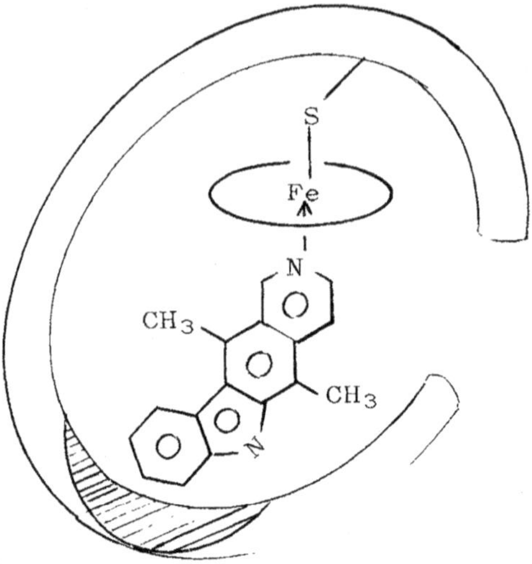

Figure 4. Schematic view of ellipticine binding to cytochrome P450.

4. CYTOCHROME P450-IRON-METABOLITE COMPLEXES

4.1. Complexes formed upon reduction of substrates

Cytochrome P450-Fe(II) catalyses the reduction by NADPH of C-halogen into C-H bonds, allowing microsomal reduction of CCl_4, CCl_3F and DDT($(pClC_6H_4)_2$ CH-CCl_3) respectively to $CHCl_3$, $CHCl_2F$ and DDD ($(pClC_6H_4)_2$CH-$CHCl_2$). We found very recently that cytochrome P450 from rat liver microsomes is also able to catalyse the reduction of benzylic halides such as pNO_2- $C_6H_4CH_2Cl$ and $C_6H_5CH_2Br$, by NADPH, in anaerobic conditions. During these reductions, new cytochrome P450 complexes appear with very characteristic Soret peaks around 478 nm (16). They are formed in steady-state concentrations, their decomposition leading to $ArCH_3$ (or $ArCH_2D$ in D_2OpH7.4). A study of these benzyl halides reduction by a heme model system using Fe(II)(TPP) as catalyst and ascorbic acid as reducing agent (17), demonstrates the following mechanism involving the intermediate formation of a ferric σ-alkyl complex (18) :

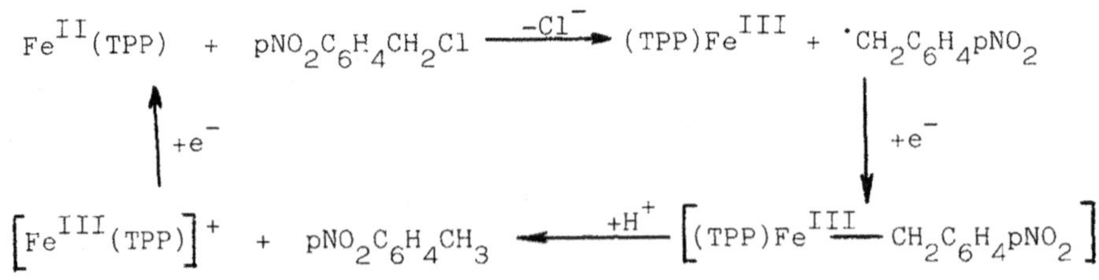

The σ-alkyl (TPP)Fe^{III}—CH_2Ar complexes formed upon reduction of $ArCH_2X$ by Fe^{II}(TPP) were found to exhibit visible and ^1H NMR characteristics quite similar to those of σ-alkyl Fe^{III}(TPP)(R)(R=CH_3 or nBu) complexes prepared by other described techniques (18). These results support that the 478 nm-absorbing complexes formed upon reduction of benzylhalides by cytochrome P450 are also σ-alkyl complexes :

$$P450Fe^{II} + ArCH_2X \xrightarrow{-X^-} P450Fe^{III} + {}^{\bullet}CH_2Ar \xrightarrow{+e^-} P450Fe^{III}\text{—}CH_2Ar \quad \text{(eq.2)}$$

Metabolic reduction of compounds containing at least two halogen atoms on the same carbon are more complex, from the coordination chemistry point of view. For instance, halothane, $CF_3CHClBr$, a volatile anaesthetic widely used in clinics, is reduced by hepatic cytochrome P450 in hypoxic conditions leading to a 470 nm-absorbing complex, whose reactivity (19) and EPR spectrum (20) are in favor of a σ-alkyl-Fe^{III} structure. Model reactions between Fe(II)(TPP) and halothane in the presence of an excess of reducing agent lead to the quantitative formation of a crystalline complex, the elemental analysis and spectral studies of which establish the σ-alkyl structure Fe(III)(TPP)(CF_3CHCl)(21) :

$$Fe^{II}(TPP) + CF_3CHClBr \xrightarrow{-Br^-} (TPP)Fe^{III} + CF_3\dot{C}HCl \xrightarrow{+e^-} \left[(TPP)Fe^{III}\text{—}CHClCF_3\right] \quad \text{(eq. 3)}$$

Addition of $nBuS^-$ to this complex at -70°C leads to the hexacoordinate complex $Fe(III)(TPP)(CF_3CHCl)(nBuS)^-$, the electronic and EPR spectra of which are very similar to those of the cytochrome P450-halothane metabolite complex (20). These model studies together with those performed on cytochrome P450 (19) point to the following mechanism for microsomal halothane reduction :

$$P450Fe^{II} + CF_3CHClBr \xrightarrow{-Br^-} P450Fe^{III} + CF_3\dot{C}HCl \xrightarrow{+e^-}$$

$$\left[P450Fe^{III}\text{—}CHClCF_3\right] \longrightarrow P450Fe^{III} + F^- + CF_2{=}CHCl \quad \text{(eq. 4)}$$

Most often, the σ-alkyl ferric complexes formed upon reduction of halogenated compounds by cytochrome P450 or iron-porphyrins, in the presence of an excess of reducing agent, are obtained in steady state concentrations, which depend upon their respective rates of formation and evolution of their iron-carbon bond by protonation, or β-elimination as in eq. 4. Another possible evolution is the α-elimination of an halogen atom occasionally present on the carbon bound to iron : this occurs upon microsomal reduction of some polyhalogenated compounds such as CCl_4, CBr_4 or CCl_3F, leading to cytochrome P450 complexes characterized by Soret peaks between 450 and 460 nm (22). Model reductions of these compounds by iron(II)-porphyrins, in the presence of a reducing agent in excess, led to a general preparation method of porphyrin-iron-carbene complexes (23) :

$$(P)Fe^{II} + RR'CX_2 \xrightarrow[-2X^-]{+2e^-} \left[(P)Fe^{II} \leftarrow CRR'\right] \leftrightarrow \left[(P)Fe^{IV}{=}CRR'\right] \quad \text{(eq.5)}$$

Crystalline complexes of Fe(TPP) with the carbenes CFCl, CFBr, CBr_2, CI_2, CClCN, CClCOOEt, $CClCH_3$, $CClCHOHCH_3$, $CClCHOHC_6H_5$, $CClSC_6H_5$, $CHSC_6H_5$ have thus been prepared in high yields and well-characterized. Moreover, an X-ray structure has been obtained for the $Fe(TPP)(CCl_2)(H_2O)$, 2DMF carbene complex derived from CCl_4 reduction (24). It thus seems highly probable that the complexes formed upon cytochrome P450-dependent reduction of CCl_4, CBr_4, $CFCl_3$ are ferrous-carbene complexes. Actually, upon slow hydrolysis, they are transformed into CO-cytochrome P450 complexes (22) :

$$P450Fe^{II} + CX_4 \xrightarrow[-2X^-]{+2e^-} \left[P450Fe^{II} \leftarrow CX_2\right] \xrightarrow[-2X^-]{+H_2O} P450Fe^{II}\text{-}CO \quad \text{(eq.6)}$$

All these results are in favor of the following general mechanism for reductions of halogenated compounds by cytochrome P450 and iron-porphyrins :

$$(P)Fe^{II} + RR'CX_2 \xrightarrow{-X^-} (P)Fe^{III} + {}^{\bullet}CXRR' \xrightarrow{+e^-} (P)Fe^{II} + {}^{\bullet}CRR'X$$

$$\longrightarrow \left[(P)Fe^{III}\text{—}CRR'X\right] \xrightarrow{+e^-} \left[(P)\overset{\ominus}{Fe}{}^{II}\text{—}\underset{X}{C}RR'\right] \xrightarrow{-X^-} \left[(P)Fe^{II} \leftarrow CRR'\right] \quad \text{(eq.7)}$$

This mechanism underlines the existence of an important organometallic chemistry of cytochrome P450. Depending upon the relative rates of the different steps involved in eq. 7 and the evolutions of the σ-alkyl (protonation, β-elimination) and carbene complexes (hydrolysis for instance), these complexes are or are not detectable, and, in some cases, can be isolated. In that respect, it is noteworthy that, in the porphyrin series, the iron-carbene complexes appear most often more stable than the iron-σ-alkyl complexes.

4.2. Complexes formed upon oxidation of 1,3-benzodioxole derivatives : a possible mechanism for C-H bond hydroxylation.

Some derivatives of 1,3-benzodioxole are oxidatively metabolized by cytochrome P450-dependent monooxygenases with the formation of very stable complexes which are characterized in the ferric and ferrous state respectively by Soret peaks around 439 and 455 nm (25). Since these complexes are very stable and resistant to ligand exchange, benzodioxole derivatives act as strong inhibitors of cytochromes P450. Some 1,3-benzodioxole derivatives, such as piperonyl butoxide, are used as insecticide synergists, since they inhibit insect monooxygenases which are responsible for detoxification of the insecticide. A $[Fe(TPP)(1,3\text{-benzodioxol-2-carben})(nBuS)]^-$ complex has been prepared by the following method (26) :

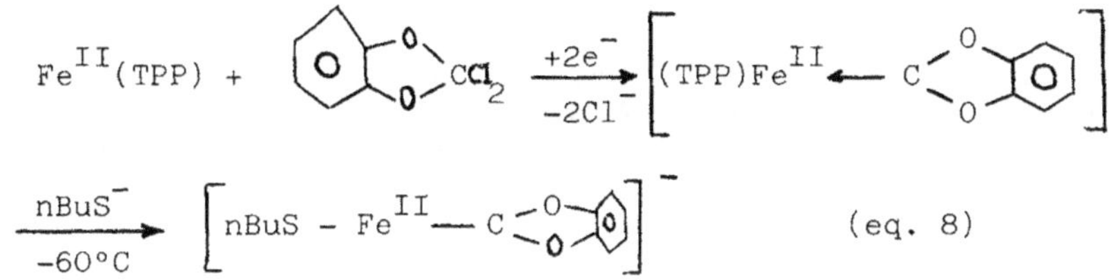

(eq. 8)

This complex exhibits a visible spectrum very similar to that of the microsomal cytochrome P450-Fe(II)benzodioxole metabolite complex (26), indicating a 1,3-benzodioxol-2-carbene nature for the cytochrome P450 exogenous ligand. If one admits an iron-oxo structure for the active oxygen cytochrome P450 complex (Fig. 3), the formation of the iron-carbene bond upon 1,3-benzodioxole oxidation corresponds formally to the replacement of the oxo ligand by its carbon analog, the carbene ligand :

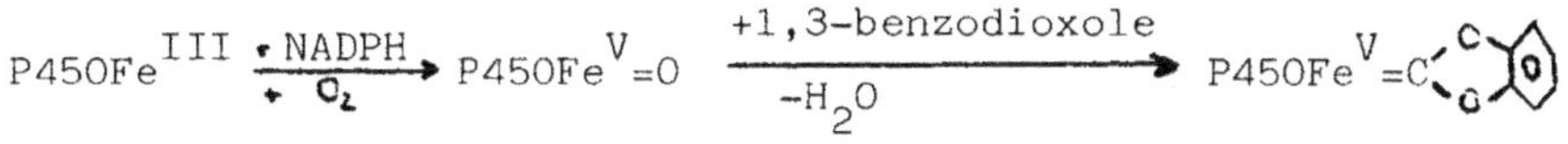

Taking into account the previously described free radical reactivity of the active oxygen cytochrome P450 complex (27), the following detailed mechanism has been proposed for this reaction (28) :

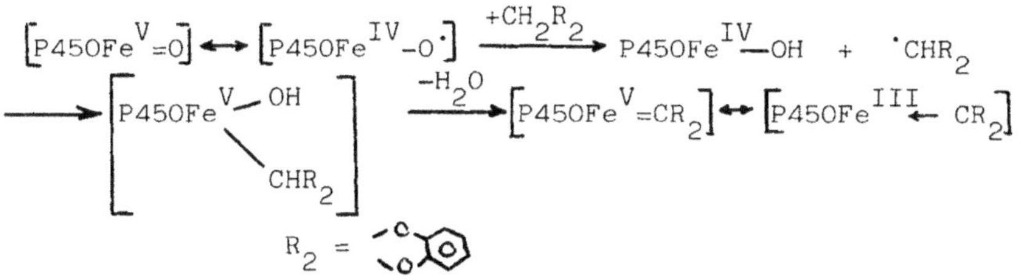

(eq.9)

In a more general manner, a similar mechanism could be proposed for the cytochrome P450-dependent hydroxylation of compounds containing a methylene group, CH_2R_2. The very unstable intermediate σ complex then undergoes a reductive elimination of its OH and σ CHR_2 cis ligands :

$$\left[P450Fe^{V} \begin{matrix} -OH \\ -CHR_2 \end{matrix} \right] \longrightarrow P450Fe^{III} + R_2CHOH \quad \text{(eq. 10)}$$

leading to the hydroxylated product R_2CHOH, and regenerating P450 Fe^{III} (28).

4.3. Complexes formed upon oxidation of amines.

Several exogenous compounds containing an amine function, like amphetamines or various other drugs, are able to form, after microsomal oxidative metabolism in vivo and in vitro, stable cytochrome P450-iron(II)-metabolite complexes that exhibit Soret peaks around 455 nm. This leads very often to a severe inhibition of the hydroxylating functions of these cytochromes (25). Since such a complex is equally formed upon cytochrome P450-dependent oxidation of an amine RNH_2 or of the hydroxylamine RNHOH as well as upon reduction of the corresponding nitroalkane RNO_2, it has been proposed that this complex derives from the binding of the nitrosoalkane RNO to the iron(II) of cytochrome P450 (29,30). Nitrosoalkanes are unstable as monomers, but they are formed in close proximity to the iron inside the active site of cytochrome P450 and are stabilized upon coordination to iron. In order to prove the existence of these iron-RNO bonds and to know their properties, model iron-porphyrin-RNO complexes have been synthesized. Actually, reactions of primary alkylhydroxylamines with Fe(III)(porphyrins) lead to the quantitative formation of very stable porphyrin-Fe(II)-nitrosoalkane complexes. Several such complexes have been isolated in a crystalline form and their structure established by various spectroscopic techniques (31).

$$(P)Fe^{III} \xrightarrow{+RNHOH} \left[(P)Fe^{II} \leftarrow N{\scriptstyle \lessgtr^{O}_{R}}\right] \xleftarrow[+S_2O_4^{2-}]{+RNO_2} Fe^{II}\ (P) \qquad \text{(eq. 11)}$$

An X-ray analysis of one of these complexes, Fe(TPP)(iPrNO)($iPrNH_2$), definitely proved the existence of the Fe-RNO bond and showed that the nitrosoalkane is bound to iron by its nitrogen atom, the Fe-N distance of the Fe-RNO bond (1.83 Å) being considerably shorter than the Fe-N distance of the Fe-RNH_2 bond (2.10 Å)(32). The strength of the Fe-RNO bond is also demonstrated by the lack of exchange of RNO upon exposure to CO, a strong ligand of iron(II)-porphyrins (31). Subsequent studies have revealed that nitrosoalkanes constitute, in a more general manner, a new class of good ligands for hemoproteins and iron-porphyrins. For instance, several hemoglobin-Fe(II)-RNO complexes have been obtained upon in situ oxidation of RNHOH or reduction of RNO_2 (33). For all nitrosoalkane-hemoprotein or -iron-porphyrin complexes studies so far, a particular strength of the Fe-RNO was noticed. This strength explains the severe inhibition of several monooxygenase activities when cytochrome P450 is engaged in such nitrosoalkane complexes (25). Since any nitrosoalkane not too hindered around nitrogen is able to bind strongly to cytochrome P450-iron (34), one should expect that any xenobiotic able to be metabolized to such a nitrosoalkane may exert some inhibitory effects on cytochrome P450-dependent monooxygenase activities. This could be of great importance in the case of therapeutic drug associations when one drug is able to form a nitrosoalkane-cytochrome P450 complex, thus inhibiting the metabolism of another drug and modifying its elimination rate and therapeutic response. Actually, it was recently found that an antibiotic widely used in human therapeutics, troleandomycin (TAO), (Fig. 5), is able to form in vivo an inhibitory nitrosoalkane complex with hepatic cytochrome P450, blocking up to 70-80% of these cytochromes in rat liver (35,36). This complex formation is a first molecular mechanism explaining some adverse effects observed after therapeutic associations of TAO with drugs such as ergotamine, warfarin or theophylline (37).

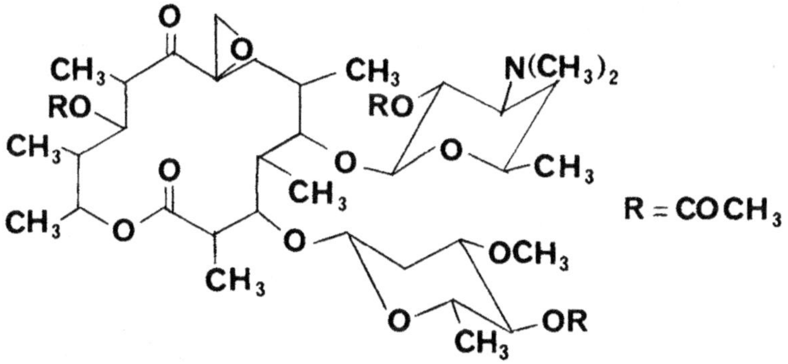

Figure 5. Formula of the antibiotic TAO : troleandomycin (oleandomycin triacetate).

4.4. Complexes formed upon oxidation of dialkylhydrazines

Certain 1,1-dialkylhydrazines have been found recently to form, upon metabolic oxidation by hepatic microsomes, cytochrome P450-metabolite complexes with Soret peaks at 438 and 455 nm for the ferric and ferrous state respectively (38,39). The nature of the hydrazines-derived ligands in these complexes is not yet established ; however, a nitrene structure has been proposed for these ligands (39). Very recently, we made experiments aiming at the isolation of model nitrene-iron-porphyrin complexes and obtained preliminary results on successful preparations of such complexes either by dioxygen-dependent oxidation of a hindered 1,1-dialkylhydrazine in the presence of Fe(III)(TPP)(Cl), or by direct interaction of the corresponding dialkylamino-nitrene with Fe(II)(TPP) at -70°C (40) :

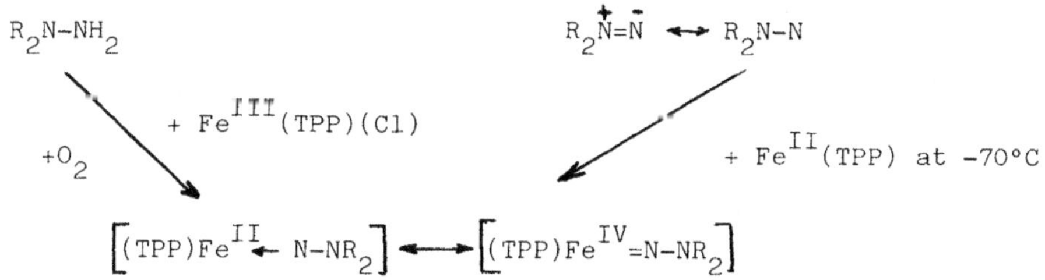

When starting with 1-amino-2,2,6,6-tetramethyl-piperidine, and Fe(III)(TPP)(Cl), the corresponding nitrene complex is obtained as a stable crystalline solid in a nearly quantitative yield. Its structure has been established by elemental analysis, ^{1}H NMR and IR spectroscopy, and from its reaction with pyridine and phosphines (40) :

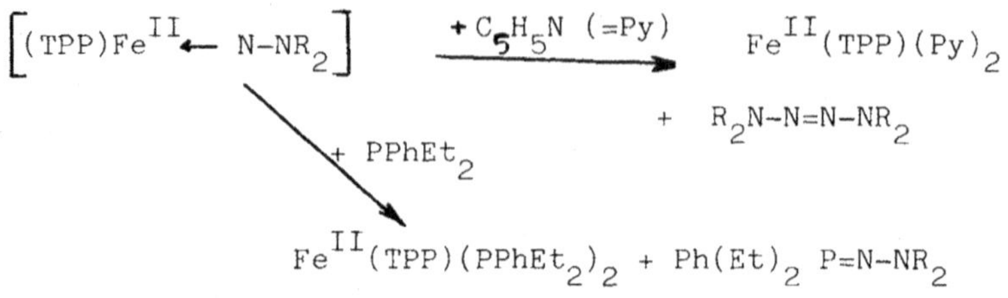

Contrary to other pentacoordinate Fe(TPP)(L) complexes with L=carbene or nitrosoalkane which are diamagnetic (Fe(II), low spin), it exhibits a magnetic susceptibility corresponding to S=2 in agreement either with a Fe(II)high spin or Fe(IV)high spin state (40).

The formation of this nitrene complex upon dioxygen-dependent oxidation of a 1,1-dialkylhydrazine in the presence of an iron-porphyrin is a strong argument in favor of the proposed nitrenic structure of the complexes formed upon dioxygen-dependent oxidation of 1,1-dialkylhydrazines in the presence of cytochrome P450.

5. CONCLUSION : THE UNIQUE RICHNESS OF CYTOCHROME P450 COORDINATION CHEMISTRY

Because of the diversity of the iron oxidation, spin and coordination states involved in their catalytic cycle, and, above all, because of the easy access to iron allowed by their active site for a wide range of organic compounds, cytochromes P450 are especially prone to form complexes with a variety of ligands (1,2).

Very often, results from molecular studies on cytochrome P450-dependent metabolic oxidation or reduction of organic compounds have been at the origin of the discovery of new types of complexes in hemoprotein and iron-porphyrin coordination chemistry. However, the structure of these complexes has been established most often after isolation of model iron-porphyrin complexes. This is the case for complexes involving an iron-carbon bond (σ-alkyl-iron(III) and carbene-iron(II)), and for nitrosoalkane-iron(II) and nitrene complexes. Actually, the nitrene and carbene complexes which are respectively the nitrogen and carbon analogs of the iron-oxo complexes presumably involved as reactive intermediates in the catalytic cycle of certain hemoproteins, have been discovered and prepared for the first time by taking into account results from studies on cytochrome P450.

$$[Fe^{n} = O] \updownarrow [Fe^{n-2} \leftarrow O] \qquad [Fe^{n} = N\text{-}R] \updownarrow [Fe^{n-2} \leftarrow N\text{-}R] \qquad [Fe^{n} = CRR'] \updownarrow [Fe^{n-2} \leftarrow CRR'] \quad n = IV \text{ or } V$$

It is noteworthy that these nitrene and carbene complexes can be formed by reaction of the iron-oxo active oxygen cytochrome P450 complex with certain substrates. The following figure gives a very schematic illustration of the present knowledge on cytochrome P450 coordination chemistry.

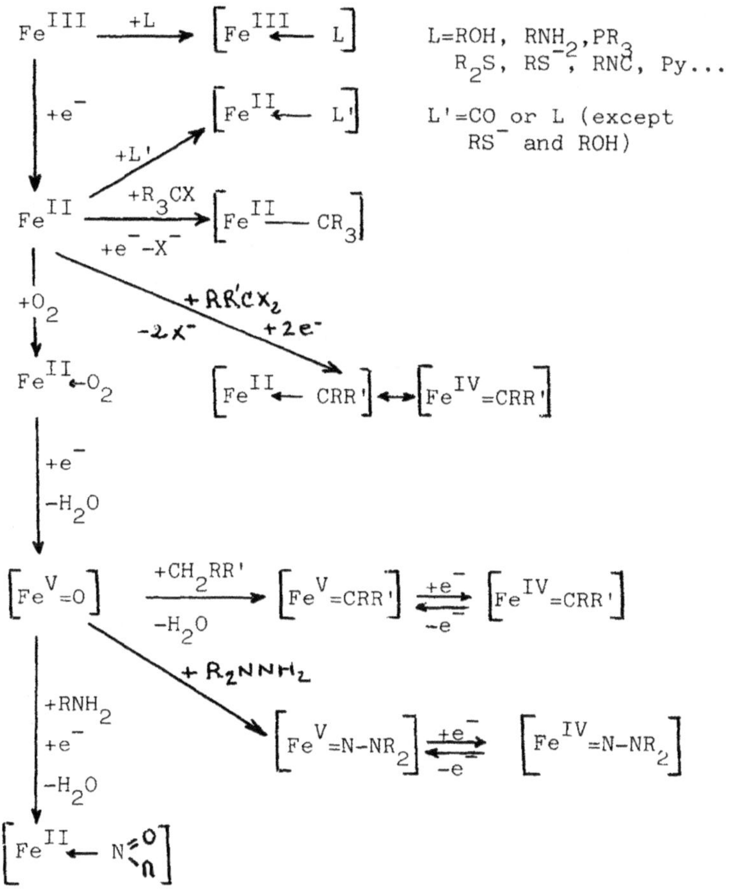

REFERENCES

(1) Ullrich, V., 1979, Topics in Current Chemistry, 83, pp. 68-104.
(2) Mansuy, D., 1981, Reviews in Biochemical Toxicology, 3, pp. 283-320.
(3) Sligar S.G., Cinti D.L., Gibson G.G. and Schenkman J.B., 1979, Biochem. Biophys. Res. Comm., 90, p. 925.
(4) a) Gunsalus I.C., Meeks, J.R., Lipscomb J.D., Debrunner P. and Münck E., 1974, Molecular Mechanisms of oxygen activation, Hayaishi Edit., Academic Press Inc. p. 561-608 ; b) Dolphin D., James B.R. and Curtis-Welborn H., 1980, J. Molec. Catal., 7, pp. 201-214.

(5) Tang S.C., Koch S., Papaefthymiou G.C., Foner S., Frankel R.B., Ibers J.A. and Holm R.H., 1976, J. Amer. Chem. Soc., 98, p. 2414.
(6) Chang C.K. and Dolphin D., 1975, J. Amer. Chem. Soc., 97, p. 5948.
(7) Schappacher M., Ricard L., Weiss R., Montiel-Montoya R., Bill E., Gonser U. and Trautwein A., 1981, J. Amer. Chem. Soc., 103, pp. 7646-7648.
(8) Collman J.P., Sorrell T.N., Hodgson K.O., Kulshrestha A.K. and Strouse C.E., 1977, J. Amer. Chem. Soc., 99, p. 5180.
(9) Caron C., Mitschler A., Riviere G., Ricard L., Schappacher M., and Weiss R., 1979, J. Amer. Chem. Soc., 101, p. 7401.
(10) Ullrich V., Sakurai H. and Ruf H.H., 1979, Acta Biol. Med. Germ., 38, p. 287.
(11) Yoshida Y. and Kumaoka H., 1975, J. Biochem. Tokyo, 78, p. 455.
(12) Lesca P., Lecointe P., Paoletti C. and Mansuy D., 1978, Biochem. Pharmacol., 27, pp. 1203-1209.
(13) Lesca P., Rafidinarivo E., Lecointe P. and Mansuy D., 1979, Chem. Biol. Interactions, 24, pp. 189-198.
(14) Lesca P., Lecointe P., Paoletti C. and Mansuy D., 1979, Chem. Biol. Interactions, 25, pp. 279-287.
(15) Lesca P. and Mansuy D., 1980, Chem. Biol. Interactions, 30, pp. 181-187.
(16) Mansuy D. and Fontecave M., to be published.
(17) Mansuy D. and Fontecave M., 1982, Biochem. Biophys. Res. Comm., 104, pp. 1651-1657.
(18) Mansuy D., Fontecave M. and Battioni J.P., 1982, J.C.S. Chem. Comm., pp. 317-319.
(19) Ahr H.J., King L.J., Nastainczyk W. and Ullrich V., 1982, Biochem. Pharmacol., 31, pp. 383-392.
(20) Ruf. H.H., Nastainczyk W., Ullrich V., Battioni J.P. and Mansuy D., to be published.
(21) Mansuy D. and Battioni J.P., 1982, J.C.S. Chem. Comm., pp. 638-639.
(22) Wolf C.R., Mansuy D., Nastainczyk W., Deutschmann G. and Ullrich V., 1977, Molec. Pharmacol., 13, pp. 698-705.
(23) Mansuy D., 1980, Pure and Appl. Chem., 52, pp. 681-690.
(24) Mansuy D., Lange M., Chottard J.C., Bartoli J.F., Chevrier B. and Weiss R., 1978, Angew. Chem. Int. Edit., 17, pp. 781-782.
(25) Franklin M.R., 1977, Pharmacol. Ther. A, 2, p. 227.
(26) Mansuy D., Battioni J.P., Chottard J.C., and Ullrich V., 1979, J. Amer. Chem. Soc., 101, pp. 3971-3972.
(27) a) Groves J.T., Mc Clusky G.A., White R.E. and Coon M.J., 1978, Biochem. Biophys. Res. Comm., 81, pp. 154-161. b) White R.E., Coon M.J., 1980, Annu. Rev. Biochem., 49, pp. 315-356.
(28) Mansuy D., Chottard J.C., Lange M. and Battioni J.P., 1980, J. Molec. Catalysis, 7, 215-226.
(29) Mansuy D., Beaune P. Chottard J.C., Bartoli J.F. and Gans P. 1976, Biochem. Pharmacol., 25, pp. 609-613.
(30) Jonsson J. and Lindeke B., 1976, Acta Pharm. Suecica, 13, pp. 313-320.
(31) Mansuy D., Battioni P., Chottard J.C. and Lange M., 1977, J.

Amer. Chem. Soc., 99, pp. 6441-6443.
(32) Mansuy D., Battioni P., Chottard J.C., Riche C. and Chiaroni A., submitted for publication.
(33) Mansuy D., Chottard J.C. and Chottard G., 1977, Eur. J. Biochem., 76, pp. 617-625.
(34) Mansuy D., Gans P., Chottard J.C. and Bartoli J.F., 1977, Eur. J. Biochem., 76, pp. 607-616.
(35) Pessayre D., Descatoire V., Mitcheva K.M., Wandscheer J.C., Cobert B., Level R., Benhamou J.P., Jaouen M. and Mansuy D., 1981, Biochem. Pharmacol., 30, pp. 553-560.
(36) Mansuy D., Delaforge M., Le Provost E., Flinois J.P., Columelli S. and Beaune P., 1981, Biochem. Biophys. Res. Comm., 103, pp. 1201-1208.
(37) Pessayre D., Mitcheva M.K., Descatoire V., Cobert B., Wandscheer J.C., Level R., Feldmann G., Mansuy D. and Benhamou J.P., 1981, Biochem. Pharmacol., 30, pp. 559-570.
(38) Hines R.N. and Prough R.A., 1980, J. Pharmacol. Exp. Ther., 214, pp. 80-86.
(39) Prough R.A., Freeman P.C. and Hines R.N., 1981, J. Biol. Chem., 256, pp. 4178-4184.
(40) Mansuy D., Battioni P. and Mahy J.P., 1982, in press.

MÖSSBAUER STUDIES OF SYNTHETIC ANALOGUES FOR THE ACTIVE SITE IN CYTOCHROMES P450.

R. Montiel-Montoya[+], E. Bill[++], U. Gonser[+], S. Lauer[++], A.X. Trautwein[++], M. Schappacher[§], L. Ricard[§], R. Weiss[§]

[+] Angewandte Physik, Universität des Saarlandes, 6600 Saarbrücken 11, FRG
[++] Physik, Medizinische Hochschule Lübeck, 2400 Lübeck 1, FRG and
[§] Laboratoire de Cristallochimie, (ERA 08) Institut Le Bel Université Louis Pasteur, 67070 Strasbourg Cedex, France

Abstract: Different $TP_{piv}P$ model compounds for the active site of cytochromes P450 have been prepared and the structure were analyzed by x-rays. Of particular interest were Mössbauer investigations on five coordinated $[TP_{piv}P(SC_6HF_4)]^-$ and its Co- and O_2-adducts with different cations, namely $[Na\text{-}222]^+$ and $[Na\text{-}18C6]^+$. In addition molecular orbital calculations have been performed in order to derive the temperature dependence of the quadrupole splittings.

I. INTRODUCTION

Following our former spectroscopic investigation of TPP (1) and TPpivP (2) synthetic analogues for the active site in cytochromes P450 we extensively studied ligand and cation effects in TPpivP compounds.

II. PREPARATION OF THE COMPOUNDS.

1. Picket fence porphyrin tetrafluorophenylthiolate ferrous anions.

a) Iron(III)ClTPpivP (0.088mmol) was reduced on Zn/Hg in C_6H_6. Addition of a five-fold excess of crowned $NaSC_6HF_4$ led to the formation of $FeTPpivPSC_6HF_4$ $NaC_{12}H_{24}O_6$. The compound is isolated by addition of pentane.

b) The analogous 222-cryptated product is formed when 222-cryptand is substituted to 18-crown6 to solubilize the sodium thiolate. Care must be taken to remove all chlorinated impurities from the reagents. The complex is insoluble in C_6H_6. It is soluble in C_6H_5Cl.

I. Bertini, R. S. Drago, and C. Luchinat (eds.), The Coordination Chemistry of Metalloenzymes, 363–367.

2. Oxygenation and Carbonylation.

The oxygen adducts were obtained in solution when the dry gas was admitted and precipitated with pentane. The CO adducts are similarly obtained.

In the case of the 18-crown6 derivative, complete oxygenation is observed in the solid state when the product is exposed to dry oxygen at 273 K for one hour.

3. Ferrous chloro derivative.

The iron(III)porphyrin is reduced with Zn/Hg and filtered into a C_6H_6 solution of the 222-cryptated KCl. The product precipitates upon formation.

The synthesis of $[FeTPPSC_2H_5]^-$ $[Na\text{-}222]^+$ has been reported in (1). All high-spin complexes must be handled under an inert atmosphere. All solvents must be dried and distilled prior to use.

From x-ray diffraction measurements the molecular structures of the following analogues have been derived:

III. X-RAY STRUCTURES

Two of the structures are shown in Fig. 1. From these Ortep plots it is obvious that the Na-18crown6 cation is specific in the sense that it strongly interacts via 046 with one of the pickets while the Na-222 cation is in ionic contact only.

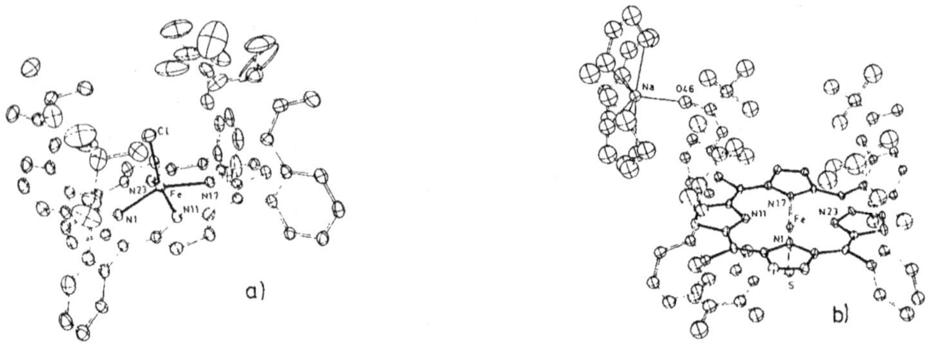

Fig. 1 Ortep plot of $[TPpivPCl]^-[K222]^+$ (a) and $[TPpivPSC_6HF_4]^-$ $[Na18C6]^+$ (b).

IV. MÖSSBAUER STUDIES

From a variety of TPpivP compounds which differ in preparation (precipitated from solution of solid state), in ligandation (pentacoordinated, oxygenated, CO form and Cl form) and in the cation used $[Na-222]^+$ and $[Na-18C6]^+$ we have performed Mössbauer Spectra in a temperature range between 4.2K and 215K. Some representative spectra are shown in Fig. 2. From a least squares fit of these data using Lorentzian lines we derived isomer shifts δ (with respect to α-iron at room temperature), quadrupole splittings ΔE_Q and linewidths Γ. The temperature dependence of ΔE_Q for oxygenated and pentacoordinated $TPpivP(SC_6HF_4)^-$ with different cations $[Na-222]^+$ and $[Na18C6]^+$ respectively, are shown (Fig. 3).

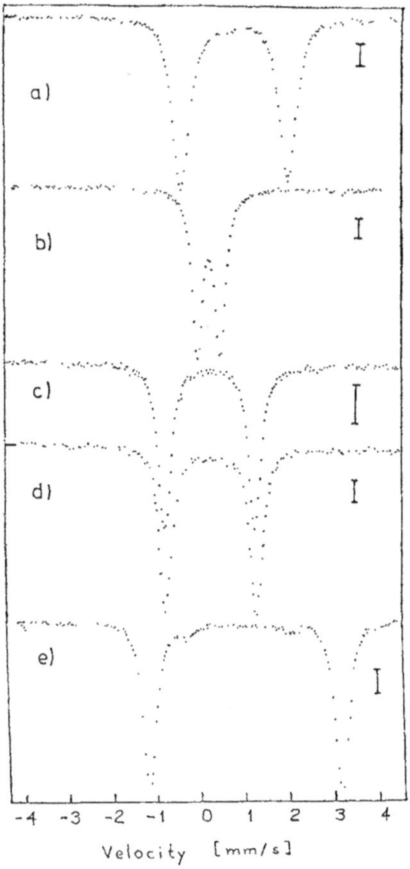

Fig. 2 Experimental Mössbauer spectra of synthetics analogs for the active site in cytochromes P450 at 77K.

(a) $[TPpivP(SC_6HF_4)]^-$ $[Na-18C6]^+$

(b) $[TPpivP(SC_6HF_4)$ $(CO)]^-[Na-222]^+$

(c) $[TPpivP(SC_6HF_4)$ $(O_2)]^-[K-222]^+$ precipitated from chlorobenzene

(d) $[TPpivP(SC_6HF_4)$ $(O_2)]^-$ $[Na18C6]^+$ and

(e) for an analogous high-spin ferrous chloro complex $[TPpivPCl]^-$ $[K-222]^+$.

Bars represent one percent effect.

The strong interaction of $[Na-18C6]^+$ with TPpivP obviously affects the electronic structure of the heme iron in both the oxygenated and pentacoordinated case, probably by increasing the rhombicity of the t_{2g}-like molecular orbitals compared with the case where $[Na-222]^+$ is the cation. We believe that this is the reason for the smooth temperature dependence of ΔE_Q of $[TPpivP(SC_6HF_4)]^-[Na-18C6]^+$.

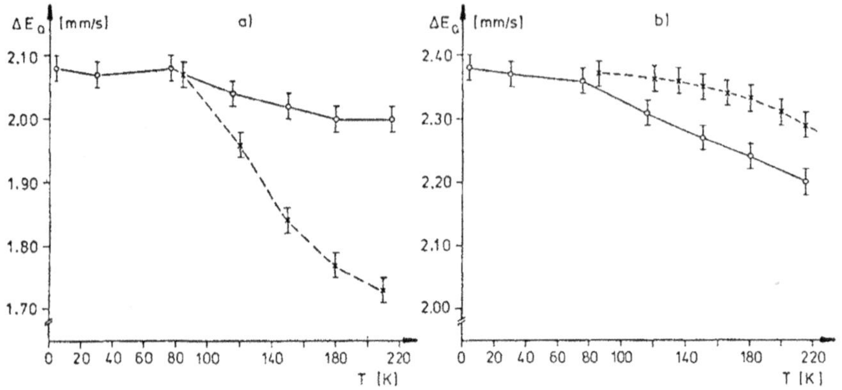

Fig. 3 Temperature dependence of quadrupole splittings for oxygenated (a) and pentacoordinated (b)[TppivP(Sc$_6$HF$_4$)]$^-$ with different cations, [Na-222]$^+$ (solid line) and [Na18C6]$^+$ (dashed line).

V. MOLECULAR ORBITAL CALCULATIONS AND ΔE_Q - INTERPRETATION

X-ray structures from [TPP(SC$_2$H$_5$)]$^-$[Na-222]$^+$ which have been derived recently (1), and from [TPpivPCl]$^-$[K-222]$^+$ were used to perform semiempirical molecular orbital (MO) calculations, including spin-orbit interaction among the MO's with mainly iron-character (3). From the MO electronic structure we have derived temperature dependent electric field gradients and quadrupole splittings. For [TPpivPCl]$^-$[K-222]$^+$ we find that due to the relatively large energy spacing between low-lying MO's spin-orbit interaction is negligible, while for TPPSC$_2$H$_5$$^-$ Na-222$^+$ it is not negligible at all. In Fig. 4 we show the experimental T-dependence for both compounds and in Fig. 4b their highest occupied MO's.

This work has been supported by Deutsche Forschungsgemeinschaft and by CNRS (ATP No. 0182). One of us R.M.-M. thanks Conacyt-Mexico for a partial fellowship.

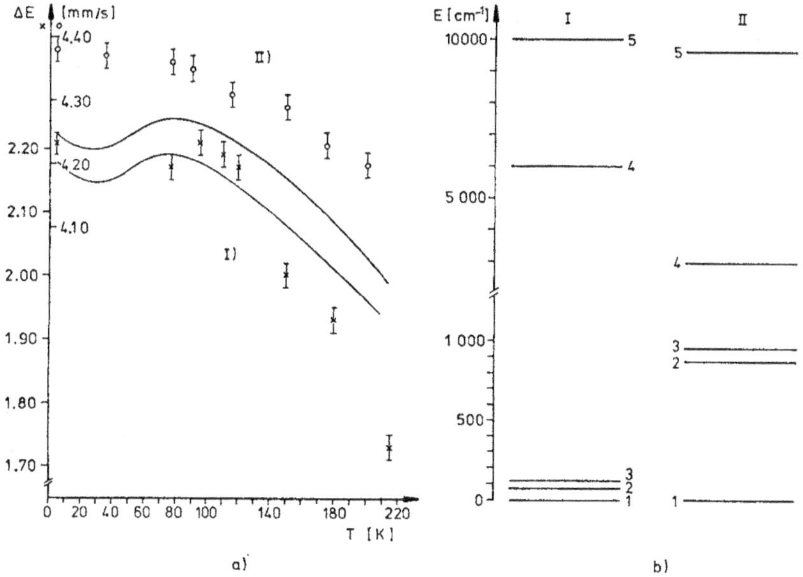

Fig. 4 Comparison of quadrupole splittings (a) and electronic structures (b) as derived from semiempirical MO calculations including spin-orbit coupling for (I)[TPP(SC_2H_5)]$^-$[Na·222]$^+$ and (II) [TPpivPCl]$^-$[K-222]$^+$. The Molecular Orbitals with increasing energies have the following wave functions:

(I) 1. $0.99\, d^{Fe}_{xy} + \cdots$

2. $0.64 d^{Fe}_{yz} - 0.58 d^{Fe}_{xz} + 0.14 d^{Fe}_{xy} - 0.1 P^{S}_{y} + 0.12 P^{N11}_{z} +$

3. $-0.56 d^{Fe}_{xz} - 0.42 d^{Fe}_{yz} + 0.33 P^{S}_{y} + 0.34 P^{S}_{x} - 0.11 P^{N1}_{z} - 0.14 P^{N11}_{z} + 0.12 P^{N17}_{z} + 0.14 P^{N23}_{z} + \cdots$

4. $-0.71 d^{Fe}_{z^2} + 0.18 d^{Fe}_{yz} + 0.13 d^{Fe}_{xz} + 0.18 S^{Fe} + 0.3 P^{S}_{z} + \cdots$

5. $-0.82 d^{Fe}_{x^2-y^2} - 0.31 P^{N1}_{x} + 0.32 P^{N11}_{y} + 0.31 P^{N17}_{x} - 0.32 P^{N23}_{y} + \cdots$

(II) 1. $0.99 d^{Fe}_{xy} + \cdots$ (d^{Fe}_{xy} means a 3dxy iron atomic orbital)

2. $0.84 d^{Fe}_{yz} - 0.29 P^{Cl}_{y} - 0.11 P^{N11}_{y} + 0.13 P^{N11}_{z} - 0.11 P^{N23}_{y} + \cdots$

3. $0.86\, d^{Fe}_{xz} - 0.29 P^{Cl}_{x} + 0.1 P^{N1}_{z} - 0.11 P^{N1}_{x} - 0.12 P^{N17}_{z} + \cdots$

4. $-0.85 d^{Fe}_{z^2} + 0.18 P^{Fe}_{z} - 0.25 P^{Cl}_{z}\ 0.32 P^{N1}_{z} - 0.2 P^{N11}_{z} - 0.14 P^{N17}_{z} - 0.16 P^{N23}_{z} + \cdots$

5. $0.83 d^{Fe}_{x^2-y^2} + 0.3 P^{N1}_{x} - 0.31 P^{N11}_{y} - 0.3 P^{N17}_{x} + 0.3 P^{N23}_{y} + \cdots$

Solid lines are calculated ΔE_Q from spin-orbit coupling, upper line with the following orbital energies: 1. 0 cm^{-1}; 2. 75 cm^{-1}; 3. 120 cm^{-1}; 4. 6000 cm^{-1}; 5. 10000 cm^{-1}, the lower line with - -orbital 3 at 150 cm^{-1}.

REFERENCES

(1) Caron, C., Mitschler, A., Riviere, G., Ricard, L., Schappacher, M., & Weiss, R. (1979) J. Am. Chem. Soc. 101, pp. 7401-7402.

(2) Schappacher, M., Ricard, L., Weiss, R., Montiel-Montoya, R., Bill, E., Gonser, U., Trautwein, A. (1981) J. Am. Chem. Soc.103, pp.7646-7648.

(3) Trautwein, A., Harris, F.E., (1973) Theoret. Chim. Acta 30, pp. 45-58.

MAGNETIC CIRCULAR DICHROISM SPECTROSCOPY AS A PROBE OF FERRIC CYTOCHROME P-450 AND ITS LIGAND COMPLEXES

John H. Dawson, Laura A. Andersson, Masanori Sono, Susan E. Gadecki, Ian M. Davis, Joseph V. Nardo and Edmund W. Svastits

Department of Chemistry, University of South Carolina, Columbia, South Carolina, 29208, U.S.A.

ABSTRACT

In order to identify the ligand *trans* to cysteinate in resting ferric cytochrome P-450 and to investigate the roles of both ligands in the reaction cycle of the enzyme, we have examined numerous ferric P-450 ligand complexes by UV-visible absorption and magnetic circular dichroism spectroscopy. The sixth ligand to the native enzyme has been found to be an oxygen donor such as an endogenous alcohol-containing amino acid, or possibly water; this ligand may play an important role in the catalytic mechanism. Comparison of the properties of P-450 and myoglobin ligand complexes suggests that the heme iron of P-450 is measurably more electron-rich, presumably due to its cysteinate fifth ligand. This ligand may aid in the formation and subsequent stabilization of reactive metal-bound oxygen intermediates that occur in P-450 catalyzed transformations.

INTRODUCTION

Cytochrome P-450 is a heme iron mono-oxygenase noted both for its ability to activate molecular oxygen for insertion into organic substrate molecules and for its unusual spectroscopic properties (1). Intense interest has been focussed on the catalytic pathway of P-450 because the enzyme has both beneficial (drug and steroid metabolism) and harmful (carcinogen activation) physiological roles (see 1, and references therein). Since accurate mechanistic interpretations require a thorough understanding of the basic structural features of an enzyme, attempts to define the non-porphyrin, axial ligand(s) in the various reaction states of P-450 have been of particular importance. Cysteinate sulfur has been well-established as the fifth ligand to the heme iron in states 1-5 (Scheme 1).[1] However, although histidine (4,5) and oxygen donors such as alcohol, water or an amide (6-9) have been

[1]Species 6 and 7 in Scheme 1 are postulated intermediates. For extensive discussions of the active-site environment of P-450 and its characterization, see References 2 and 3.

I. Bertini, R. S. Drago, and C. Luchinat (eds.), The Coordination Chemistry of Metalloenzymes, 369–376.

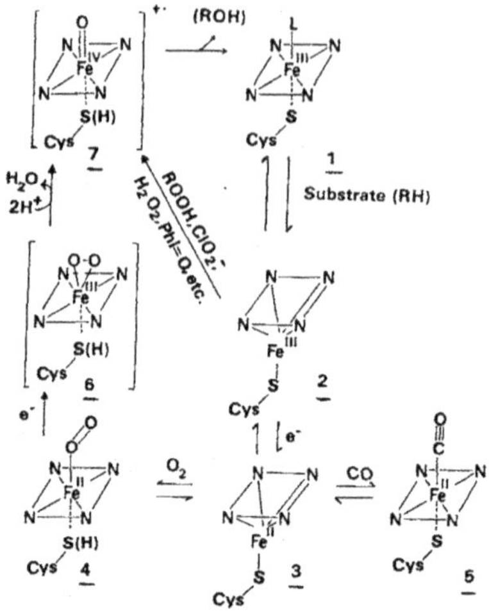

Scheme 1

suggested as the ligand *trans* to cysteinate in 1, the identity of this crucial ligand has not been conclusively determined. Thus, the goals of the work reported herein have been to identify the sixth ligand of resting ferric P-450, 1, and to obtain evidence for the roles of both this ligand and the cysteinate fifth ligand during P-450 catalyzed mono-oxygenase (10) reactions. To that end an extensive variety of ferric P-450 ligand complexes with biomimetic nitrogen, oxygen and sulfur donor adducts has been examined by both UV-visible absorption and magnetic circular dichroism (MCD) spectroscopy and compared to P-450 state 1 (2). Also, anionic and sulfur donor complexes of ferric P-450 and myoglobin have been studied in order to observe the effect of the cysteinate fifth ligand on adduct formation (11,12).

MCD[2] spectroscopy is a technique of particular utility in the study of heme protein complexes. It has been shown to be capable of distinguishing between species having highly similar absorption spectra (17-18) and can be diagnostic of both spin state (19) and ligand type (20). Further, since MCD is a differential spectroscopic method, both sign and intensity are plotted versus wavelength, thus providing twice the information content of more conventional UV-visible absorption spectroscopy. It is this "finger-printing" power that makes MCD spectroscopy such a useful technique with which to make comparisons between generally similar chromophores. Finally, due to the exquisite sensitivity of MCD spectroscopy to heme iron systems, it is possible to obtain spectra at low concentrations of protein (~10 μM).

[2]For reviews of MCD theory, see References 13-16.

MATERIALS AND METHODS

Cytochrome P-450-CAM was purified from *Ps. putida* to electrophoretic homogeneity (2,21). Substrate removal utilized the procedure of Peterson and co-workers (22). Myoglobin samples were prepared as described by Sono et al. (12). Ligand purity and complex generation were as reported elsewhere (2,11,12,23). UV-visible absorption and MCD/CD spectroscopic procedures were as previously reported (2). All P-450 ligand complexes were examined for the presence of the inactive form of P-450, known as P-420 (1), by bubbling with carbon monoxide and treatment with $Na_2S_2O_4$; none of the species reported exhibited more than a few percent P-420.

RESULTS AND DISCUSSION

While cysteine anion has been well established as the fifth ligand of cytochrome P-450 (Scheme 1), previous investigators have not concurred on the identity of the ligand *trans* to cysteinate in P-450 state 1. Imidazole has been suggested as the sixth ligand on the basis of both EPR (4) and electron spin echo (5) studies; the latter experiments provide evidence for a nitrogen atom (presumed to be the non-coordinated nitrogen of imidazole) weakly coupled to the heme iron, at a distance of a few angstroms. On the other hand, histidine (imidazole) can be ruled out as the ligand *trans* to cysteinate by electron nuclear double resonance (7) and NMR (8,9) investigations, which suggest the presence of a dissociable proton approximately 2.6-2.9 Å from the heme iron of P-450, and are thus most consistent with amine, alcohol, thiol or water ligation. Further studies by Ruf et al. (6) of thiolate·Fe(III)porphyrin ligand adducts support alcohols as possible candidates for the sixth ligand to P-450 state 1.

Our approach to this problem as been based on the assumption that a P-450 ligand adduct with an exogenous ligand of the same type as the sixth ligand in native ferric P-450 (Scheme 1, 1) would exhibit UV-visible absorption and MCD spectral properties that are very similar to those of the resting enzyme. Such six-coordinate ligand complexes of ferric P-450 can be readily formed from either P-450 state 1 or 2 and their integrity verified via competition experiments (2). Fig. 1 depicts the MCD spectrum of homogeneous low-spin P-450-CAM, and the key points of this spectrum which should be reproduced by a ligand complex. Numerous biomimetic ligands were examined as representative models for all reasonable amino acid coordinating types (2). These included numerous imidazoles (histidine), amines (lysine) and indole (tryptophan) as nitrogen donors; numerous thioethers (methionine), disulfide (cystine), thiols (cysteine) and thiolates (cysteinate) as sulfur donors; and numerous alcohols (serine, threonine, tyrosine), amides (asparagine, glutamine) and carboxylates (glutamate, aspartate) as oxygen donors.

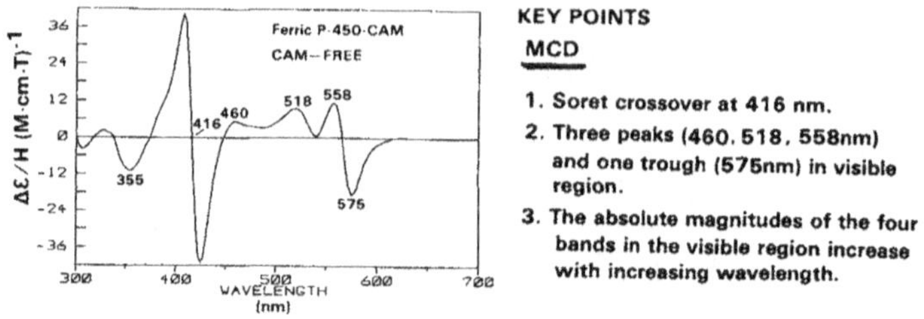

Figure 1: MCD Spectrum of Ferric P-450 State 1 and Key Points

Fig. 2 is the MCD spectra of low-spin ferric P-450 complexes with imidazole and N-octylamine. Direct comparison to Fig. 1 readily reveals a lack of overall similarity, particularly with respect to the key points of the native enzyme spectrum, and thus suggests that the sixth ligand to low-spin resting P-450 is not a nitrogen donor. Of the various sulfur

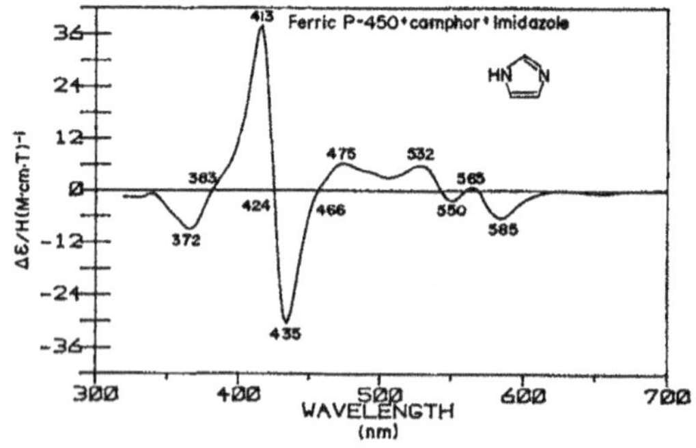

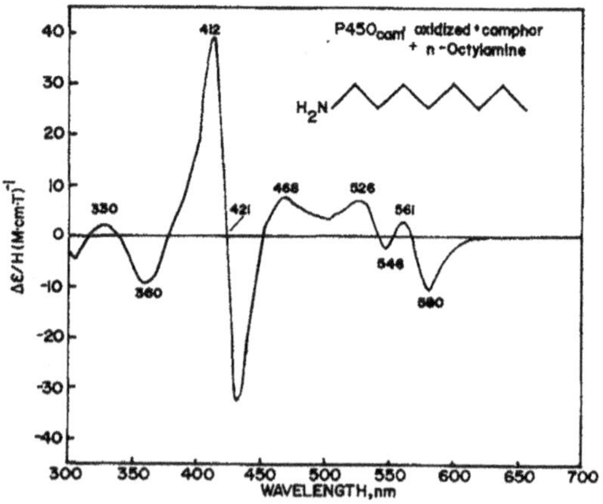

Figure 2: MCD Spectra of Ferric P-450 Nitrogen Donor Complexes

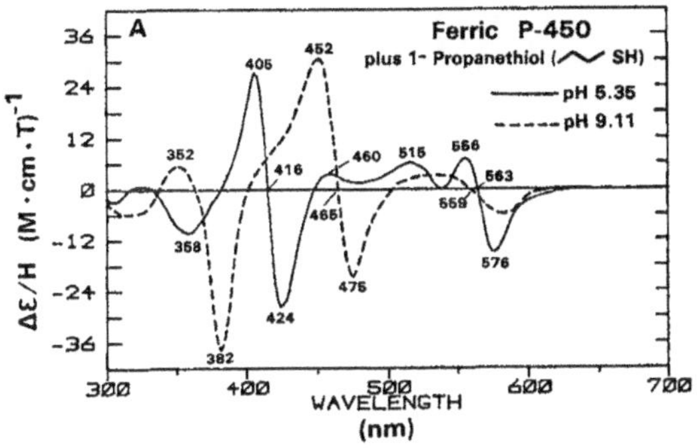

Figure 3: MCD Spectra of Thiol and Thiolate Adducts of Ferric P-450

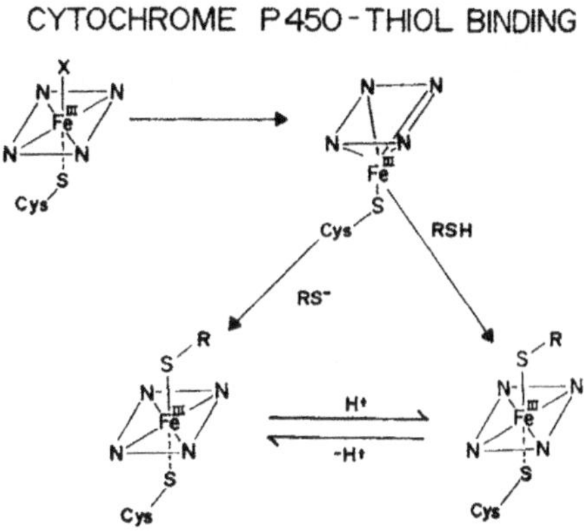

Scheme 2

donor adducts of P-450, the thiol species (Fig. 3, solid line) has an MCD spectrum closely analogous to that of the native enzyme (Fig. 1). However, as shown in Scheme 2, the thiol adduct of P-450 undergoes a pH-dependent conversion to a thiolate species which has spectral properties distinct from those of P-450 State 1 (compare Fig. 3, dotted line, and Fig. 1). Further, the native enzyme does not undergo such pH-dependent spectral changes; thus, the sixth ligand to low-spin ferric P-450 cannot be a thiol sulfur donor. Finally, Fig. 4 is the MCD spectrum of a typical oxygen donor complex; comparison to Fig. 1 readily reveals close analogies, particularly in the pattern of transitions in the 450-700 nm region. The UV-visible absorption and EPR

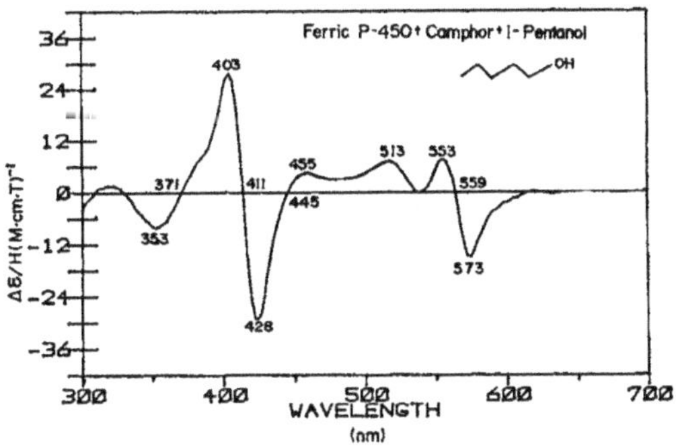

Figure 4: MCD Spectrum of Ferric P-450 Oxygen Donor Complex

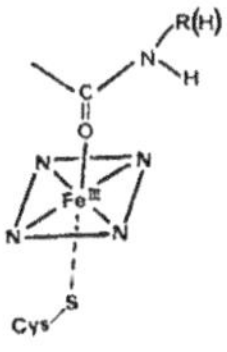

8

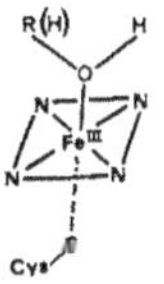

9

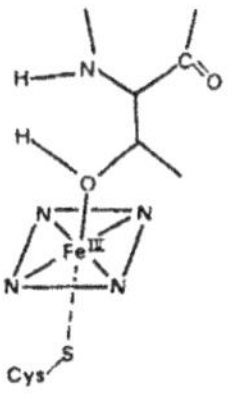

10

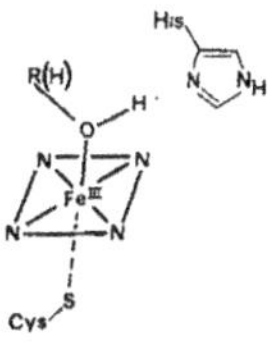

11

spectroscopic parameters are also highly similar to those of native low-spin P-450 (2,23), thus suggesting that the sixth ligand is an oxygen donor. Further, the fact that spectra of neutral oxygen donors such as ethers, esters and ketones (2,23) are more analogous to the spectra of P-450 State 1 than those of carboxylate oxygen donors (2) suggests that the oxygen donor sixth ligand of P-450 is protonated. Structures 8-11 depict coordination to the heme iron of P-450 of an amide[3], 8, an alcohol, 9, and further hypothetical bonding patterns (10 and 11) that would be consistent with all of the experimental evidence reported concerning the structure of low-spin ferric P-450 (5,7-9, 24). Since an amide ligand would lack the dissociable proton in close proximity to the metal as required by magnetic resonance studies (7-9), the sixth ligand to native P-450, 1, is most likely an endogenous alcohol-containing amino acid, or possibly water. Structures 10 and 11 incorporate the weakly coupled nitrogen seen in the spin echo study (5) in the form of a peptide nitrogen of serine or threonine and of a histidine nitrogen hydrogen bonded to the sixth ligand, respectively. The sixth ligand may have a critical role in the catalytic cycle of P-450 through chemical, electronic or steric interactions between bound dioxygen and the sixth ligand; possibly positioning the bound substrate for interaction with activated oxygen or hydrogen-bonding to O_2 or subsequent metal-oxo species.

One point of concern regarding the possibility that the sixth ligand is an alcohol or water pertains to the fact that for myoglobin, when water is coordinated to the ferric state, the resulting complex is primarily high-spin in contrast to the low-spin nature of the resting state of P-450. Evidence to overcome this concern has been obtained in that all P-450 ligand complexes examined to date have been found to be exclusively low-spin. This includes weak ligand field donors such as esters, ethers and ketones (2,23) as well as all of the anions that would bind to P-450 (12) which normally give high-spin complexes with myoglobin (Table I). These data indicate that the ligand field strength of the cysteine thiolate sulfur is sufficiently large that all six-coordinate P-450 adducts are enforced to be low-spin, regardless of how weak the ligand field strength of the *trans* ligand.

Indirect support for a ligand such as an oxygen donor amino acid or water as the sixth ligand to resting ferric P-450 comes from their low affinity for ferric P-450. Thus, such a ligand would be readily displaced by substrate (Scheme 1, 1→2) to give the five-coordinate substrate complex. In addition, oxygen donor ligands other than dioxygen, itself, are not known to coordinate to ferrous heme proteins, consistent with the known five-coordinate structure (3) of ferrous P-450 (3).

[3]Amide coordination is assumed to be through oxygen since amide adducts display spectra closely similar to those of oxygen donor complexes and because N,N-disubstituted amides coordinate to P-450 even though analogous tertiary amines do not.

Table I[a]

Anionic Ligand	P-450-CAM % Low-Spin	Myoglobin % Low-Spin
CN^-	∿100	∿100
SH^-	∿100	∿100
$CH_3(CH_2)_2S^-$	∿100	∿100
$SeCN^-$	∿100	> 80
N_3^-	∿100	79-83
OH^-	not confirmed	31-35
NO_2^-	not confirmed	24-28
SCN^-	∿100	16-18
CH_3COO^-	∿100	15
$CH_3CH_2COO^-$	∿100	not formed
$HCOO^-$	∿100	9-14
OCN^-	∿100	10
F^-	not formed	4

[a]Adapted from Table II, Reference 12.

In order to define the role of the cysteinate fifth ligand to the heme iron of P-450 in the catalytic cycle, we have examined the binding of thiol(ate) ligands to P-450 and to myoglobin (11). The pH-dependent interconversion of P-450-bound thiols to thiolates (Scheme 2) reflects the fact that the pKa of thiols is lowered by about 4 units (from 10.7 to 6.7) upon ligation to P-450. Only the thiolate-bound form is seen for myoglobin regardless of thiol acidity or solution pH (5.5-11.0); thus the pKa's of added thiols are lowered by greater than 6 units (from 10.7 to less than 4.5) upon binding to myoglobin. This quantitatively demonstrates that the heme iron of P-450 is appreciably more electron-rich than that of myoglobin (11). This increased electron density presumably results from the endogenous cysteinate ligand of P-450 and may help to explain the ability of P-450 to cleave dioxygen (Scheme 1, 6→7) and ultimately incorporate an oxygen atom into its substrate. Once the oxygen-oxygen bond is cleaved, the cysteinate ligand could also serve to stabilize the high-valent metal-oxo species that may result. Both roles for the cysteinate ligand in the P-450 mechanism have previously been suggested (17).

CONCLUSION

The data reported herein favor an endogenous oxygen donor, such as an alcohol-containing amino acid or possibly water, as the sixth ligand to resting ferric P-450. Such a ligand may play an important role in the subsequent reaction cycle. The cysteinate fifth ligand of P-450 has been shown to substantially increase the electron density at the heme iron, as compared to that of myoglobin, thus suggesting a possible role for the cysteinate fifth ligand in stabilization of active metal-bound oxygen during the process of P-450 mono-oxygenase reactions.

REFERENCES

(1) Sato, R., and Omura, T., eds.: 1978, "Cytochrome P-450", Academic Press, New York.
(2) Dawson, J.H., Andersson, L.A., and Sono, M.: 1982, J. Biol. Chem. 257, pp. 3606-3617.
(3) Hahn, J.E., Hodgson, K.O., Andersson, L.A., and Dawson, J.H.: 1982, J. Biol. Chem., in press.
(4) Chevion, M., Peisach, J., and Blumberg, W.E.: 1977, J. Biol. Chem. 252, pp. 3637-3645.
(5) Peisach, J., Mims, W., and Davis, J.: 1979, J. Biol. Chem. 254, pp. 12379-12387.
(6) Ruf, H.H., Wende, P., and Ullrich, V.: 1979, J. Inorg. Biochem. 11, pp. 189-204.
(7) LoBrutto, R., Scholes, C.P., Wagner, G.C., Gunsalus, I.C., and DeBrunner, P.G.: 1980, J. Amer. Chem. Soc. 102, pp. 1167-1170.
(8) Griffin, B.W., and Peterson, J.A.: 1975, J. Biol. Chem. 250, pp. 6445-6451.
(9) Philson, S.B., DeBrunner, P.G., Schmidt, P.G., and Gunsalus, I.C.: 1979, J. Biol. Chem. 254, pp. 10173-10179.
(10) Hayaishi, O.: 1974, in "Molecular Mechanisms of Oxygen Activation" (Hayaishi, O., ed.) pp. 1-28, Academic Press, New York.
(11) Sono, M., Andersson, L.A., and Dawson, J.H.: 1982, J. Biol. Chem., in press.
(12) Sono, M., and Dawson, J.H.: 1982, J. Biol. Chem. 257, pp. 5496-5502.
(13) Vickery, L.E.: 1978, Methods Enzymol. 54, pp. 284-302.
(14) Stephens, P.J.: 1974, Ann. Rev. Phys. Chem. 25, pp. 201-232.
(15) Holmquist, B.: 1978, in "The Porphyrins" (Dolphin, D., ed.) Vol. III pp. 249-270, Academic Press, New York.
(16) Sutherland, J.C.: 1978, in "The Porphyrins" (Dolphin, D., ed.) Vol. III, pp. 225-248, Academic Press, New York.
(17) Dawson, J.H., Holm, R.H., Trudell, J.R., Barth, G., Linder, R.E., Brunnenberg, E., Djerassi, C., and Tang, S.C.: 1976, J. Amer. Chem. Soc. 98, pp. 3707-3709.
(18) Dawson, J.H., and Cramer, S.P.: 1978, FEBS Lett. 88, pp. 127-130.
(19) Vickery, L.E., Nozawa, T., and Sauer, K.: 1976, J. Amer. Chem. Soc. 98, pp. 343-350.
(20) Vickery, L.E., Nozawa, T., and Sauer, K.: 1976, J. Amer. Chem. Soc. 98, pp. 351-357.
(21) Andersson, Laura A.: 1982, Ph.D. Thesis, to be submitted to the University of South Carolina.
(22) O'Keeffe, D.H., Ebel, R.E., and Peterson, J.A.: 1978, Methods Enzymol. 52, pp. 151-157.
(23) Andersson, L.A., and Dawson, J.H.: submitted for publication.
(24) White, R.E., and Coon, M.J.: 1982, J. Biol. Chem. 257, pp. 3073-3083.

SUBJECT INDEX